P9-CKN-521

TRANSPORT AND ACCUMULATION
IN BIOLOGICAL SYSTEMS

TRANSPORT AND ACCUMULATION IN BIOLOGICAL SYSTEMS

By

E. J. Harris, Ph.D., D.I.C., D.Sc.

Department of Biophysics
University College, London

with contributions from other authors

BUTTERWORTHS LONDON
UNIVERSITY PARK PRESS BALTIMORE

THE BUTTERWORTH GROUP

ENGLAND
Butterworth & Co (Publishers) Ltd
London: 88 Kingsway, WC2B 6AB

AUSTRALIA
Butterworth & Co (Australia) Ltd
Sydney: 586 Pacific Highway Chatswood, NSW 2067
Melbourne: 343 Little Collins Street, 3000
Brisbane: 240 Queen Street, 4000

NEW ZEALAND
Butterworth & Co (New Zealand) Ltd
Wellington: 26–28 Waring Taylor Street, 1

SOUTH AFRICA
Butterworth & Co (South Africa) (Pty) Ltd
Durban: 152–154 Gale Street

First published 1956
Second Edition 1960
Third Edition 1972

© Butterworth & Co. (Publishers) Ltd, 1972

ISBN 0 408 37901 4

Published 1972 jointly by
BUTTERWORTH & CO (PUBLISHERS) LTD, LONDON
and
UNIVERSITY PARK PRESS, BALTIMORE

Library of Congress Cataloging in Publication Data

Harris, Eric James, 1915-
   Transport and accumulation in biological systems.

   Includes bibliographical references.
   1. Biological transport. 2. Cells—Permeability.
   I. Title
QH601.H3   1972   574.8'75   73-38633
ISBN 0 8391 0633 5

Filmset and printed in England by Page Bros (Norwich) Ltd

PREFACE

This volume is a successor to *Transport and Accumulation in Biological Systems* last published in 1960. However, it differs in several important respects. The publication explosion, like all excesses, has not led to greater ease in gleaning information. Nowadays, the scientific literature is so diluted with accounts of repetitious, unoriginal and unrevealing work that it takes much time to pick out useful 'bits' of information. The accretion of information on all subjects in the field of biological membranology means that a specialist one can produce a useful summary only on any one topic; one needs a panel of obliging authors and a great deal of space to accommodate their contributions. In preparing this volume I have been fortunate in obtaining chapters from a number of colleagues. My own diversified and consequently attenuated sections will by their omissions illustrate how difficult it is to maintain any breadth of knowledge. It will be important to emphasise here that many topics briefly mentioned in the earlier editions are now entirely omitted. It will be so easy for a critic to complain of there being no treatment of amphibian skin, kidney secretion, plant cells, bacteria, heart muscle and so on but perhaps he will consider whether he will not himself write a chapter or a book to fill in the regrettable gap.

Advances in the techniques and materials for the study of artificial membranes have struck me as the most important factors in helping to understand the processes going on in biological membranes. The first chapter is devoted to some model systems. For the last five years I have been interested in mitochondria and the correspondence obtainable with some agents applied to mitochondrial and to artificial membranes is impressive. Much earlier work on muscle seems to be clouded by the complexity of the membrane system with paths involving diffusion in tubules in series with permeation through internal membranes. As for nerve, since the work of Hodgkin and Huxley, the squid axon has been perfused by an astonishing variety of chemicals but the phenomena described by Rudin and Müller using artificial membranes doped with alamethicin seem more relevant to our understanding of the action potential. In some fields such as gastric secretion it is humbling to hear from life-long investigators that their earlier conclusions were based on oversimplifications and that the explanation has to be generated anew.

E. J. H.

ACKNOWLEDGEMENTS

As editor I have to thank my colleagues who have kindly contributed chapters to this volume. Only by the concerted efforts of other workers in the field of membranology can even a fraction of its compass be covered. Also, a general thanks for the use of published material for the figures and tables must be given. It is appropriate to mention here my indebtedness to the Medical Research Council, the Wellcome Trust, and the Muscular Dystrophy Association of America, Inc. for their support for my own research by provision of assistance and apparatus. It is only be receiving such help in a sustained way that research in the universities can proceed.

E. J. H.

CONTENTS

1

MODELS OF MEMBRANES AND CARRIERS

E. J. HARRIS

Department of Biophysics, University College, London

INTRODUCTION

It is profitable to commence this volume on 'membranology' with a description of some models that can illustrate the applicability of particular laws or processes, but we must remember that the mere choice of a model involves selection, and the range of properties displayed is thereby limited. The living membrane is much more labile and complex than the product of our attempts to mimic it, so it will be unwise to draw too facile conclusions from such parallels as are found. However, at the present time it is true that many results obtained with models can be used to predict the responses of natural membranes. For this reason, a fairly detailed discussion of the models and of the substances active upon them is justified. We must include consideration of membrane materials, particularly those that provide fixed charges to give the membrane an ion-exchange character, and the 'ionophores'[1], a group of substances that can profoundly alter the permeability of lipid membranes to charged substances.

NATURAL MEMBRANE MATERIALS

Models can be built using membranes of porous glass, cellophane, Millipore filter and the like, impregnated, if desired, with lipid. Alternative systems use water-immiscible fluids that act as the link between two aqueous solutions in separated compartments. The 'oil'–water interface can then be given sufficient lipid to spread over it to form a monolayer, a process that usually occurs spontaneously. What lipid is used in these cases? Most work has been concerned with those lipids that occur as major components in the makeup of the cell or other membrane. Examples are listed in Table 1.1. Some authors have preferred to use the natural mixtures as extracted from brain, mitochondria, and so on, while others take purified fractions of lecithin, cephalin, or sphingomyelin, to which they may add a proportion of cholesterol or a long-chain hydrocarbon, e.g. tetradecane. Whether the lipid component is a two-dimensional liquid, or is solid, or alternates between both conditions, is a subject for enquiry. Besides the phospholipid component of the biological membrane, there is always associated protein, and in special cases mucopolysaccharides (in gram-positive bacteria) and polyhydroxy-compounds mannan and glucan in some yeast cells. How the protein and lipid components are arranged in the membrane is still a matter of speculation[2]. Many electronmicroscope pictures of membranes show two dark bands separated by about 7·0 nm (Robertson's unit membrane[3–5]) and about 0·5 nm outside the dark

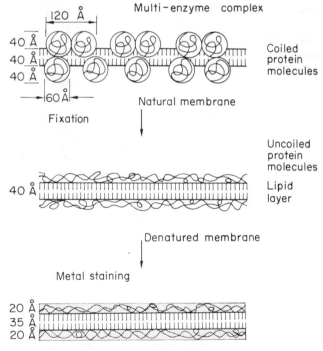

Figure 1.1. Scheme to explain the formation during fixation of a protein-coated bilayer from the natural membrane. The latter is supposed to have localised protein molecules on its surface (see Figure 1.2) before exposure to fixative (After Mühlethaler, Moor, and Szarkowski[7]; courtesy Planta)

2

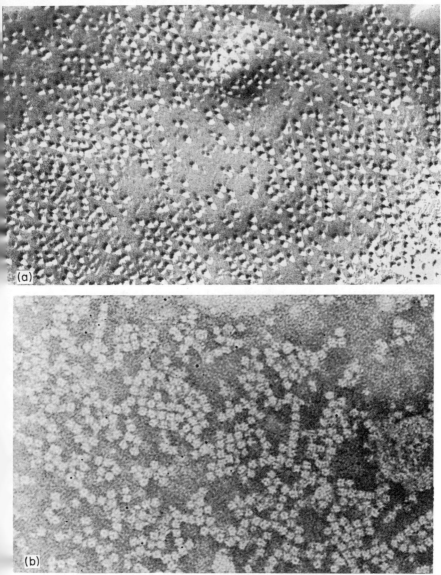

Figure 1.2. Ultrastructure of the thylakoid lamellae that make up the grana (green pigmented areas) and stroma of the chloroplast. The photographs show particles covering the membrane surface. At the higher magnification, they appear to be composed of four subunits arranged round a central hole. Some of the particles lie along straight lines. (a × 160 000 × ⅔; b × 320 000 × ⅔) (From Mühlethaler and Wehrli[8]; courtesy Progr. Photosynthesis Res.)

3

bands[6] are diffuse stained bands attributed to protein layers. The unit membrane seen in fixed material is considered by some (e.g. Mühlethaler, Moor, and Szarkowski[7]) to arise during the fixation process (Figure 1.1) and their illustrations show particles with central holes situated on the membrane (Figure 1.2). The particles seem to be made from four subunits of 6·0 nm diameter. They may be the sites of the membrane ATPase and other membrane enzymes. It has been proposed[9] that specific chemicals plug up the central hole and so give rise to toxic effects by excluding the substrates from the enzyme sites.

An interwoven protein–lipid structure has been discussed by Benson[10]. It is possible to extract a specific protein from the membranes of *E. Coli*, adapted to galactoside permeability,[11] that is lacking from the untrained strain. Pardee[12] was able to obtain a protein from a *Salmonella* that would endow sulphate permeability upon a mutant lacking the property. Accordingly, it seems correct to think that proteins are an integral part of the natural 'bilayer' membrane. Certainly, it is attractive to be able to invoke allosteric modification of the protein to account for effects of specific agents on permeability. This picture would not exclude the possession of a parallel system for moving substances on mobile carriers into and through the lipid, which is suggested by the action of the ionophores on ion movements in both natural and artificial membranes (see pp. 12 and 306).

The inner membrane of the mitochondrion is exceptional in having only 30% phospholipid in its composition. The outer membrane has more phosphatidylcholine as a major component; while the inner membrane has considerable cardiolipid (Table 1.1). The low lipid content of the inner membrane accounts for the observation that in the electronmicroscope the appearance is unchanged by lipid extraction. Its enzymic properties do, however, depend on the presence of lipid[13, 14]. Other presentations of composition of mitochondrial membranes and the respective enzymic activities of outer and inner components have been given by Ernster and Kuylenstierna[15] and Lévy *et al.*[16]. The latter topic will find some mention in a later chapter.

The chemistry and mode of formation of membrane components has been reviewed by Dawson[17]. Van Deenen and his collaborators[18–20] have explored the changes in composition of cell membranes in response to different conditions of growth.

The adaption of *Bacillus megatherium* to acid conditions occurs by glucosamine becoming linked to phosphatidyl glycerol in the membrane phospholipid. In this way, the membrane gains fixed cationic sites, so that anion transfer is facilitated to compensate for

4

Table 1.1 PHOSPHOLIPID COMPOSITIONS OF SOME MEMBRANES

Component	Erythrocyte (strongly bound)	Membrane	
		Outer mitochondrial	Inner mitochondrial[21]
Lecithin	40·5	55·2	44·5
Cephalin	20	25·3	27·7
Cardiolipin	?	3·2	21·5
Phosphatidyl inositol	?	13·5	4·2
Sphingomyelin	28	?	?
Lysolecithin	8	?	?

Cholesterol is present in the weakly bound fraction of the red cell membrane lipids and in the outer membrane of the mitochondrion. Values are as percentages of the alkali labile phosphorus.

there being few anions from the weak substrate acids when the pH is low. Adaption of yeast to low temperature requires that the membrane becomes more permeable. This is effected by a change from saturated to unsaturated acid esterifying the glycerol to form phospholipid. Since the unsaturated acids are held by their double bonds in a bent conformation, they pack less tightly and the membrane is more permeable to small molecules.

The osmotic fragility of erythrocyte membranes is increased if part of their cholesterol is removed and at the same time the glycerol permeability is raised. Glycerol permeability is enhanced by the addition of 3-ketosteroids[22].

MEMBRANE MODELS OF GROSS DIMENSIONS

Models made using collodion, which after oxidation has fixed anionic charges, have a preferential cation permeability.

If protamine is incorporated in the collodion, the product has fixed basic groups and is preferentially anion-permeable. These materials have been much investigated by Sollner (for a review and list of references see reference 23). He measured what are known as bi-ionic potentials between solutions having one common ion (e.g. Cl). With an anionic membrane, the potential difference depended on the nature and concentrations of the cations on the respective sides. The converse was true with a cationic membrane. Unequal concentration of the same salt give rise to a cationic or anionic potential difference, depending on the sign of the fixed charges. A general expression for the p.d. between the membrane boundaries is:

$$E = -\frac{RT}{F} \int_1^2 \frac{t_i}{z_i} \, \mathrm{d} \ln a_i$$

This is the theoretical expression; t_i, a_i, and z_i are the transport number, the activity, and the valency of ion species i and contributions from all mobile ions have to be included. The total p.d. across a fixed-charge membrane also includes the two contributions from Donnan potentials existing between each membrane boundary and the respective solutions[24, 25]. Only when equal concentrations of equally charged electrolytes are present on each side do the Donnan potentials cancel out[26]. Teorell[27] has described an interesting model using Sephadex as a membrane. Oscillations of electrical potential and hydrostatic pressure can be obtained when a current is passed through the membrane.

Biologists always sought for a model system mimicking the cell's selection of K in preference to Na. Osterhout[28] had some success using guiacol as a soluble carrier in nitrobenzene for K ions, which would then carry current through the non-aqueous phase. More recently, ion-selective glasses having an Na/K selectivity of 1000 or a K/Na selectivity of 20 have been made. They operate by having an ordered array of adsorption sites at certain distances from each other; with a suitable spacing, cations of a given size require less energy to migrate between sites than do larger or smaller ones. Eisenman[29, 30] has discussed the dependence of rank order of cation on the composition of the glass. Modification of the structure, which tends to take place during hydration, alters the surface selectivity. This change can give rise to transient responses to alterations in composition of the medium, and may well mimic some biological phenomena. In nature, the zeolites, which like the glasses are aluminosilicates, also have marked ion-selective properties[31-33]. Some zeolites show a transition in structure and ion selectivity when exposed to sufficiently high concentrations of a given ion. Since the transition can be reversed with high enough concentrations of another ion, one can obtain a hysteresis curve relating properties to the ion composition in the solution applied to the material. The movement of ions in zeolites and glasses is from site to site and it has a high temperature coefficient (54–113 kJ per g ion). This high value serves to emphasise that a physical process involving no net chemical change is not necessarily distinguishable from a chemically mediated process by its temperature dependence.

Selective glass electrodes, besides serving as models of ion-selective membranes, have also been used to measure the ion activities resulting from the operation of the selectivity of natural membranes. In the interiors of various animal and plant cells it seems that K^+ is held at the activity expected for the prevailing ionic strength, the activity factor being about 0·7. On the other hand, Na^+ has an activity factor of only about 0·4[34–36]. It is difficult to allow for effects of compartmentation within the cell; it is possible that Na^+ is excluded from some of the total water, so that the microelectrode's tip is exposed to a non-uniform region, so leaving its response ill-defined. However, investigation by NMR of the Na of cells indicates that 70% does not give the signal characteristic of free Na^+[37].

A convenient gross model system is obtained by impregnating a microporous membrane (such as Millipore) with phospholipid with or without cholesterol. When interposed between salt solutions, the impregnated membrane acts as a cation exchanger. In the Ca^{2+} form, the conductance, water permeability, and water content are less than in the K^+ or Na^+ form[38, 39]. The model has 340 mequiv fixed anion per kg water or per 500 g lipid; its behaviour in response to ions is rather similar to that of red cell membrane fragments, which have only 92 mequiv fixed charges per kg. A further study showed that penetration by Ca^{2+} is optimal between pH 3 and 7. At lower pH, Ca^{2+} is displaced by protons, and at high pH Ca^{2+} is firmly bound. The Ca^{2+} impregnated membrane allows Ca^{2+} for Ca^{2+} exchange more readily than Ca^{2+} for Na^+ exchange. A marked inhibition of Ca movement is obtained by the more firmly bound La^{3+} ion[40]. This property resembles that obtained with mitochondria.

The K permeability of lipid-impregnated membranes is increased by the ionophorous antibiotics such as valinomycin and the actins. This response can most readily be shown by resistance measurements (Table 1.4), though movements of the ions can be measured.

BILAYER OR 'BLACK' MEMBRANE MODELS

When a solution of lipid, phospholipid, or oxidised cholesterol dissolved in hydrocarbon or chloroform–methanol is spread over an aperture immersed under water, it forms a film that will spontaneously thin down to the dimensions of an ordered bilayer of the lipid[41–44]. Some of the decane remains in the bilayer[45]. The thickness according to interference studies is 5·5 nm[46, 47]. An alternative method is to form a hanging drop of aqueous medium at the end of a tube projecting down into a density gradient with a film of lipid enclosing the drop. When shaken, the drop is released as a spherule and finds a position in the gradient just higher, because of the low density of the lipid, than the equilibrium level for the internal solution against the external solution[48]. There is a residual difference of water activity across the membrane, giving rise to a slow entry of water.

The electrical resistance of the film supported on an aperture is measured between reversible electrodes inserted in the two compartments linked by the film. When the spherule technique is used the film has to be impaled with microelectrodes to find the transmembrane potential difference set up by passing a measured current. Either method leads to a value of about 10^8–10^9 Ω cm^2 associated with a capacitance of about 1 mF cm^{-2}. The water permeability can be measured, either by using an isotopically labelled water on one side to find the rate of interdiffusion, or by setting up an osmotic gradient and measuring the volume of water moving per unit time.

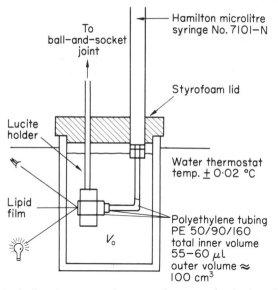

Figure 1.3. Sketch of cell used to measure the osmotic flux across thin, lipid membranes (After Cass and Finkelstein[49]; courtesy J. gen. Physiol.)

For the latter method, the compartment is made to be the end of a modified microsyringe; the membrane is observed optically to detect distortion and bulging or concavity is corrected by adjusting the syringe setting (Figure 1.3). Osmosis into the spherules can also be followed by watching the displacement upwards in a density gradient as water enters the slightly heavier internal solution that is

8

held above its equilibrium concentration in the gradient by the buoyancy of the lipid film. Earlier measurements of tracer and osmotic methods indicated a slower interdiffusion permeability than osmotic permeability, but the origin of the discrepancy was shown by Cass and Finkelstein[49] to reside in inadequate stirring. Their figure by either method was about 10^{-3} cm s^{-1}. With poor stirring, the effect on the tracer interdiffusion is more marked[50] than on osmosis, in part because of the convective stirring osmosis itself provides. The lack of stirring in cells that are of biological interest probably explains the very high factors (as much as 100) by which tracer movements can be slower than osmotic movements under equal gradients of water[51]. Dick[52] pointed out that the factor increased in proportion to the cell radius, as would be expected for a diffusion delay. The mathematical treatment of the effect of an unstirred layer has been given by Mackay and Meares[53] in the context of an ion-exchange membrane. They measured rates of diffusion for different salt gradients and rates of stirring, and found an irreducible minimum thickness of unstirred layer at each boundary of 50 μm. In certain bilayer experiments, the arrangement of the film left at least 500 μm of unstirred fluid, so it is not surprising that diffusion delays were significant.

The lipid capacity, water permeability, and breakdown potential of the bilayers approximate those of cell membranes, but their resistance is some 10^5 times higher. It is here that the striking effect of small quantities of various additives is brought out. Historically, the first compound used was the still ill-defined product of bacterial action on egg white, called EIM by Rudin and Müller. The material, added in small quantity, led to the appearance of a cation permeability, so that in salt solutions the resistance would drop to 10^4 Ω cm^2 or less. A remarkable property of the membrane so modified was the non-linear and time-dependent current p.d. relation[42] (Figure 1.4). If the current fed in causes the p.d. to exceed some threshold value, the resistance falls after a brief delay and the transmembrane p.d., after its initially high value, eventually decreases. Later[44], Rudin and Müller found that the natural cyclic polypeptide alamethacin would give the same effects (Figures 1.4 and 1.5). The membranes modified with either EIM or alamethacin differ from natural membranes in having a rather unspecific cation permeability and a low anion permeability. By adding protamine in carefully adjusted amounts, a parallel anion permeability could be provided. This recalls Sollner's use of protamine in collodion membranes. Alternative substances for providing anion permeability are histone or spermine. The bilayer treated with both protamine and alamethacin has two parallel conductances, for

9

anions and for cations. Both depend on the p.d. and concentrations of salt, and, since the ions move into and out of some part of the membrane (perhaps akin to the barrier layer of semiconductor rectifiers), there is a time dependence. By holding a current on, in the right sense, a delayed increase in resistance can be obtained; that is, just the contrary to the delayed decrease mentioned above. If set up across a salt gradient, the treated membranes can give rise to repeated oscillations of potential (see Figure 1.5); basically, they are then switching between being anionic electrodes, as a membrane with fixed positive charges would be, and being cation electrodes as a membrane having fixed negative charges would be. To achieve this, some part of the membrane thickness would have to have

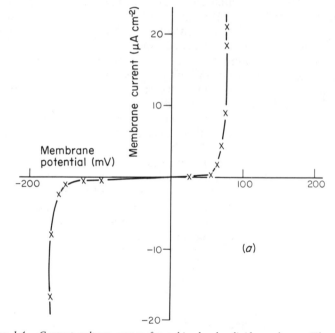

Figure 1.4. Current–voltage curves for a bimolecular lipid membrane. The inside aqueous compartment contained alamethicin at $10^{-7} g\ cm^{-3}$ and there was 5 mmol histidine buffer at pH 7 in each side. The outward current is shown as positive and the potential is of the inside minus the outside. The membrane was made from 2·5 % sphingo-myelin dissolved in tocopherol : chloroform : methanol 5 : 3 : 2. Curve (a) was obtained with 0·1 mol NaCl both sides; the asymmetry is due to alamethicin having been added only to one side. Curve (b) was obtained with 50 mmol NaCl outside and none inside. The resting potential with zero current is due to the ion gradient. Part of curve (a) is repeated for comparison on the same scale (After Rudin and Müller[44]; courtesy Nature, Lond.)

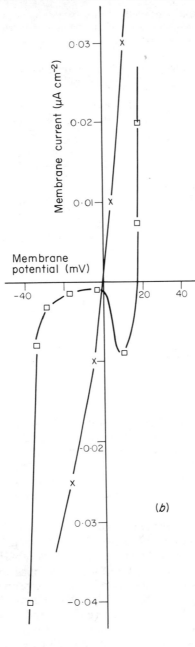

Table 1.2 SOME SUBSTANCES CONFERRING CATION PERMEABILITY ON
NATURAL AND ARTIFICIAL MEMBRANES—IONOPHORES

Group 1
Neutral compounds which form positively charged complexes
with cations

Substance	Mol. wt.	Rank order of cation selectivity induced	Method of ranking (below)
Valinomycin	740·8	Rb $\geqslant$ K > Cs $\gg$ Na	1, 2, 3
Nonactin	736	K > Cs > Rb $\gg$ Na	4
Monactin	750	Rb = K > Cs = Na	4
Dinactin	764	Rb > K > Cs $\gg$ Na	1, 3, 4
Trinacitin	778	Cs $\geqslant$ Rb$\geqslant$ K > Na	4
Gramicidin B	3725	Rb > K > Cs > Na	2, 3

Group 2
Carboxylic acids which form neutral salt complexes. They
promote exchanges between cations and protons

Substance	Mol. wt.	Rank order of cation complexing[60]
Nigericin	736	K > Rb > Na
Dianemycin	956	Na > K > Rb = Cs > Li
Monensin	670	Na > K > Rb

Methods used for ranking
1. Respiratory rate of mitochondria when cycling of the cation has been induced[54, 55]
2. Conductance of bilayer[43]
3. Mitochondrial ion flux measurement[54]
4. ATPase activity induced[56].

alternately a surplus of bound positive and bound negative charges. Most excitable cell membranes seem to possess the ability to undergo an alternation of Na and K permeabilities, so a much more specific change has to occur. It is to the understanding of this phenomenon that the ion-specific substances obtained from the fungi of the *Streptomyces* family can be helpful. Some of the compounds are listed in Table 1.2.

Valinomycin has been extensively investigated. This compound is a cyclic depsipeptide with 36 atoms in the ring. There are three repeats of a set of four acid residues that are notable in including the uncommon D-stereoisomers. Valinomycin applied to artifical membranes, to red cells, or to mitochondria has the property of inducing a high permeability to Rb, K, and, to a lesser extent, Cs ions. Variants on the structure have been synthesised[57–59] and tested[55, 60]. Few of these approach the parent compound in activity. The specificity of the permeability induced in the bilayers can be shown by setting up cells having different alkali chloride solutions separated by the membrane and reading the resulting bi-ionic potential difference. The membrane is impermeable to Cl ions, so the p.d. gives approximately the ratio of the cation mobilities[43, 61] (see Table 1.3).

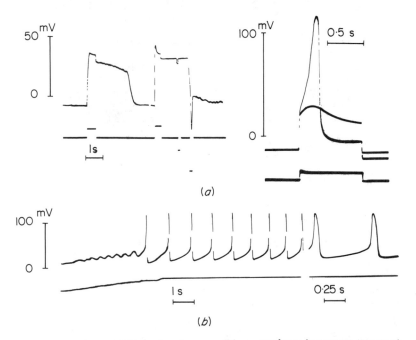

Figure 1.5. (a) *Potentials developed across a bilayer membrane in response to current pulses just under and just over a critical value. When the current exceeds the critical value the membrane resistance first rises, so the p.d. increases, but then the resistance falls to a low value and the p.d. is low. The system had 40 mmol KCl on one side, together with a trace of alamethicin (0·1 µg cm⁻³) and protamine (100 µg cm⁻³). (b) Rhythmic fluctuations of the p.d. produced by passing a constant current just above the critical value through the membrane (After Rudin and Müller[44]; courtesy Nature, Lond.)*

Alternatively, by titrating in a salt into one side the concentration required to give zero p.d. against a standard salt on the other side can be found by interpolation. Then, the respective cation activities on the two sides are related inversely as the products (mobility in the lipid) times (lipid:water distribution ratio). The lipid:water distribution ratios have to be found in separate experiments.

It is revealing to carry out a titration for a cation in order to relate the conductance it provides to the activity in the lipid. Using double reciprocal plots familiar in enzyme and adsorption studies,

Table 1.3 BI-IONIC POTENTIALS ACROSS LIPID BILAYERS*

1. From Lev and Buzhinsky[61]. Valinomycin at 10^{-11} M

Salt conc.	Cation tested				
	Rb^+	K^+	Cs^+	Na^+	Li^+
10 mM p.d. (mV)	0	−24	−35	−110	−110
1 mM p.d. (mV)	0	−12	−34	−85	−80
Selectivity, referred to Li^+ as standard	164	85	35	1·2	1·0

2. From Rudin and Müller[43]. Valinomycin at 3×10^{-9} M

Salt conc.	Cation tested				
	Rb^+	K^+	Cs^+	Na^+	Li^+
50 mM p.d. (mV)	0	−21	−37	−164	−172
Selectivity referred to Li^+ as standard	920	395	210	1·4	1·0

* Lipid bilayers interposed between solutions of RbCl and other alkali chlorides in the presence of valinomycin. Selectivity ratios have been deducted that depend on both lipid solubility and mobility of cation complex in the lipid.

a limiting conductance attainable at infinite cation concentration applied to the water in contact with the lipid can be deduced. For a given ionophore concentration in the lipid, there is a stoichiometric amount of cation and the conductivity depends on the mobility of the cation–ionophore complex[60, 62].

Table 1.4 COMPARISON BETWEEN RELATIVE CONDUCTIVITIES INDUCED BY VALINOMYCIN APPLIED TO LIPID MEMBRANES[43] AND ION UPTAKE RATES INDUCED IN RAT LIVER MITOCHONDRIA

Cation	Relative conductivity of bilayer	Ion flux (nmol mg^{-1} min^{-1})
Li^+	1	0
Na^+	1·2	0
Cs^+	50	345
K^+	200	470
Rb^+	300	530

Conditions of mitochondrial experiments: temperature 22 °C, pH 6·8. Medium: sucrose 250 mM, tris acetate 20 mM, tris glutamate and succinate 3 mM each, alkali chloride 5 mM. Protein 2·5 mg cm^{-3}. Valinomycin 20 μg per g protein.

Tables 1.2, 1.3, and 1.4 show that the chemically dissimilar compounds valinomycin and the actins endow membranes with rather similar cation permeability and selectivity. The gramicidins, which are obtained as fractions from the commercial mixture (so-called gramicidin D) are as effective on a mole-for-mole basis as the former compounds, but provide less inter-cation selectivity. These compounds are all neutral, having peptide and lactone linkages, so when associated with a cation the complex formed must bear the cation's charge. The complex then provides a current carrier in the lipid. The amount of such a charged complex that can enter the lipid from water will be limited by the counter-potential developed between water and lipid; only when lipid soluble anions can accompany the cation complex will chemically important quantities of the ions be found in the lipid. Iodide is an example of a comparatively lipophilic anion[63] but thiocyanate is still better[60].

Anion conductivity in lipid bilayers is increased by the presence of Fe^{3+} ions at 10^{-5} mol; this may perhaps be due to production of uncharged complexes to facilitate transfer of the ions into the lipid. The increase in conductance can be by a factor of 10^3 [64]. Cardiolipin reduces anion permeability of lipid films, presumably by providing negative fixed charges[65].

Carboxylic acids having lipid solubility can confer lipid solubility on cations, but the effect becomes most pronounced when the acids also clathrate the cations. Examples of such compounds are monensin[66], nigericin, and dianemycin (see latter part of Table 1.2). Since the compounds are undissociated salts, they do not confer electrical conductance on the lipid, yet they act as effective transporters of the cations across concentration gradients. The cations either will exchange for another when the aqueous layer is reached or will be replaced by a proton drawn from the water. Using a concentration gradient of protons between two aqueous media separated by a lipid containing a dissolved ionophore of this class, one can obtain a counter-movement of cations driven by dissipation of the proton gradient. Given a supply of protons from metabolism, this exchange could be the basis for cellular accumulation of selected cations, depending on the properties of an ionophore present in the cell membrane.

X-ray diffraction studies of complexes formed between the ionophores and cations have shown that the cation becomes enveloped by coordination with oxygen atoms leaving a hydrocarbon-like outer shell to confer the lipophilic property. The nonactin complex is shown in Figure 1.6 and the valinomycin complex in Figure 1.7[67, 59]. The latter compound appears to exist in two conformations[68]. The complex formation alters the positions

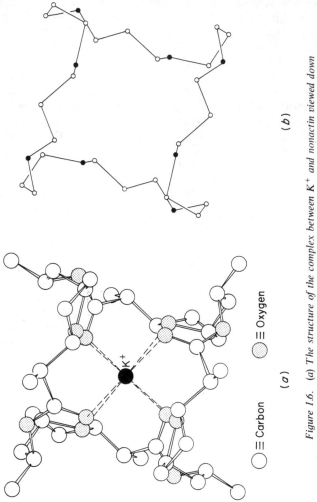

Figure 1.6. (a) The structure of the complex between K^+ and nonactin viewed down the α-axis. (b) The outline of the ring (After Kilbourne et al.[67]: courtesy J. molec. Biol.)

≡ Carbon   ≡ Oxygen

(a)

(b)

of the peaks of the H atoms in the nuclear magnetic resonance spectrum[62] (Figure 1.8).

Another class of compound capable of complexing cations, but leaving a net charge, is the cyclic polyethers[69], which Pressman[62] has shown to assist the entry of malonate into a lipid layer.

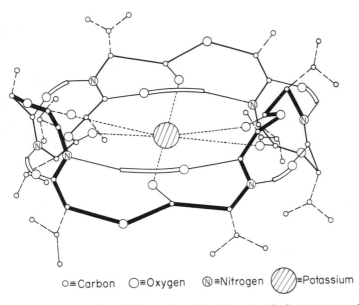

○≡Carbon   ◯≡Oxygen   Ⓝ≡Nitrogen   ▨≡Potassium

Figure 1.7. Representation of the structure of the K complex of valinomycin according to Ivanov et al.[59] (courtesy Biochem. biophys. Res. Comm.)

The classical uncouplers of mitochondrial phosporylation, such as dinitrophenol, dicoumarol, trifluoromethylcarbonyl cyanide phenylhydrazone, and the like, are also active on lipid bilayers, giving them an effective conductance to protons and probably hydroxyl ions also. It has been suggested that the mobile charge carrier is a complex $[A^-.HA]$ (where HA represents the uncoupler)[70]. Various kinds of experiment to show that the proton conductance given by an uncoupler plus the cation conductance given by valinomycin will promote a rapid K-for-H exchange across the membranes of mitochondria, red cells, and liposomes have been described.[71, 48, 72]. Bielawski, Thompson, and Lehninger[73] drew the parallel between H^+ conductance and uncoupling action. The result is consistent with the thesis that a vital part of the phosphorylation process is a charge separation because provision

17

of mobile ions would evidently short-circuit this and lead to dissipation of the energy. The potency of the different uncouplers towards membrane conductance and as uncoupler are well correlated[74, 75]. Energy dissipation by K^+-for-H^+ exchange also has been demonstrated in chloroplasts[76].

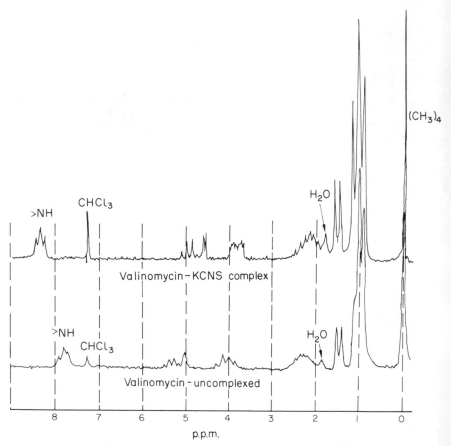

Figure 1.8. The nuclear magnetic resonance spectra of valinomycin dissolved in deuterochloroform to show the changes brought about when K is complexed (After Pressman[60, 62]; courtesy Fedn Proc. Fedn Am. Socs exp. Biol. and Academic Press)

Maintenance of electroneutrality in a system demands either that like-charged particles exchange, or that unlike-charged particles move together. For this reason, endowing a high cation permeability on a membrane will show up as a raised conductance or raised

18

ease of interchange of the cation with one sufficiently similar to be carried by the ionophore. In the red cell, valinomycin accelerates K^+ for K^{42} exchange (e.g. see reference 77) but there is little change in rates of net K^+ movement unless Rb^+ or Cs^+ is offered in exchange. To obtain K^+-for-H^+ exchange, either nigericin alone can be used or valinomycin and uncoupler together. To obtain K^+-for-Na^+ exchange, gramicidin is sufficient because of its broad spectrum of selectivity.

Attempts to solve the mechanism by which cells concentrate K^+ and reject Na^+, at the expense of ATP hydrolysis, are being made with artificial membranes to which ATPase proteins are added. The protein adsorbs on the lipid film and the resistance could be made to fall by between 10^2 and 10^3 times, provided Na and ATP were present[78]. Since the change was inhibited by the specific blocker of Na^+ movement—ouabaine—the model resembles the cell membrane in several ways. To cancel the p.d., a Na and ATP-dependent current was needed.

The structure of bilayers has been investigated by Branton[79] and Deamer and Branton[80], using a freeze-etch method. With ice formed on each side, the bilayer can be split at the junction between the opposed lipid chains. This was used as a way to test for migration of tracer; there is a slow exchange of labelled lipid from one of the opposed monolayers to the other with time constant about 1 h. Adding Ca^{2+} ions slowed the exchange.

OIL–WATER DISTRIBUTION

Although measurements of the partition of substances between oil and water phases have been made in attempts to relate the distribution to the permeation of biological membranes[81], until recently only a qualitative correlation for a number of uncharged molecules could be established. A major problem in such work is to choose a suitable non-aqueous solvent; olive oil has often been used. But natural membranes are mainly phospholipids, and it is likely that the compounds are spread in an ordered way over a skeleton of protein. The act of penetration involves transfers from water to lipid, diffusion through the lipid, and transfer back to water. The energy barrier involved in escaping from the water can be a limiting factor to the permeation, and this is particularly the case when ions are involved. The hydration energies of the cations are concerned with a statistical degree of association of water with the ion. To change this, as happens in varying degrees on transfer of the ion to the non-aqueous phase, requires energy. The energetic requirement can be modified if electrostatic forces between the ions and the fixed charges in the membrane assist transfer. For example, in a model water–oil–water system, the rates of transfer of ions from one to the other aqueous phase was accelerated by spreading between the oil and the water a monolayer of phospholipid[82, 83].

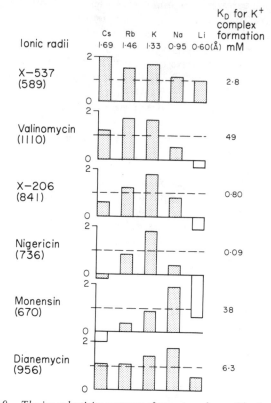

Figure 1.9. *The ion selectivity patterns of some ionophores. The data were primarily based on the ability of various ions to displace* [86]*Rb from complexes with the ionophores. The ordinate is the logarithm of the selectivity ratio. The absolute affinities are indicated by the composite dissociation constant for K, indicated by* K_D, *which represents the concentration of K required in the aqueous phase to half-saturate the organic solution of the ionophores. (After Pressman*[60, 62]*; courtesy Fedn Proc. Fedn Am. Socs exp. Biol. and Academic Press). The ionic radii are for the unhydrated ions.*

Examination of the temperature coefficient for the transfer process to obtain the change in heat content, coupled with determination of the absolute reaction rate to find the free energy change, indicates that a major obstacle to putting an ion into lipid is one of an entropy change[84]. The same conclusion was reached by Johnson and Bangham[85] as a result of study of permeation of K^+ through lecithin spherules with and without valinomycin. The valinomycin accelerated the rate without altering the temperature dependence, and only when a local anaesthetic was added, supposedly to induce disorder in the boundary structure, did the presence of valinomycin lessen the energy requirement for the transfer.

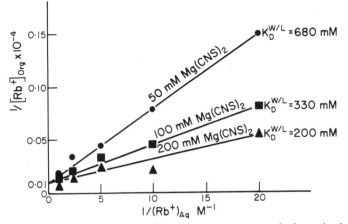

Figure 1.10. Double reciprocal plot of the Rb concentration in the butanol–tolune phase against the concentration of Rb in the aqueous phase in the presence of the cyclic polyether dicyclohexyl-18-crown-6 with different thiocyanate concentrations. Transfer of the complex to the oil is greater as the thiocyanate concentration is raised, because the anion must accompany the complex to maintain electroneutrality. The extrapolated limiting concentration of Rb at infinite aqueous concentration is independent of the anion and depends only on the concentration of the complexing agent. (After Press-man[60, 62]; courtesy Fedn Proc. Fedn Am. Socs exp. Biol. and Academic Press)

Reference has been made to studies of oil–water partition. Figure 1.9 summarises results of Pressman's for the rank order of cation lipophilisation by some ionophores. The concentration of cation in a lipid obtainable with a neutral ionophore in presence of lipophilic anion (thiocyanate) is increased as the concentration of anion is increased, but the limiting concentration is unaffected (Figure 1.10). The limiting concentration does depend, however, on the concentration of ionophore present and in fact is related thereto stoichiometrically (Figure 1.11 shows a 1 : 1 relation between nigericin and rubidium). Cs seems alternatively to associate with two valinomycin molecules, or there can be two Cs per valinomycin, depending on the concentration ratio. This dual behaviour is

21

consistent with the large size of unhydrated Cs that will not fit (as do K and Rb) in the assembly of O atoms (Figure 1.7).

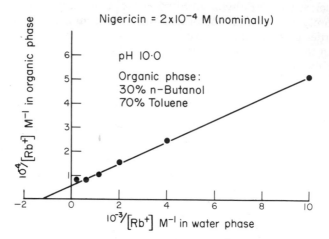

Figure 1.11. Double reciprocal plot of the saturation by nigericin of a butanol (30 %)–toluene (70 %) solution containing nigericin at 20 mol. Note that the limiting quantity of Rb found at the intercept is 1.5×10^{-4} mol, close to the concentration of nigericin (After Pressman[60, 62]; courtesy Fedn Proc. Fedn Am. Socs exp. Biol. and Academic Press)

A model experiment to show the carriage of Rb through carbon tetrachloride is shown in Figure 1.12. There has been shown to be a practical and theoretical correlation between the results of ion distribution studies in presence of the ionophores and the effects of the latter on ion conductance in bilayers. The investigation extended to use of lipids having different charges, so as to set up different interface potentials between the bilayers and the water[86–88]. It is notable that the size of the mobile species is little affected by the cation, so the mobility in the lipid is nearly independent of the presence of the cation.

PARTITION OF WEAK ACIDS

The distribution of weak acids between water and lipid is of special importance because of its relation to permeation of metabolites through cell and mitochondrial membranes. Weak acids are usually relatively lipophilic at pH sufficiently acid to stop their being ionised. This, however, would require pH below 5, and this is

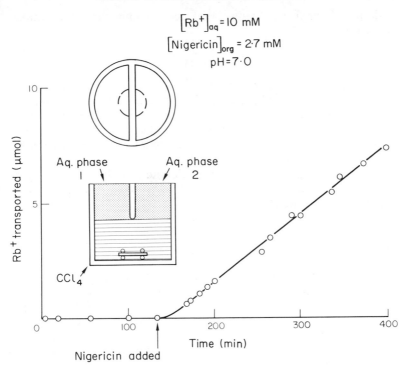

$$\left[Rb^+\right]_{aq} = 10 \text{ mM}$$
$$\left[Nigericin\right]_{org} = 2\cdot7 \text{ mM}$$
$$pH = 7\cdot0$$

Figure 1.12. Sketch of a test cell used to measure transport of Rb through a CCl₄ layer under the influence of nigericin as carrier (After Pressman[60, 62]; courtesy Fedn Proc. Fedn Am. Socs exp. Biol. and Academic Press)

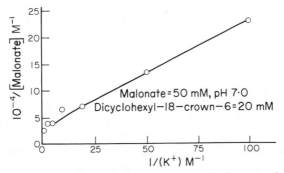

Figure 1.13. Double reciprocal plot of concentration of malonate in a butanol–toluene phase versus K concentration in the aqueous phase in the presence of the complex forming polyether dicyclohexyl-18-crown-6 at 20 mmol. Malonate is carried into the oil with the K complex (After Pressman[60, 62]; courtesy Fedn Proc. Fedn Am. Socs exp. Biol. and Academic Press)

23

seldom tolerated by living organisms. Using the dissociation equation:

$$K_a = \frac{[H^+][A^-]}{[HA]}$$

we see that the major change from the acid being predominantly undissociated (HA) to ionised (H^+ and A^-) occurs in a span of 10 in H^+ or 1 pH unit centred on the pK_a value. Since, experimentally, most penetration processes undergo a gradual change over a range of several pH units, it may be inferred that more than mere dissociation of the acid is involved.

Figure 1.13 shows an increasing malonate transfer to an oil layer as the amount of a cation complex is increased by increasing the cation concentration.

MONOLAYERS

When lipids possessing polar groups are spread on water, for example by letting a solution in decane evaporate on a clean water surface in a trough, a unimolecular film can be obtained. Since the molecule has a definite area, according to its size and orientation, there is a relation between the number of molecules spread and the limiting area they occupy without crumpling the film. Generally the technique is to reduce the free area of the surface by moving a sweep along the trough. By monitoring either the pressure acting on a fixed barrier, or by observing the surface tension acting on a foil dipping into the liquid (the Wilhelmy balance), the completion of the monolayer is signalled. The surface tension provided by uncovered water exceeds that found when there is a complete film.

There is an extensive literature on monolayer films and several useful books can profitably be consulted[89,90]. Here, only a few observations on special features will be mentioned.

It has been shown that lecithin monolayers correspond to a limiting area per molecule of $0.7\ nm^2$. If cholesterol is added, the limiting area falls to $0.55\ nm^2$[91]. It has been proposed[92] that the cholesterol fits in between the hydrocarbon chains so that hydrogen bonds pull them together. It is relevant that many hormones are steroids, as are the potent inhibitors of ion or sugar transport such as ouabaine and phlorizin.

A possible parallel between an effect seen with monolayers and a mitochondrial phenomenon is afforded by the sensitivity of cardiolipin to Ca^{2+} ions. The area of the spread molecule is reduced by Ca because it tends to draw together the phosphate groups.

Spherules made with a proportion of cardiolipin become anion impermeable when Ca^{2+} ions are added[93].

On account of the properties of the weak β-particle emission from Ca^{45}, it is possible to measure the adsorbed Ca^{2+} on surface layers floating on solutions containing a little of the radioactive ion. The technique has been applied to show that local anaesthetics displace Ca^{2+} from adsorption sites on lipid films[94]. Univalent ions compete with the Ca^{2+}[39]. Phosphatidyl serine was shown by the latter authors to find one Ca^{2+} per molecule. Application of pressure to the layer led to displacement of part of the Ca^{2+}.

There is still controversy about the mode of action of insulin in making cells more permeable to sugars and aminoacids. It is possible that model experiments may be revealing; Kafka and Pak[95] have described a displacement of Ca^{2+} from an octadecyl phosphate monolayer by insulin. If insulin is added before Ca^{2+}, it inhibits attachment of Ca^{2+} to the film. These observations suggest that the permeability of the cell membrane may be changed by insulin because of a Ca^{2+} shift. Analogues of insulin are less effective. Cyclic AMP, a stimulant of many enzymes, acts to increase Ca^{2+} adsorption.

Monolayers formed from sterols are disorganised and penetrated by the polyene antibiotics (e.g., filipin). Bilayer films made from the steroids are ruptured by the same agents. The results are paralled by these antibiotics being effective only against organisms whose membranes include a proportion of sterol[96]. On the other hand, certain psychoactive drugs specifically alter the force–area curves of monolayers of the gangliosides, which are a major constituent of nerve membranes[18]. Tetrodotoxin, an agent potent on nerve receptors, causes an expansion of cholesterol films[97].

LIPOSOMES

When a phospholipid is exposed, either as dry powder or as a solution in a volatile solvent to an aqueous solution, with shaking to disperse it and remove solvent, it usually consists of a multilayer film of the lipid enclosing a core of the aqueous solution[98]. By careful sonication, most of the vesicles can be reduced to have a single bilayer membrane (Figure 1.14[99]). In this condition, the vesicles provide an effective test system to examine lipid permeability and effects of agents on it. Spherules prepared from lecithin have a high anion permeability, on account of their having the fixed cationic groups provided by choline. The activation energy for anion penetration is low. The cation permeability of lecithin spherules is low and its energy of activation is high[100]. Ca increases the cation permeability, which can also be increased if the lipid is given acidic groups by incorporating dicetyl phosphate (Figure 1.15). The rate of ion transfer can be related to the surface charge density as given by the zeta. potential (Figure 1.16). An increased potential difference with respect to the solution tends to favour ions of the opposite sign reaching the membrane; this is one of the Debye effects.

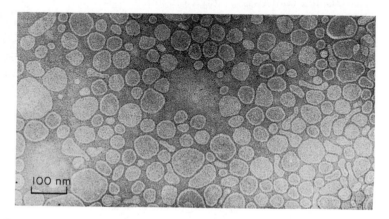

Figure 1.14. Electronmicrograph of single-compartment liposomes produced by long sonication of swollen lecithin micelles. These are used as a model system to study lipid permeability (From Johnson and Bangham[99]; courtesy Biochim. biophys. Acta)

Liposomes have been used to show that most sterols, when incorporated in the film, increase the cation permeability; only hydrocortisone diminishes the permeability and counters the effects of the other sterols. Effects of ionophores[93] (Figure 1.17) have been shown to resemble those seen on bilayers and mitochondria, and the effect of uncouplers in producing proton permeability can be demonstrated.

The effect of cholesterol on the permeability of lecithin films to water was investigated. using swelling in an osmotic gradient as the index. The change can be followed readily using light scattering to indicate the volume change[101]. At higher temperatures (30–40 °C) cholesterol increased permeability, perhaps due to disordering; at lower temperatures, the membrane becomes less permeable on

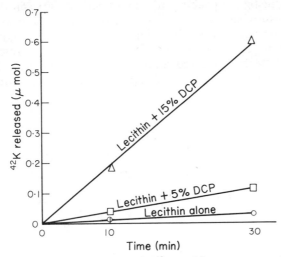

Figure 1.15. Initial rates of leakage of K^{42} out of liposomes prepared from lecithin alone (—○—), lecithin with 5% dicetyl phosphate (—□—) or lecithin with 15% dicetyl phosphate (—△—) into KCl solution 145 mmol at 37 °C. The dicetyl phosphate provides anionic sites to pass the K ions through the lipid (After Bangham, Standish Watkins[98]; courtesy J. molec. Biol.)

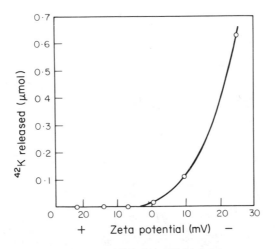

Figure 1.16. The amount of release of ^{42}K from K^{42} loaded liposomes prepared from various lecithin–long hydrocarbon chain ions mixtures to obtain different zeta potentials. Time 30 min, temp. 37 °C. Dialysis was against unlabelled KCl at 145 mmol. The anionic membrane, with negative zeta potential, is more permeable (After Bangham, and Watkins[98]; courtesy J. molec. Biol.)

27

account of the tighter packing. It is convenient to use easily detected ions when studying the permeability of the spherules to ions. For this chromate has been used[102]. This shows that filipin induces an anion leakage through lecithin, irrespective of the presence of cholesterol, but nystatin and amphotericin only caused leakage if cholesterol was included in the compositon of the membrane.

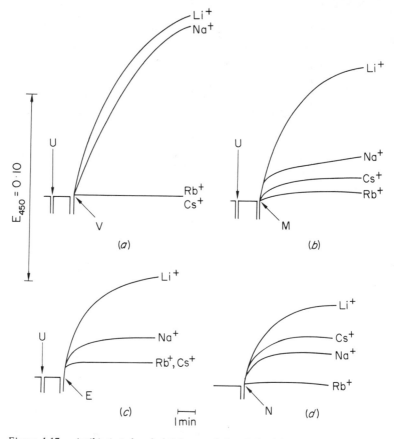

Figure 1.17. Antibiotic-induced shrinkages of phospholipid liquid crystals, prepared in KCl (40 mM) and suspended in salts of other cations (40 mM). These were then rendered permeable to K^+ by the agents: U, carbonyl cyanide p-chlorophenylhydrazone (0·2 mol); V, valinomycin (0·2 µg cm^{-3}); M, monactin (1 µg cm^{-3}); E. enniatin (8 µg cm^{-3}); N, nigericin (1 µg cm^{-3}). When K^+ is exactly replaced by its equivalent of another cation, there is no net volume change, e.g. Rb^+, Cs^+ in (a). When KCl emerges but the other cation only penetrates poorly, there is a net loss of water, shrinkage, and the O.D. rises (After Henderson, McGivan, and Chappell[93]; courtesy Biochem. J.)

28

LABELLING OF MEMBRANES

There is a growing interest in the idea that changes in membrane conformation bring about ion movements or phosphorylation. It then becomes useful to have available probes of membrane structure. Two types of compound have been applied; one is a fluorescent substance, exemplified by anilinonaphthalene sulphonic acid, whose fluorescence yield depends on the dielectric constant of the material in which the molecule is situated (see, e.g. references 103, 104). With this substance, changes in enzyme conformation resulting from cofactor additions can be demonstrated, and changes in membrane vesicles derived from mitochondria have also been obtained in response to substrates[105]. The other class of compound used for labelling membranes have a −NO group in their composition. These nitroxyl compounds give a signal on the electron spin resonance detector by virtue of their unpaired electron. Changes in the environment of the unpaired electron are detected by a shift in frequency and broadening of the band of the resonance[106, 107].

CONCLUSIONS

Developments in search for and discovery of compounds active in conferring specific ionic conductivity on membranes, presumably by acting as carriers, coupled with ways of making model bilayer membranes allow many properties of natural membranes to be mimicked. When proteins are added to these systems, they will become still more like the natural objects and eventually should be capable of using chemical energy to move ions against electrochemical gradients.

REFERENCES

1. PRESSMAN, B. C., HARRIS, E. J., JAGGER, W. S., and JOHNSON, J. H., 'Antibiotic-mediated transport of alkali ions across lipid barriers', *Proc. natn. Acad. Sci. U.S.A.*, **58**, 1949 (1967)
2. BRANTON, D., 'Membrane structure', *A. Rev. Pl. Physiol.*, **20**, 209 (1969)
3. KORN, E. D., 'Structure of biological membranes', *Science*, **153**, 1491 (1966)
4. FLEISCHER, S., FLEISCHER, B., and STOECKENIUS, W., 'Fine structure of lipid-depleted mitochondria', *J. Cell Biol.*, **32**, 193 (1967)
5. STOECKENIUS, W., in *Membranes of Mitochondria and Chloroplasts* (ed. E. RACKER), Van Nostrand, New York, 53 (1970)
6. HECHTER, O., 'Intracellular water structure and mechanism of cellular transport', *Ann. N.Y. Acad. Sci.*, **125**, 625 (1965)
7. MÜHLETHALER, K., MOOR, H., and SZARKOWSKI, J. W., 'The ultrastructure of the chloroplast lamellae', *Planta*, **67**, 305 (1965)
8. MÜHLETHALER, K., and WEHRLI, E., *Progress in Photosynthesis Res., Proceedings Int. Congr. of Photosynthetic Res.*, Freudenstadt (1968)

9. HOLAN, G., 'New halocyclopropane insecticides and the mode of action of D.D.T.', *Nature, Lond.,* **221**, 1023 (1969)
10. BENSON, A. A., 'The cell membrane—A lipoprotein monolayer', in *Membrane Models and the Formation of Biological Membranes.* (eds. L. BOLIS and B. A. PETHICA) North Holland Publishing Co., Amsterdam, 190 (1968)
11. FOX, C. F., and KENNEDY, E. P., 'Specific labeling and partial purification of the M protein, a component of the beta-galactoside transport system of Escherichia coli', *Proc. natn. Acad. Sci. U.S.A.,* **54**, 891 (1965)
12. PARDEE, A. B., 'Purification and properties of a sulfate-binding protein from Salmonella typhimurium', *J. biol. Chem.* **241**, 5886 (1966)
13. RACKER, E., 'The membrane of the mitochondrion', *Scient. Am.,* **218**, (2) 32 (1968)
14. RACKER, E., and BRUNI, A., 'The role of phospholipids and proteins in structure and function of the mitochondrial membrane', in *Membrane Models and the Formation of Biological Membranes* (eds L. BOLIS and B. A. PETHICA). North Holland Publishing Co., Amsterdam, 138 (1968)
15. ERNSTER, L., and KUYLENSTIERNA, B., in *Mitochondria Structure and Function* (eds L. ERNSTER and Z. DRAHOTA), Academic, London, 5 (1969)
16. LÉVY, M., TOURY, R., SAUNER, M. T., and ANDRÉ, J. in *Mitochondria—Structure and Function* (eds L. ERNSTER and Z. DRAHOTA), Academic, London, 33 (1969)
17. DAWSON, R. M. C., 'The metabolism of animal phospholipids and their turnover in cell membranes', *Essays Biochem. (U.S.),* **2**, 69 (1966)
18. VAN DEENEN, L. L. M., and DEMEL, R. A., 'A study of the arsenate uptake by yeast cells compared with phosphate uptake', *Biochim. biophys. Acta,* **94**, 312 (1965)
19. VAN DEENEN, L. L. M., 'Some structural and dynamic aspects of lipids in biological membranes', *Ann. N.Y. Acad. Sci.,* **137**, 717 (1966)
20. VAN DEENEN, L. L. M., in *Membrane Models and the Formation of Biological Membranes* (eds L. BOLIS and B. A. PETHICA), North Holland Publishing Co., Amsterdam, 98 (1968)
21. PARSONS, D. O., WILLIAMS, G. R., THOMPSON, W., WILSON, D., and CHANCE, B. in *Mitochondria Structure and Compartmentation* (ed. QUAGLIARIELLO et al.), Adriatica, Bari, 29 (1967)
22. BRUCKDORFER, K. R., DEMEL, R. A., DE GIER, J., and VAN DEENEN, L. L. M., 'The effect of partial replacements of membrane cholesterol by other steroids on the osmotic fragility and glycerol permeability of erythrocytes', *Biochim. biophys. Acta,* **183**, 334 (1969)
23 SOLLNER, K., 'Membrane Electrodes', *Ann. N.Y. Acad. Sci.,* **148**, 154 (1968)
24. MEYER, K. H., and SIEVERS, J. F., 'La permeabilité des membranes', *Helv. chim. Acta,* **19**, 649, 665, 987 (1936)
25. TEORELL, T., 'Transport processes and electrical phenomena in ionic membranes', *Prog. Biophys. biophys Chem.,* **3**, 305 (1953)
26. FINKELSTEIN, A., and MAURO, A., 'Equivalent circuits as related to ionic systems', *Biophys. J.,* **3**, 215 (1963)
27. TEORELL, T., *Ark. Kemi,* **18**, 401 (1961)
28. OSTERHOUT, W. J. V., 'Physiological studies of single plant cells', *Biol. Rev.,* **6**, 369 (1931)
29. EISENMAN, G., 'Cation selective glass electrodes and their mode of operation', *Biophys. J.,* **2**, 259 (1962)
30. EISENMAN, G., 'Electrochemistry of cation sensitive glass electrodes', *Adv. anal. Chem. & Instrum.,* **4**, 213 (1965)
31. BARRER, R. M., and FALCONER, J. D., 'Ion exchanges in feldspathoids as a solid state reaction', *Proc. R. Soc.* (A), **236**, 227 (1956)
32. ASH, R., PALMER, D. G., and BARRER, R. M., 'Diffusional conductances and time lags in laminated hollow cylinders', *Trans. Faraday Soc.,* **65**, (1) 121 (1968)

33. BARRER, R. M., and DAVIES, J. A., 'Thermodynamics and thermochemistry of cation exchanges in chabazite', *J. inorg. nucl. Chem.,* **31**, 290 (1968)
34. HINKE, J. A. M., in *Glass Microelectrodes* (eds M. LAVALLEE, O. F. SCHANNE, and N. C. HEBERT), Wiley, New York, 349 (1969)
35. KURELLA, G. A., in *Glass Microelectrodes* (eds M. LAVALLEE, O. F. SCHANNE, and N. C. HEBERT), Wiley, New York, 154 (1969)
36. KOSTYUK. P. G., SOROKINA, Z. A., and KHOLODOVA, YU. D., in *Glass Microelectrodes,* (eds M. LAVALLEE, O. F. SCHANNE, and N. C. HEBERT), Wiley, New York, 322 (1969)
37. COPE, F. W., 'NMR evidence for complexing of Na^+ in muscle, kidney, and by actomyosin. The relation of cellular complexing of Na^+ to water structure and to transport kinetics', *J. gen. Physiol.,* **50**, 1353 (1967)
38. MIKULECKY, D. G., and TOBIAS, J. M., 'Phospholipid-cholesterol membrane model. I. Correlation of resistance with ion content. II. Cation exchange properties. III. Effect of Ca on salt permeability. IV. Ca—K uptake by sonically fragmented erythrocyte ghosts', *J. cell. comp. Physiol.,* **64**, 151 (1964)
39. ROJAS, E., TOBIAS, J. M., 'Membrane model association of inorganic cations with phopholipid monolayers', *Biochim. biophys. Acta.* **94**. 394 (1965)
40. VAN BREEMEN, D., and VAN BREEMEN, C., 'Calcium exchange diffusion in a porous phospholipid ion-exchange membrane', *Nature, Lond.,* **223**, 898 (1969)
41. RUDIN, D. O., and MÜLLER, P., 'Induced excitability in reconstituted cell membrane structure', *J. Theoret. Biol.,* **4**, 268 (1963)
42. MÜLLER, P., and RUDIN, D. O., 'Action potential phenomena in experimental bimolecular lipid membranes', *Nature, Lond.,* **213**, 603 (1967)
43. MÜLLER, P., and RUDIN, D. O., 'Development of K^+—Na^+ discrimination in experimental bimolecular lipid membranes by macrocyclic antibiotics', *Biochem. Biophys. Res. Commun.,* **26**, 398 (1967)
44. RUDIN, D. O., and MÜLLER, P., 'Action potentials induced in bimolecular lipid membranes', *Nature, Lond.,* **217**, 713 (1968)
45. HENN, F. A., and THOMPSON, T. E., 'Properties of lipid bilayer membranes separating two aqueous phases: composition studies', *J. molec. Biol.,* **31**, 227 (1968)
46. TIEN, H. T., 'Thickness and molecular organization of bimolecular lipid membranes in aqueous media', *J. molec. Biol.,* **16**, 577 (1966)
47. TIEN, H. T., and DIANA, A. L., Bimolecular lipid membranes: a review and a summary of some recent studies, *Chem. & Phys. Lipids,* **2**, 55 (1968)
48. THOMPSON, T. E., and HENN, F. A., in *Membranes of Mitochondria and Chloroplasts* (ed. E. RACKER), Van Nostrand, New York, 1 (1970)
49. CASS, A., and FINKELSTEIN, A., 'Water permeability of thin lipid membranes', *J. gen. Physiol.,* **50**, 1765 (1967)
50. DAINTY, J., *Adv. bot. Res.,* **1**, 279 (1963)
51. PRESCOTT, O. M., and ZEUTHEN, E., 'Water diffusion and water filtration across cell surfaces', *Acta physiol. scand.,* **28**, 77 (1953)
52. DICK, D. A. T., 'Osmotic properties of living cells', *Internat. Rev. Cytol.,* **8**, 387 (1959)
53. MACKAY, D., and MEARES, P., 'The correction for unstirred solution films in ion exchange membrane cells', *Kolloid Z.,* **167**, 31 (1959)
54. HARRIS, E. J., in *Membrane Models and the Formation of Biological Membranes* (eds L. BOLIS and B. A. PETHICA), North Holland Publishing Co., Amsterdam, 247 (1968)
55. PRESSMAN, B. C., 'Induced active transport of ions in mitochondria', *Proc. natn. Acad. Sci. U.S.A.,* **53**, 1076 (1965)
56. GRAVEN, S. N., LARDY, H., JOHNSON, D., and RUTTER, A., 'Antibiotics as tools for metabolic studies. V. Effect of nonactin, monactin, dinactin, and trinactin on oxidative phosphorylation and adenosine triphosphatase induction', *Biochemistry,* **5**, 1729 (1966)

31

57. SHEMYAKIN, M. M., VINOGRADOVA, E. I., FEIGINA, M. YU., ALDANOVA, N. A., LOGINOVA, N. F., RYABOVA, I. D., and PAVLENKO, I. A., 'The structure—antimicrobial relation for valinomycin depsipeptides', *Experientia*, 21, 548 (1965)
58. SHEMYAKIN, M. M., OVCHINNIKOV, YU. A., IVANOV, V. T., ANTONOV, V. K., SHKROB, A. M., MIKHALEVA, I. I., EVSTRATOV, A. V., and MALENKOV, G. G., 'The physicochemical basis of the functioning of biological membranes: conformational specificity of the interaction of cyclodepsipeptides with membranes and of their complexation with alkali metal ions', *Biochem. biophys. Res. Commun.*, 29, 834 (1967)
59. IVANOV, V. T., LAINE, I. A., ABDULAEV, N. D., SENYAVINA, L. B., POPOV, E. M., ORCHIMIKOV, YU. A., and SHEMYAKIN, M. M., 'The physicochemical basis of the functioning of biological membranes: the conformation of valinomycin and its K^+ complex in solution', *Biochem. biophys. Res. Commun.*, 34, 803 (1969)
60. PRESSMAN, B. C., 'Ionophorous antibiotics as models for biological transport', *Fedn. Proc. Fedn. Am. Socs exp. Biol.*, 27, 1283 (1968)
61. LEV, A. A., and BUZHINSKY, E. P., 'Cation specificity of model membranes treated with valinomycin, *Tsitologiya*, 9, 102 (1967) (in Russian)
62. PRESSMAN, B. C., in 'Mitochondria Structure and Function', (eds L. ERNSTER, and Z. DRAHOTA), Academic, London, 315 (1969)
63. LÄUGER, P., LESSLAUER, W., MARTI, E., and RICHTER, J., 'Electrical properties of bimolecular phospholipid membranes, *Biochim. biophys. Acta*, 135, 20 (1967)
64. MIYAMOTO, V. K., and THOMPSON, T. E., 'Some electrical properties of lipid bilayer membranes', *J. colloid & interface Sci.*, 25, 16 (1967)
65. MCGIVAN, J. D., and CHAPPELL, J. B., 'The effect of cardiolipin on the anion permeability of artificial phospholipid micelles, *Biochem. J.*, 105, 15 (1967)
66. AGTARAP, A., CHAMBERLIN, J. W., PINKERTON, M., and STEINRAUF, I., 'The structure of monensic acid, a new biologically active compound', *J. Am. Chem. Soc.*, 89, 5737 (1967)
67. KILBOURNE, B. T., DUNITZ, J. D., PIODA, L. A. R., and SIMON, W., 'Structure of the K^+ complex with nonactin, a macrotetralide antibiotic possessing highly specific K^+ transport properties', *J. Molec. Biol.*, 30, 559 (1967)
68. OHNISHI, M., and URRY, D. W., 'Temperature dependence of amide proton chemical shifts: the secondary structures of gramicidin S and valinomycin', *Biochem. biophys. Res. Commun.*, 36, 194 (1969)
69. PEDERSON, C. J., 'Cyclic polyethers and their complexes with metal salts', *J. Am. chem. Soc.*, 89, 7017 (1967)
70. FINKELSTEIN, A., 'Weak acid incouplers of oxidative phosphorylation. Mechanism of action of thin lipid membranes', *Biochim. biophys. Acta*, 205, 1 (1970)
71. CHAPPELL, B., and CROFTS, A. R., in *Regulation of Metabolic processes in Mitochondria* (eds J. M. TAGER *et al.*), Elsevier, London, 307 (1966)
72. HARRIS, E. J., and PRESSMAN, B. C., 'Obligate cation exchanges in red cells, *Nature, Lond.*, 216, 918 (1967)
73. BIELAWSKI, J., THOMPSON, T. E., and LEHNINGER, A. L., 'The effect of 2,4-dinitrophenol on the electrical resistance of phospholipid bilayer membranes', *Biochem. biophys. Res. Commun.*, 24, 948 (1966)
74. SKULACHEV, V. P., SHARAF, A. A., YAGUZHINSKY, L. S., JASAITIS, A. A., LIBERMAN, E. A., and TOPALI, V. P., *Currents in Modern Biology*, 2, 96 (1968)
75. SKULACHEV, V. P., JASAITIS, A. A., NAVICKAITE, V., and YAGUZHINSKY, L., in *Mitochondria—Structure and Function* (eds L. ERNSTER and Z. DRAHOTA). Academic, London, 275 (1969)
76. KARLISH, S. H., SHAVIT, N., and AVRON, M., 'The mechanism of uncoupling in chloroplasts by ion-permeability-inducing agents', *Eur. J. Biochem.*, 9, 291 (1969)
77. TOSTESON, D. C., ANDREOLI, T. E., TIEFFENBERG, M., and COOK, P., 'The effect of

macrocyclic compounds in cation transport in sheep red cells and thin and thick lipid membranes, *J. gen. Physiol.*, **51**, 3735 (1968)

78. JAIN, M. K., STRICKHOLM, A., and CORDES, E. H., 'Reconstitution of an ATP-mediated active transport system across black lipid membranes', *Nature, Lond.*, **222**, 871 (1969)

79. BRANTON, D., 'Fracture Faces of frozen membranes', *Proc. natn. Acad. Sci. U.S.A.*, **55**, 1048 (1966)

80. DEAMER, D. W., and BRANTON, D., 'Fracture planes in an ice-bilayer model membrane system', *Science*, **158**, 655 (1967)

81. NAYLER, W. G., 'The effect of pronethalol and propranolol on the facilitated transport of Ca ions', *J. Pharmac. exp. Ther.*, **153**, 479 (1966)

82. COLLANDER, R., and BARLAND, H., 'Permeabilitätsstudien an *Chara Ceratophylla.*, *Acta bot. fenn.*, **11**, 1 (1933)

83. SHAH, D. O., and SCHULMAN, J. H., 'Binding of metal ions to monolayers of lecithins, plasmalogen, cardiolipin and dicetyl phosphate, *J. Lipid Res.*, **6**, 341 (1965)

84. TING, H. P., BERTRAND, D. L., and SEARS, D. F., 'Diffusion of salts across a butanol–water interface', *Biophys. J.*, **6**, 813 (1966)

85. JOHNSON, S. M., and BANGHAM, A. D., 'Potassium permeability of single compartment liposomes with and without valinomycin', *Biochim. biophys. Acta*, **193**, 82 (1969)

86. CIANI, S., EISENMAN, G., and SZABO, G., 'A theory for the effects of neutral carriers such as the macrotetralide actin antibiotics on the electric properties of bilayer membranes', *J. Membrane Biol.*, **1**, 1 (1969)

87. EISENMAN, G., CIANI, S., and SZABO, G., 'The effects of the macrotetralide actin antibiotics on the equilibrium extraction of alkali metal salts into organic solvents', *J. Membrane Biol.*, **1**, 294 (1969)

88. SZABO, G., EISENMAN, G., and CIANI, S., 'The effects of the macrotetralide actin antibiotics on the electrical properties of phospholipid bilayer membranes, *J. Membrane Biol.*, **1**, 346 (1969)

89. ADAM, N. K., *The physics and chemistry of surfaces*, Oxford University Press, Oxford (1941)

90. DAVIES, J. T., and RIDEAL, E. K., *Interfacial Phenomena*, Academic Press, Loncon (1961)

91. DEMEL, R. A., VAN DEENEN, L. L. M., and PETHICA, B. A., 'Monolayer interactions of phospholipids and cholesterol', *Biochim. biophys. Acta*, **135**, 11 (1967)

92. VAN DEENEN, L. L. M., HOUTSMULLER, V. M. T., DE HAAS, G. H., and MULDER, E., 'Monomolecular layers of synthetic phosphatides', *J. Pharm. Pharmac.*, **14**, 429 (1962)

93. HENDERSON, P. J. F., MCGIVAN, J. D., and CHAPPELL, J. B., 'The action of certain antibiotics on mitochondrial erythrocyte and artificial phospholipid membranes. The role of induced proton permeability', *Biochem. J.*, **111**, 521 (1969)

94. HAUSER, H., and DAWSON, R. M. C., 'The displacement of calcium ions from phospholipid monolayers by pharmacologically active and other organic bases', *Biochem. J.*, **109**, 909 (1968)

95. KAFKA, M. S., and PAK, CH. Y. C., 'The effect of polypeptide hormones on lipid monolayers. II. The effect on insulin analogues, vasopressin, oxytocin, thyrocalcitonin, adrenocorticotropin, and $3'5'$-cyclic AMP on the uptake of Ca^{2+} by monomolecular films of monooctadecyl phosphate', *J. gen. Physiol.*, **54**, 134 (1969); *Biochim. biophys. Acta*, **193**, 117 (1969)

96. KINSKY, S. C., LUSE, S. A., and VAN DEENEN, L. L. M., 'Interaction of polyene antibiotics with natural and artificial membrane systems', *Fed. Proc. Fedn. Am. Socs exp. Biol.*, **25**, 1503 (1966)

C

97. CAMEJO, G., and VILLEGAS, R., 'Tetrodotoxin interaction with cholesterol', *Biochim. biophys. Acta.*, **173**, 351 (1969)
98. BANGHAM, A. D., STANDISH, M. M., and WATKINS, J. C., 'Diffusion of univalent ions across the lamellae of swollen phospholipids', *J. Molec. Biol.*, **13**, 238 (1965)
99. JOHNSON, S. M., and BANGHAM, A. D., 'The action of anesthetics on phospholipid membranes', *Biochim. biophys. Acta*, **193**, 92 (1969)
100. BANGHAM, A. D., and PAPAHADJOPOULOS, D., 'Biophysical properties of phospholipids. I. Interaction of phosphatidylserine monolayers with metal ions'. *Biochim. Biophys. Acta*, **126**, 181 (1966)
101. DE GIER, J., MANDERSLOOT, J. G., and VAN DEENEN, L. L. M., 'Lipid composition and permeability of liposomes', *Biochim. biophys. Acta*, **150**, 666 (1968)
102. SESSA, G., and WEISSMAN, G., 'Effect of polyene antibiotics on phospholipid spherules containing varying amounts of charged components', *Biochim. biophys. Acta*, **135**, 416 (1967)
103. LYNN, J., and FASMAN, G. D., 'A fluorescent probe, toluidinylnaphthalene-sulfonate, specific for the beta structure of poly-L-lysine', *Biochem. biophys. Res. Commun.*, **33**, 327 (1968)
104. DODD, G. H., and RADDA, G. K., 'Interaction of glutamate dehydrogenase with fluorescent dyes', *Biochem. biophys. Res. Commun.*, **27**, 500 (1967)
105. CHANCE, B., LEE, C. P., and YONETANI, T., in *Probes of Structure and Function of Macromolecules and Membranes*, Academic, New York (1971)
106. CALVIN, M., WANG, H. H., ENTINE, G., GILL, D., FERRUTI, P., HARPOLD, M. A., and KLEIN, M. P., 'Biradical spin labeling for nerve membranes', *Proc. natn. Acad. Sci. U.S.A.*, **63**, 1 (1969)
107. CHAPMAN, D., BARRATT, M. D., and KAMAT, V. B., 'A spin label study of erythrocyte membranes', *Biochim. biophys. Acta*, **173**, 154 (1969)

FACTORS DETERMINING PERMEATION RATES; ENERGETICS OF PERMEATION

E. J. HARRIS

Department of Biophysics, University College, London

INTRODUCTION

Cells and their internal organelles can be seen to be bounded by membranes. It is the differential properties of the membranes that are invoked to account for the discontinuities of composition encountered when the extracellular fluid, the cytosol, and the contents of the organelles, respectively, are examined. Most cell saps have K as the major cation; the extracellular fluid is usually Na-rich and K-poor, though for plant cells the external fluid often has a low salt concentration. Mitochondria have a high K content, like cell sap, but they include specific enzymes, and both adenine and pyridine nucleotides that do not freely pass into the cytosol.

Interest in the permeability of the cell membrane to specific substances arose because of theories seeking to account for the electrical responses of excitable tissues to stimulation[1]. Now attention encompasses the whole spectrum of metabolites, as well as the ions, because membrane permeability plays an important role in metabolic control. Most recent ideas about membrane structure invoke a complex of phospholipid and protein; the components carry fixed charges of both signs. Spatial changes can occur both consequent upon cross-linking and due to allosteric effects of small molecules attaching to the protein. There may be mobile, dissolved, carriers and there may be a dynamic state between a liquid surface phase and a solid one. The inference is that most attempts to fit permeability measurements have been referred to too simple a model. That the observations usually require a series of parameters, rather than a simple membrane resistance, to describe them is not surprising.

ELECTROCHEMICAL ACTIVITY

The rate of a non-energised redistribution process depends on the gradient of electrochemical activity and on the diffusivity of the moving particles. For generality we shall have to consider the effects of concentration (c), the Debye–Hückel chemical activity coefficient (f), the electrical potential (E), and the hydrostatic pressure (P). One part of the system is chosen as reference and the electrical potentials and hydrostatic pressures are measured with respect to it. The factors can be combined into a quantity called the electrochemical activity (denoted here by $\mathscr{A}$), given by

$$\mathscr{A} = cf \exp{(P\bar{v}/RT)} \exp{(zEF/RT)} \qquad (2.1)$$

where $\bar{v}$ is the partial molar volume and zF is the charge per g ion. For a molecule, z is zero and the electrical term disappears.

The electrochemical activity can be used to evaluate another quantity, the electrochemical potential (denoted μ) where

$$\mu = RT \ln{\mathscr{A}} + \mu_0 \qquad (2.2)$$

The quantity μ_0 is the measure of the amount of energy that can be extracted from 1 mol of the substance by allowing it to run down to infinite dilution at our chosen zeros of electrical and hydrostatic pressure; it is a constant depending on the substance and the reference conditions, and disappears when the expression is differentiated to obtain the gradient of μ.

The basic equation relating the rate of diffusion in the x-direction to μ or $\mathscr{A}$ is[2, 3]

$$\mathrm{d}n/\mathrm{d}t = (-Dc/RT)(\mathrm{d}\mu/\mathrm{d}x) = -Dc \, \mathrm{d} \ln{\mathscr{A}}/\mathrm{d}x \qquad (2.3)$$

where D is diffusivity. The number of moles (n) crossing a boundary normal to the x-direction in unit time gives $\mathrm{d}n/\mathrm{d}t$. Evidently, if this number is zero, there is no net movement and the gradient of μ is also zero. This corresponds to constant μ and $\mathscr{A}$ along the x-direction. For no net diffusive movement of a substance between two regions its electrochemical activity must be the same in each. It is to be noticed that f, $\mathscr{A}$, and μ vary with temperature. In a system containing temperature gradients, thermal diffusion takes place; this may be of particular importance in plants.

DIFFUSION

The spontaneous mixing in a system that initially included gradients of electrochemical potential occurs by diffusion. It may be exemplified by the familiar Brownian movement seen under the microscope when small particles are observed. In a bulk of material, be it solution or membrane material, diffusion tends to reduce the gradient with time. For this reason, the extent of the movement is related not to the time directly but rather to the square root of the time. Only when the gradient is maintained, as, for example, when a membrane separates two well-stirred reservoirs of infinite size, does the transfer become linearly time dependent.

KINETICS OF DIFFUSION

Trans-membrane diffusion follows equation (2.3) in its various integral forms. When we are more concerned with the rate of change of content of material in a certain volume than with the rate of passage across a boundary we can, with certain limitations, use another equation. If the concentration need only be considered as the variable, the diffusion equation can be written $\partial c/\partial t = D\nabla^2 c$, where ∇^2 represents the second differential of c with respect to a length and takes various forms according to the coordinate system chosen. The solutions of the equation for a number of examples involving simple geometrical shapes are given by Jacobs[4]. Combinations of diffusion and chemical reaction have been considered by Hill[5] and Roughton[6]. A useful reference work on the subject is *The Mathematics of Diffusion*[7].

The formulae relating the equilibration of three shapes to time and the quotient D/r^2 (written as 'B' for compactness) are given below and have been plotted in Figure 2.1.

For the sheet, exposed both sides and thickness $2r$,

$$\text{fractional equilibration} = \frac{8}{\pi^2} \sum_{0}^{n=\infty} \frac{\exp[-(2n+1)^2\pi^2 Bt/4]}{(2n+1)^2}$$

For the cylinder, radius r
fractional equilibration =

$$4\left[\frac{\exp(-v_1^2 Bt)}{v_1^2} + \frac{\exp(-v_2^2 Bt)}{v_2^2} + \ldots\right]$$

Here numbers v_1, v_2, etc., are the zeros of the Bessel function $J(0)$.
For the sphere, radius r,

$$\text{fractional equilibration} = \frac{6}{\pi^2} \sum_{0}^{n=\infty} \frac{\exp(-n^2\pi^2 Bt)}{n^2}$$

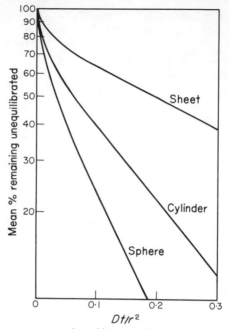

Figure 2.1. The time course of equilibration of some simple geometrical shapes by diffusion. The mean value of the fraction that has still to equilibrate is given against the parameter Dt/r^2. D = *diffusion constant,* r = *radius or half thickness*

DIFFUSION IN PRESENCE OF HYDROSTATIC PRESSURE

This problem is an example of the pressure-dependent electro-chemical activity gradient determining the diffusion; for this to hold, the size of the channels connecting the two compartments between which redistribution occurs must be small enough to preclude bulk flow that is dependent only on the hydrostatic pressure difference. Diffusion depends on the net pressure (p) given by hydrostatic pressure minus osmotic pressure. Using equation (2.6) on p. 40 to find the activity gradient, we have:

$$\frac{d \ln \mathscr{A}}{dx} = \frac{\bar{v}}{RT}\frac{dp}{dx}$$

and, putting in equation (2.3),

$$\frac{dn}{dt} = -A'\frac{D\bar{v}}{RT}\,c\,\frac{dp}{dx} = -\frac{A'}{RT}\frac{Dn}{N}\frac{dp}{dx}$$

where A' is the diffusion area and n is the number of solute particles

38

dissolved in N molecules of water. To proceed, one has to have information on the dependence of n/N and A' on the distance x.

OSMOSIS

Movement of water by diffusion caused by a gradient of electrochemical activity set up by a concentration difference is called osmosis. If the gradient is set up as the result of an electrical field acting on a net charge borne by the water, the movement is known as electro-osmosis.

Osmosis represents the movement of water towards an equality of electrochemical activity; therefore, it depends on the water concentration and activity factor, not on the nature of solute present. Application of hydrostatic pressure increases the water activity: to see how this comes about, we shall use the vapour pressure as a measure of the activity.

If we have two compartments containing solution and one is under hydrostatic pressure P, we can imagine two alternative ways water could be moved from the low-pressure compartment to the other; (1) by evaporating water, compressing the vapour to pressure P and then condensing it, or (2) by forcing the water through a semi-permeable membrane into the high-pressure compartment by applying a pressure just exceeding P. If these processes are carried out in a reversible way, they must both require the same energy, for otherwise one could be run in reverse and a surplus of energy drawn indefinitely from the system. Hence, one can equate the respective works for (say) 1 mol of water transferred. The evaporation process requires heat, but this is recovered from the condensation; the only work (W_d) required is that to compress a mole of water vapour from pressure p_1 (its vapour pressure at zero applied pressure) to pressure p_2 (its vapour pressure when under hydrostatic pressure P),

$$W_d = \int_{p_1}^{p_2} p \, dv$$

and from the gas laws $dv = -RT \, dp/p$, so

$$W_d = RT \ln p_2/p_1 \text{ per mol}$$

To force a mole of water occupying volume $\bar{v}$ through the membrane reversibly, a series of small increments of pressure δp are applied to build up to a final value just exceeding P. Since

$$\text{work done} = \text{force} \times \text{distance}$$
$$= \text{force/area} \times \text{volume moved}$$
$$= \text{excess pressure} \times \text{volume moved}$$
$$= \Sigma \bar{v} \delta P$$

in the limit where W_c is the work of compression

$$W_c = \int_0^P \bar{v} \, dP$$

Strictly, $\bar{v}$ has to be defined as $\partial V/\partial N$, where V is the volume of solution and N is the molar fraction of water; this is the partial molar volume of water that varies with P. However, the variation is slight, about 0·5% per 1 M concn., and may usually be ignored, so that

$$W_c = P\bar{v} \text{ per mol}$$

Equating W_c and W_d, one obtains

$$P\bar{v} = RT \ln p_2/p_1$$

or
$$p_2 = p_1 \exp P\bar{v}/RT \tag{2.4}$$

The factor by which the hydrostatic pressure P has raised the vapour pressure and the water activity is $\exp P\bar{v}/RT$[8, 9]. Given this result we can use it to find the relation between water activity and salt concentration in a solution.

Suppose an osmometer tube containing a solution, bounded by a semipermeable membrane at its lower end, is immersed in pure water. After equilibration, the height of the column is π cm, where π is the osmotic pressure of the solution. The solution at the base of the column is under hydrostatic pressure π cm. Since the system is then stable, the water activity on both sides of the membrane must be equal. If on the side with pure water we call the activity unity then on the solution side it is $c_w f_w = \exp \pi\bar{v}/RT$ from the general expression for activity and using the previous result. For this product to equal unity we have:

$$c_w f_w = \exp - \pi\bar{v}/RT \tag{2.5}$$

The solute has reduced the water activity by the factor $\exp - \pi\bar{v}/RT$, where π is the 'osmotic pressure' of the solution. The combined effect of a solute whose osmotic pressure is π and application of hydrostatic pressure P can be expressed in the simple form:

$$\text{water activity } \mathscr{A} = \exp (P - \pi) \bar{v}/RT \tag{2.6}$$

This relation leads to what seems the most meaningful definition of the osmotic pressure: it is equal to the hydrostatic pressure necessary to bring the activity of the water in the solution up to unity. The operation of equation (2.6) is exemplified by those plant cells whose tough walls permit a high internal hydrostatic pressure. These cells (e.g., *Nitella*) have at 100 mM or more salt solution in their vacuoles,

40

yet they grow in pond water having less than 0.5 mM salts. The internal pressure brings the water activity in the interior up to that in the outside solutions.

OSMOTIC PRESSURE AND VAPOUR PRESSURE

To relate the osmotic effect to the reduction of vapour pressure, imagine an osmometer tube has a small bubble trapped in it at the level of the membrane. In this case, the movement between solution and water must occur by isothermal distillation. The fact that equilibrium can be attained shows again that equality of water activity, this time measured by vapour pressure, subsists at the level of the membrane. Raoult's relation equates the fractional reduction of the vapour pressure $\Delta p/p$ and the ratio: number of dissolved particles/number solute particles $= n/N$. At the top of the osmotic column, the vapour pressure is less than that of pure water by Δp, so if in equation (2.4) $p_1 = p - \Delta p$ and $p_2 = p$ then

$$\pi \bar{v} = RT \ln p/(p - \Delta p)$$

Berkeley, Hartley, and Burton[10] experimentally showed the validity of the relation between vapour pressure lowering and the 'osmotic pressure' exerted by sucrose. When p is small, the logarithm approximates to $\Delta p/p$ and so

$$\pi \bar{v} = RT\Delta p/p = RTn/N$$

so that $$\pi = n\,RT/N\bar{v}$$

$n/N\bar{v}$ is the solute concentration in N partial molar volumes of solvent. It becomes sufficiently accurate in dilute aqueous solution to set $\bar{v} = 18$ ml mol^{-1} or to put $N\bar{v}/n =$ volume (V_0) of solvent associated with 1 mol solute. Then the equation becomes $\pi V_0 = RT$.

In the literature, concentrations are expressed in several different units. The molar concentration (m) is the number of moles of solute per kg water (1 kg water $= 55.5$ mol) so $V_0 = 1000/m$ and $\pi = mRT/1000$.

The commonly used molar concentration is less convenient, because it expresses the moles of solute per litre of solution. To convert to molality or molar fraction, it is necessary to know the density (d) of the solution. Then, if $1000d$ g solution contains W g solute of molecular weight M, the molar concentration is W/M and the molal concentration is $1000W/(1000d - W)M$. The difference between the two units becomes considerable at molar levels of solutes of high molecular weight, for example, 1.27 molar sucrose is

41

1 molar. The difference, however, is slight when the concentration is of the order of $10 \mathrm{~g~l}^{-1}$ and is for this reason seldom taken into account in biological solutions.

ELECTRO-OSMOSIS AND RELATED EFFECTS

A solution of an electrolyte in motion through a pore or tube bearing fixed charges on its walls tends to carry ions of sign opposite to those on the walls because the mobile ions of like sign require more energy to bring them into the pore. This means an electrical double layer is set up at the entrance that tends to oppose motion of the solution. The total potential between the two compartments is called the streaming potential and depends on the flow rate. Analogously, the passage of an electric current through a solution distributed in a fixed-charge material causes the movement of the solution because of a slight excess of charge in the fluid due to the dipole layer formed at the fixed-charge surface. This electro-osmosis can reach large values in polymers having a high fixed-charge density. The forces that occur between the solution and the fixed charges tend to move the membrane itself in a direction opposite to the medium.

Such sets of interactions can be formulated in a consistent and logical way according to the system exemplified for three components:

$$\phi_1 = L_{1,1}F_1 + L_{1,2}F_2 + L_{1,3}F_3$$
$$\phi_2 = L_{1,2}F_1 + L_{2,2}F_2 + L_{2,3}F_3 \qquad (2.7)$$
$$\phi_3 = L_{1,3}F_1 + L_{2,3}F_3 + L_{3,3}F_3$$

where the L's are proportionality factors between forces and fluxes and the ϕ's are the fluxes of the three components.

FLOW CAUSED BY APPLICATION OF PRESSURE

So far, account has been taken of diffusion, but a solution can be forced without change of composition from one compartment to another by the application of hydrostatic pressure when the connecting holes are of large enough size compared with the molecular sizes. In biological systems, it may be difficult to decide whether this criterion is met; if it is not one has a combination of flow and diffusion.

When bulk movement of a fluid takes place, the rate of movement is limited by viscosity. When stream line flow in a cylindrical tube occurs, the flow rate dQ/dt is given by the Poiseuille equation $(\pi r^4/8\eta)(dP/dx)$ where r is the radius, η the viscosity, and dP/dx the hydrostatic pressure gradient; Q is the volume.

When the size of the holes or channels through which the liquid is being forced becomes comparable with the dimensions of solute particles, these tend to accumulate in the neighbourhood of the entrance side of the porous barrier, and the local concentration of solute is increased. A steady state can be set up with a concentration gradient of the large solute particles diminishing against the direction of flow. The rate of passage of the filtered fluid through the barrier is ultimately determined by the diffusivity of the particles that cannot pass the filter (a useful discussion of this will be found in Kuhn[11]). Pappenheimer[12] has discussed the limiting pore size in relation to filtration in the kidney.

WATER MOVEMENT THROUGH THE WALLS OF THE CAPILLARIES

An important problem in physiology is that of the movement of water between the blood plasma and the lymph. Starling[13] originally proposed that the movement of water was determined by the combined effect of the hydrostatic pressure existing between plasma and lymph space and the osmotic pressure difference. That is, the water *activity* difference between the two sides of the wall determines the rate of movement and not the hydrostatic pressure as such. The difference in the osmolar concentration on the two sides arises from the fact that normally the capillary wall is nearly impermeable to the plasma proteins. Then, although the salt concentrations can balance on the two sides, the plasma has an extra concentration of colloids that reduces the water activity.

In the frog, osmosis to the plasma from the lymph is stopped by application of about 11·5 cm water pressure to the blood. This pressure is the 'colloid osmotic pressure' of the plasma, not to be confused with a hydrostatic pressure for it is merely one way of expressing the reduction of water activity due to dissolved colloid. A hydrostatic pressure exceeding 11·5 cm water applied to the plasma in a frog will outweigh the osmotic effect and water will be ultra-filtered from plasma to lymph. This takes place at the arterial end of the capillaries. Where the hydrostatic pressure acting on the plasma is less than the colloid osmotic pressure, osmosis of water back into the plasma occurs. This occurs at the venous ends of the capillaries. The pressure equivalent to the osmotic effect of dissolved protein in the blood varies from species to species. In man the value is about 36 cm water.

Experimental demonstration that this is the correct explanation of the movement of water between plasma and the lymph was provided by Landis[14], who used blood cells trapped in a capillary as indicators of fluid movement. A small blood vessel was closed at one point by mechanical pressure, trapping a volume of plasma between a blood cell and the closure. A micropipette inserted into the vessel communicated with a pressure gauge. When the hydrostatic pressure acting on the fluid was equal to the colloid osmotic pressure difference there was no movement of the blood cell. The rate of passage of the water when the hydrostatic pressure did not balance the activity difference set up by unequal concentrations of solute was $0.03\ \mu m^3\ \mu m^{-2}$ per min at 5 cm water-pressure difference across the frog capillary wall. The rate of passage of dyes through the capillary wall was unaffected by the diameter of the vessel, and by its being dilated.

Landis[5] found the permeability was increased during anoxia by a factor of 4. The walls become permeable to the smaller of the blood colloids, so that about half the total colloid passes through, and

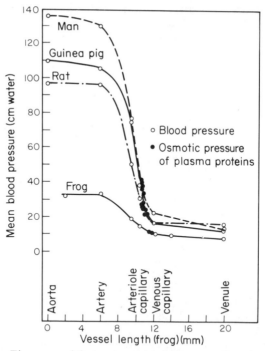

Figure 2.2. The pressure (O) at various points between arteries and venules compared with the pressure equivalent of the plasma colloid (●) for four species (After Landis[15]; courtesy Am. J. Physiol.)

44

this reduces the colloid osmotic pressure difference across the capillary wall, so that the hydrostatic pressure now causes water to move into the lymph.

In a view of the subject of water movement, Landis[16] collected figures for the pressure differences across the walls of arterioles and venules, together with the pressure equivalent of the colloid in the blood (see Figure 2.2). The pressure equivalent of the colloid is always in excess of the venous capillary hydrostatic pressure, as must be the case if water is to diffuse back into the plasma.

Pappenheimer and Soto-Rivera[17] perfused intact limbs with fluids containing different proportions of dissolved colloid. When the pressure equivalent of the colloid was very slightly above the hydrostatic pressure applied to fluid in the arterial part of the system, there was no change of weight, that is, no water moved between the vessels and the tissue. The slight excess of colloid needed was due to the capillary wall being slightly permeable to the colloid. The capillary wall is estimated to be 3000 times as permeable as cell walls. Reviews of capillary wall permeability have been made by Landis[18] and Pappenheimer[12]. The latter author has tried to apply corrections for the effect of unfavourable collisions on rate of passage through pores of size little more than that of the colliding particles (the diameters are 2–4 nm) and has also examined the case of slit-like pores[19]. Limitations to our knowledge of hydrodynamics make estimation of pore sizes from rates of movement subject to considerable uncertainty. In particular, the lower limit of diameter to which the Poiseuille law can be applied has to be considered. Pappenheimer sets this at 2 nm.

The wall of the capillary is built up of endothelial cells that lie side by side. It seems that Ca atoms provide the cross-links in the intercellular cement between the cells. Lack of calcium makes the wall much more permeable to proteins and other colloidal particles[20]. This was demonstrated by Chambers and Zweifach[21], who injected colloidal carbon into the aorta of a frog and then examined the capillary walls microscopically. With normal calcium, carbon was deposited so that it outlined the endothelial cells. In the absence of calcium, deposition was absent, since the vessels had allowed the carbon to escape into the interstitial space.

WATER MOVEMENT THROUGH THE GASTRIC MUCOSA

The gastric mucosa secretes water from the serosal to the mucosal side; that is, a net movement of water even between equi-concentrated solutions takes place.

It has been shown by Durbin, Frank, and Solomon[22] that the net water movement can be influenced by the water activity gradient across the membrane; in this respect the system behaves as if there were a metabolically potentiated movement amounting to 11 µl cm^{-2} h^{-1}, and an osmotic component of 75 µl cm^{-2} h^{-1} for a water concentration difference of 1 mol l^{-1} at 25 °C. The osmotic component could be reversed by applying a concentration difference in the reverse sense and the metabolic component could be abolished by adding 12 mmol of NaCNS to the serosal side fluid. A possible mechanism of water movement through such an organised layer of cells will be outlined below. The osmosis of water through the mucosa corresponds to its having a water permeability of 11·6 µm s^{-1}.

WATER AND ION MOVEMENTS

The energised ion movements described elsewhere are usually accompanied by movement of water; this is often in the form of a

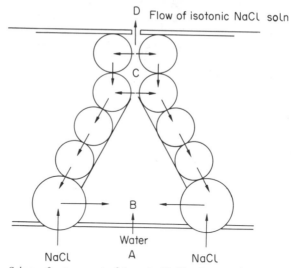

Figure 2.3. Scheme for transport of isotonic NaCl solution. There is secretion of NaCl from the lumen (A) through the adjacent cells into the space B where the NaCl concentration tends to rise. Water then diffuses from A to B down its activity gradient. A hydrostatic pressure tends to develop in B, which in turn leads to flow of the hypertonic solution towards the outside D. In its path at C the lining cells remove part of the NaCl and pass it back towards the cells on the luminal side. The resulting secreted fluid could then be isotonic and a surplus of NaCl cycles round between the cells. This proposal is elaborated from Diamond's scheme[23], since it includes the resorption of part of the NaCl at C

solution of the salt isosmotic with the fluid round the membrane (see Diamond[23] for references). Such water movement is in a sense energy-linked, because the salt secretion requires energy, but there is no suggestion that energy has to be fed to the water molecules to bring about the movement. Water movement is regarded as secondary to the energised ion secretion.

The mechanism of water transport has been investigated using the gall bladder[23]. Over an eightfold range of osmolarity, which could be adjusted by alteration of salt or sugar concentrations, the secreted fluid was isotonic, with the fluid bathing the luminal side. This disposes of several alternative postulated mechanisms involving either two membranes in series, or passive movement of water with active NaCl secretion according to Diamond. The most attractive mechanism (see Figure 2.3) involves actively transported NaCl raising the salt content in a space near the luminal side into which water diffuses from the luminal side down its activity gradient. The movement of water builds up the local pressure in the space towards that required to equalise the water activities. The raised pressure causes flow towards the serosal side of the hypertonic solution. This, however, can be depleted of salt by the cells in the path, which may return the excess NaCl to the other cells nearer the lumen. In this way, an active NaCl movement including a recycling of the excess NaCl could give rise to isotonic secretion.

WATER MOVEMENT CAUSED BY A CONTRACTILE VACUOLE

A number of unicellular organisms, including various rhizopoda and ciliates, possess the property of expelling a quantity of fluid from the cell interior at intervals. Kitching[24] has listed some species together with the frequency of expulsion and the volume of fluid expelled. The phenomenon consists in the development of a vacuole within the cytoplasm, which gradually increases in size, the phase of diastole; then at some critical size the vacuole discharges its contents to the external surroundings through a temporarily formed pore, the phase of systole.

Vacuole formation appears to be a mechanism for keeping the cell at constant volume despite the osmotic entry of water. In confirmation of this idea, Kitching found that the output depended upon the osmotic activity of the medium (Figure 2.4). When more salts were present in the solution, osmotic entry of water would be less, and, in fact, the water output was also reduced.

The control of the body volume is by no means perfect (Figure 2.5). However, if the vacuole formation is stopped by the addition of cyanide, the body volume varies much more. If, after cyanide, the swollen cell is returned to a non-poisonous solution, pumping starts again and occurs more rapidly and with greater output than usual until the normal volume is restored. Swelling in cyanide can be stopped by making the outside solution 0·05 M with sucrose, which has led to the conclusion that the internal osmolarity may be of this order.

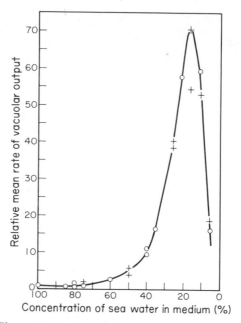

Figure 2.4.   The variation in output from a contractile vacuole as the tonicity is varied (After Kitching[25]; *courtesy* J. exp. Biol.)

The fluid that collects in the vacuole presumably contains a proportion of the waste products of the cell's metabolism. The reason for the accumulation of fluid in the vacuole is not known. It has been attributed to osmotic attraction of water by a local concentration of a soluble substance, to filtration induced by pressure, and to the operation of a secretory mechanism. Kitching favours the last proposal. The force inducing emptying is thought to be set up in the wall of the vacuole. The fact that some cells will discharge their vacuoles even when they are in a shrunken condition (Kitching[26]) indicates that it is not cytoplasmic pressure.

48

The idea that water is secreted into the vacuoles receives support from the fact that the increase of vacuolar volume is linear with respect to time up to the discharge point. Assuming an osmotic force is operative, and allowing for increasing area, Kitching showed that the rate would diminish, unless, of course, more osmotically active material were continually being fed into the vacuole.

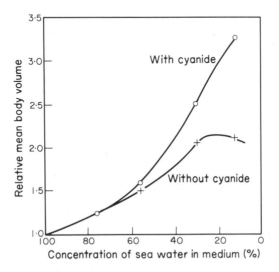

Figure 2.5. *The volume of* Cothurnia curvula *as function of the tonicity of the medium. The effect of cyanide poisoning is to make the volume more dependent upon tonicity* (*After Kitching*[25]; *courtesy* J. exp. Biol.)

Some organisms operate several vacuoles, and the fact that these do not operate synchronously has been used as an argument against vacuolar discharge being controlled solely by cytoplasmic pressure. It has been pointed out that some other factor, such as local weakening of the cell wall in readiness for pore formation, is more likely to control discharge.

Here it should be mentioned that in some cells the vacuole or vacuoles form at fixed points, whereas in other cells the position moves about, and it has been suggested that the pressure of the vacuolar membrane against the inner surface of the cell wall induces a weakening of the latter.

The influence of external factors upon vacuolar frequency and output has been investigated by Kitching[27, 26]. The frequency and output increase with temperature, but in the first 20 min after a change the frequency first rises, then falls, and subsequently rises again, suggesting that the behaviour is controlled by some chemical balance that has to be re-established following the disturbance. Application of pressure up to 2×10^7 N m^{-2} increased the frequency, but decreased the output because the maximum size attained was reduced.

To give an idea of the magnitude of the water pumping, the amoebae expel 20–250 μm^3 s^{-1}, or 0·036–0·090 μm^3 per μm^2 of surface per min and ciliates 0·36–0·7 μm^3 μm^{-2} min^{-1}. Estimates of the permeability of the surface of ciliates to water are of the order of 0·12–0·25 μm^3 μm^{-2} min^{-1} per atmosphere osmotic pressure difference, so that the pumping from the latter can cope with a water entry from a solution having as much as 1 atm osmotic pressure.

KINETICS

ANOMALIES PRODUCED BY MULTICOMPARTMENTS IN TRACER METHODS

An adsorptive membrane is, in effect, a series of compartments, and when labelled substances are used with the object of finding one-way fluxes in the presence of unlabelled substances going the other way the results can be misleading. In the ideal case without adsorption, the flux from side 1 to side 2 can be equated to the electrochemical activity on side 1 multiplied by a permeability, and the flux from 2 to 1 is equated to the electrochemical activity on side 2 multiplied by the same permeability. This leads to a test for the presence or absence of energised transport[28]. The test is made as follows: the electrochemical activities of an ion are equalised on the two sides of a barrier by making the electrical p.d. zero through feeding current in, and using equal chemical concentrations. In this condition, the fluxes in the two senses should be equal; if they are not, some metabolically energised process must be feeding energy to operate the higher flux.

When, as an example of a multicompartment system, we take a system having three compartments in series within each of which there is good mixing, and ask how the specific labelling of a substance is in the compartments, we find that it depends on the flux ratio. For equal fluxes from 1 to 2 and 2 to 1, the dilution of the labelled material occurs in steps (Figure 2.6a) to give specific activities 0·75, 0·5, and 0·25 in the compartments. The observed tracer flux is only 0·25 times the true flux, and the 'non-tracer' flux in the opposite sense is also 0·25 for the true value, but at least the values found are equal and will not mislead us into supposing that there is an asym-

50

metry. We only infer too low a permeability. When the two fluxes are unequal, the consequence of the multicompartment structure is to magnify the discrepancy between the observed fluxes and the true fluxes. Taking, for example, the 1-to-2 flux at 10 units (Figure 2.6b), and the 2-to-1 flux at 1 unit, the specific activities are, respectively, 0·999, 0·990, and 0·900, the tracer flux into 2 is 9 units not 10, and the non-tracer flux into 1 is only 0·001 instead of 1. Although the true fluxes stand in the ratio of the electrochemical activities,

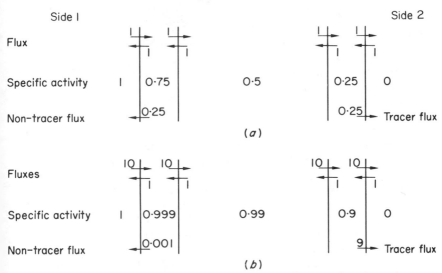

Figure 2.6. The effect of intermediate compartments on the tracer flux observed in two conditions of a two-way flux system. (a) In this condition, the true fluxes are represented as unity in each direction. All the material in the left-hand compartment is supposed to be marked (specific activity = 1) and it is kept so by a flow; similarly all the material in the extreme right-hand compartment is unmarked (specific activity = 0) and is kept so. The specific activities in the three intermediate compartments are calculated from a set of simultaneous equations and are found to be 0·75, 0·5, and 0·25. The non-tracer flux observed into the left-hand compartment is 0·25 units and the tracer flux into the right-hand compartment is 0·25. (b) In this condition, the true flux from right to left is 10 units and that from left to right is 1 unit. The specific activities in the internal compartments are 0·999, 0·990, and 0·900. The non-tracer flux is 0·001 into the left-hand compartment, the tracer flux is 9 into the right-hand compartment

namely, 10 and 1 in the case of an ideal membrane, no matter how many layers it is made of, the tracer flux from 1 to 2 and non-tracer flux from 2 to 1 stand in the ratio of 9000:1. Between the case of the electrochemical activity on one side being 1, as in our first example, and that of it being 10, as in the second, the flux of labelled material into compartment 2 rises by $9 \div 0\cdot25 = 36$ times; the observed back-flux into 2 from 1 falls from $0\cdot25$ to $0\cdot001$, that is by 250 times. With diffusion occurring along adsorption sites, the effective number of compartments is high and the anomalies greater. One may infer that observations of the movements of labelled substance or of a chemical analogue in a multicompartment system lead to grossly exaggerated measure of the changes in flux. Only when intermediate dilutions are absent can unambiguous results be obtained for the one-way flux.

ADSORPTION ON MEMBRANE MATERIAL AS A CONTROL OF KINETICS

Most membranes have a certain capacity to take up those substances that move through them. This fact in itself means there is an intermediate compartment, the membrane. Although the capacity may be low, it can give rise to anomalies and interactions. It is a curious historical fact that while the permeations of sugars and amino acids have for years been interpreted in terms of the degree of saturation of putative membrane sites, the penetration of ions has conventionally been discussed as if the membrane had no capacity for them. The passage of material from a system of sites whose equilibration with the medium builds up logarithmically has been discussed[29]; it resembles a diffusion process rather than resistance-limited permeation. Once the sites are equilibrated with the external medium, the rate depends on their adsorption properties.

Where the flux in one direction is dependent on the adsorbed quantity the relation becomes like the classical enzyme kinetic equation[30], or the Langmuir isotherm. If the saturation flux is ϕ_{max}, C_s is the concentration of substrate applied, C_i the concentration of an inhibitor, K_m the concentration of substance that provides half maximum flux, and K_i the concentration of each respective inhibitor that reduces the flux to half, then

$$\phi = \frac{\phi_{max} C_s/K_s}{1 + C_s/H_s + \sum_n C_i/K_i} \tag{2.8}$$

When the substance is present on both sides of the membrane, the net flux is supposed to be equal to the difference between two separate contrary fluxes

$$\phi_{net} = \phi_{1 \to 2} \sim \phi_{2 \to 1} \tag{2.9}$$

It has often been assumed that the substance traverses the membrane in combination with a 'carrier', and that the loaded and unloaded forms have equal mobilities (e.g. reference [31, 32]). According to the relative values of the parameters in the equation, different time dependences for equilibration of cells with a medium are obtained (e.g. reference 33). With a high K_m, there is no saturation in physiological range and the process is like the diffusion-limited case. When the K_m is low, the rate soon saturates and since all the sites or carriers are occupied the net movement may be quite slow because unoccupied carriers seldom arise and cycling of occupied carriers cannot provide net movement.

Discussion of the mechanism of active or energised transport usually includes an assumption that there exists a leak path linearly dependent upon the concentration difference. This assumption implies complete irreversibility of both components making up the transfer. Insufficient attention has been paid to 'either–or' possibilities where metabolic energy changes the K_m of the carrier for the actively moved substance and in so doing alters the amount of carrier available to move in the leakage path.

When a substance enters a cell or organelle and is converted to another substance that competes for use of the carrier, the overall process becomes one of an exchange of the entering material for its metabolic product; an example may be succinate entering mitochondria and malate (formed via fumarate) leaving. If a membrane carrier is a weak acid, it is reasonable to suppose that metabolically generated protons can displace, say, K ions entering with the carrier anion to restore the carrier to its protonated form. The latter then diffuses out and can take up more K from the exterior. If the protonated form had the higher mobility, the interior would tend to become positive and tend to accumulate anions.

CARRIER MOBILITIES IN THE OCCUPIED AND UNOCCUPIED CONDITIONS

The kinetics for cases in which the two forms of the carrier have unequal mobilities have been considered by several groups[34–38]. When the occupied carriers are the less mobile form, they tend to accumulate at the high-substrate side of the membrane, so retention of the gradient is favoured (Figure 2.7). This displacement of carrier concentration adds to the effect of their becoming saturated in causing a slow net movement. For a steady state, the product of the gradient of unoccupied carriers times their mobility equals the gradient of occupied carriers times their mobility.

There is evidence that sugar-carrier complexes move more rapidly than the free carriers. Kotyk[38] estimated a ratio of 2 between the respective mobilities in the yeast membrane for xylose, arabinose, and glucose and a ratio of 3 for fructose. The ratio for glucose in the ascites cell membrane is about 2·8 (Levine et al.[36]) and in the red cell membrane about 4 (Mawe and Hempling[37]).

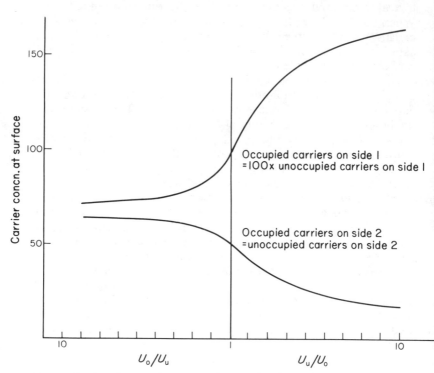

Figure 2.7. *Distributions of carriers at the two sides of a membrane separating side 1 with 100 substrate and side 2 with 1 mmol substrate when the unperturbed carrier concentration is 100 units and the ratio of the mobility of the occupied form U_o to that of the unoccupied form U_u is between 1:10 and 10:1. Note that to keep the graph expanded the abscissa scale is partly of U_o/U_u and partly of U_u/U_o. The calculation was made for a carrier K_m equal to 1 mM*

That sugar-for-sugar exchange is more rapid than the net movement gives rise to the possibility of using a gradient of one sugar to concentrate temporarily one applied at low concentration outside a cell preloaded with another at high concentration. The outside sugar enters and replaces the internally more concentrated species faster than the latter leaks out. This phenomenon is called 'countertransport' and is regarded as a good test of coupled movement on carriers[39, 40]. The same kind of coupling has been described between cations in the red cell[41].

Kotyk[38] gave equations for the total carrier concentrations on sides 1 and 2, denoted by $C_{t, 1}$ and $C_{t, 2}$, in presence of a substrate S and a competitor R having, respectively, dissociation constants for combination with carrier $K_{CS} = [C] [S]/[CS]$ and $K_{CR} = [C] [R]/[CR]$. Set $S' \equiv S/K_{CS}$ and $R' \equiv R/K_{CR}$, C_t is the unperturbed carrier concentration.

Then

$$C_{t, 1} = \frac{2C_t x_2 y_1}{x_1 y_2 + x_2 y_1} \quad \text{and} \quad C_{t, 2} = \frac{2C_t x_1 y_2}{x_1 y_2 + x_2 y_1}$$

where $x_1 = D_C + {}_{CS}S'_1 + D_{CR}R'_1$, $x_2 = D_C + D_{CS}S'_2 + D_{CR}R'_2$, $y_1 = 1 + S'_1 + R'_1$, and $y_2 = 1 + S'_2 + R'_2$. The D's are the respective diffusivities of carrier, carrier S, and carrier R. The rate of transport of S in presence of R comes to:

$$V_S = 2D_{CS}C_t \frac{(x_2 S'_1 - x_1 S'_2)}{x_1 y_2 + x_2 y_1}$$

An example has come to light of a carrier-mediated transport that appears to involve two substrate molecules per carrier. This case is proline entering the pancreas.[42] The double reciprocal plot of the square root of the difference: (concentration in cell (a_c) − concentration in medium (a_f)), plotted against the concentration, gives a straight line. The gradient in presence of an inhibitor (such as methionine) can be related to the maximum gradient by:

$$\frac{\max(a_c - a_f)}{a_c - a_f} = \left[1 + \frac{K_m}{a_f}\left(\frac{1 + I}{K_i}\right) \right]^2$$

KINETICS LIMITED BY MEMBRANE RESISTANCE

When the rate of crossing the membrane is determined by the quotient of the electrochemical activity gradient over the resistance, or when a constant active process causes the transmembrane movement, the rate with one or both compartments of finite volume decays exponentially with time. In most biological experiments, one compartment is kept open and may have sufficient capacity to allow one to ignore the change in concentration of the permeating substance; then the concentration on the closed side will build up according to the expression:

$$APt/V = \ln \frac{S_\infty - S_0}{S_\infty - S_t}$$

When already in the cell the substance is lost to a sink of infinite volume, according to the law:

$$APt/V = \ln S_0/S_t$$

in which A is the membrane area, V is the volume of the cell or closed compartment, S_0 is the initial concentration, S_t is the concentration at time t, and S_∞ is the final concentration. Slightly more cumbersome logarithmic equations arise when the volume of the second compartment is limited. Leaving aside the bizarre effects of compartmentation on tracer fluxes, the true fluxes are affected by many factors. These include changes in volume, and hence of the ratio A/V during the experiment, and distension of the membrane leading to an increase in P per unit area. Assuming constant P, the case of variable A/V can be dealt with, provided A/V has a known relation to the amount of substance transferred. Examples of integral equations applied to the red cell volume changes have been given by Jacobs and Stewart[43] and Miller[31, 32]. Another reason the resistive membrane rate law becomes distorted is on account of delays imposed by diffusion up to and away from the membrane. The same limitation to movement is met in considering the equilibration of ion exchanger particles (Helferrich[44]). The faster the permeation process, the greater will be the divergence between the local concentrations at the boundaries and the respective values in the bulk solutions. An unstirred region is inevitable within cells and artificial spherules. Since the time for diffusion to reach a given degree of completion depends on the square of the diameter, while the time of permeation varies as V/S, i.e., directly as the diameter, it follows that as the sphere becomes bigger there is an increased reason for diffusion to become the limiting factor. As already explained, the effect becomes magnified when dealing with the rate of exchange of labelled water with boundary water. The problem of unstirred regions is important, even for slow-moving substances, when there is a relatively long diffusion path. If tissues are used without perfusion, the passage round the more or less adherent cells slows down equilibration between the medium and the cell interior, the effect being maximal at the centre of the tissue. Such non-uniformity has been shown to occur in relation to oxygen tension in muscle[45, 46] and in those movements in mammalian tissue that occur relatively rapidly. An allowance for diffusion delay in a thin tissue can be made, with a number of assumptions on the path followed[47-51]. The problem is best avoided by use of either single cells or a perfusion method.

Even single cells have complex structures that can impose diffusion delays on rapid penetrants. Access to myelinated nerve seems to be only at its nodes, so diffusion along the core has to take place. An energised flow of axoplasm has been described[52]. The tubular system within striated muscle (see the electronmicrographs in Chapter 7) are in free enough communication with the medium to become filled with ferritin particles. Movement into the muscle interior occurs in part via the tubules up to the internal membranes, in which case a long diffusion path is present.

The observed time courses of equilibration of tissues and of cells commonly involves a fast initial phase that may be extracellular material and a slower, but not necessarily uniform, phase of material that is either internal or adsorbed or both. The kinetics when passage takes place through an adsorption region was discussed by Harris[53].

LIMITATION BY PORES

Some artificially prepared materials have regular pores. Biological membranes have been assigned populations of pores, often of about 4 nm diameter, largely because of wrong interpretation of the differences between tracer and osmotic water movements. Indeed, the complex of lipid and protein making up the biological membrane must have intermolecular interstices and may also have discontinuities between lipid and protein, so pores of a sort exist. These will be lined with the fixed-charge groups of the lipid (phosphate and choline) and of the protein (amino-, imino-, and carboxyl). Cross-linkages from hydrogen bonds, disulphide bridges, and ionic interaction with trapped Ca^{2+} and Mg^{2+} ions will be present. Flow of water through such a channel from which larger solutes are excluded will set up gradients of electrochemical potential tending to limit the flow; and it becomes determined by the rates of diffusion of solutes up to or away from the vicinity of pore ends, rather than by the pore dimensions.

The pore is an attractively simple way to seek to explain how selection between molecules or ions of different sizes might occur. However, the thickness of biological membranes, 5·5–7 nm, is several times the molecular dimension of common substances that can penetrate. There are then bound to be several molecules in each pathway, and interactions between them are inevitable. Models lacking a fixed-charge property are unlikely to be useful in explaining all observed phenomena.

DEPENDENCE OF KINETICS ON TEMPERATURE

The Boltzmann relation states that the proportion of particles having energy exceeding E is related to temperature by the factor $\exp(-E/kT)$. As E becomes higher, so does the temperature dependence become steeper. In assessing results from biological experiments, there has been some tendency to regard E as diagnostic between chemical processes with high E and physical ones with low E. The test is, in fact, not a certain guide, because when high physical forces are involved, as in the binding of ions to oppositely charged groups, there is a high activation energy; examples have been given by Davies and Wigsill[54], and Ting, Bertrand, and Sears[55].

The temperature dependence of a permeation process may not be linear. Changes from diffusion limitation at higher temperature to membrane limitation at lower temperature can be envisaged as a membrane becomes less permeable on cooling. On the other hand, the protein component of a membrane will denature at around 40 °C and this can lead to a sharp fall of permeation[56, 57].

KINETIC OBSERVATIONS

Movement of a permeable substance is usually associated with movement of its osmotic equivalent of water, so the cell volume changes. This can be observed by optical methods. If cells are equilibrated with a solute and the medium is suddenly diluted, the dilution artifact is followed by a transient swelling as the water moves down gradient into the cells, and then there is a prolonged shrinkage as both water and solute move together outward and into the more dilute medium. The latter phase can be used to evaluate solute permeability[58-62]. The method is adapted to rapid changes by taking observations in a flow system at various distances from a zone in which mixing occurs[63].

In evaluating such results, it is important to confirm that the changes are due to the known solute and not to other cell constituents escaping. Further, the volume change itself can be accompanied by permeability changes. Shrinkage, so that solutes become concentrated, can alter the state of internal proteins. Dick[64] showed that the haemoglobin in the erythrocyte gives rise to anomalous osmotic responses because its own osmotic coefficient varies with concentration.

The relative permeabilities of a membrane to a series of ions of the same sign can be ranked if a readily penetrating oppositely charged partner can be provided to prevent electrostatic forces limiting the movement. One way to achieve this object is to use a dissociable ion-forming agent such as ammonia or CO_2. The ammonium salt swelling method was applied by Jacobs and Stewart[43] to find anion permeabilities in the red cell membrane. Ammonia, NH_3, readily crosses the membrane and exists each side in equilibrium with ammonium ions. The transfer of NH_3 and protons (or —OH ions) separately has the same effect as a permeable cation. The method has more recently been used to find anion permeabilities of mito-chondria[65]. A readily permeating anion for red cell studies is the CO_2–bicarbonate system, and this has been used to rank cation permeabilities[66, 67]. The transfer of salt in all the examples mentioned is accompanied by its osmotic equivalent of water, and the volume changes are followed optically.

When exchanges occur so that the volume of the particle remains constant, other techniques must be used. For example, the rate of disappearance or appearance of an isotopically marked substance can be found if a rapid separation of a sample of medium can be done. By flowing a cell suspension past a port closed with a fine filter, a sample of medium can be sucked out after a short contact time. This has been applied to the red cell[68]. A flow system can be used when a spectroscopically observable change occurs with the mixed fluid passing through an observation cuvette. Either a length and the flow rate determine the time of mixing, or the flow is sud-denly stopped so that the reaction then proceeds to completion (so-called 'stopped flow'). Ion-selective electrodes can be inserted in a flow system to monitor changes in the ionic makeup of the medium. Some subsidiary reaction in the medium can be set up so that the emergent substance gives an optically detectable reaction. This technique was applied to follow mitochondrial production of citrate by adding enzymes and NADH to the medium. The citrate was split to exalacetate, which consumed NADH whose absorption was being measured. Conductance changes in the medium, when they occur, are easily measured and recorded, and afford a method for following some ionic changes.

When both sides of a membrane are accessible, as in frog skin, or bladder wall, a powerful method for measuring ion transfers is to employ Ussing's 'short-circuit' method. A feedback amplifier is set to keep the p.d. across the membrane at zero and the current necessary is recorded. This current is equivalent to the net ion movement and it is necessary to resort to analytical methods to find out what ions are contributing to it. In a number of examples, only two (e.g. Na^+ and Cl^-) or a single ion (Na^+) are involved; the system can sometimes be simplified by substituting a non-penetrat-ing ion for one of the permeants, e.g., sulphate is substituted for Cl.

A refined example of the technique of clamping the p.d., not necessarily at zero, was used in studies of nerve permeability (see Chapter 10).

ENERGY-DEPENDENT MOVEMENTS

When a substance is moved against an electrochemical gradient, the process is often called 'active transport'. This way of defining an endergonic process meets a logical difficulty when we change the conditions slightly, so that movement is no longer against the gradient, though its rate remains dependent upon energy expenditure. Hence, the more general 'energy-dependent' description is preferable. Another problem to be faced in each example is the decision about whether an exchange of a molecule or ion for a like one (marked isotopically) is part of the energy-dependent process. Part or all of such exchange, called 'exchange diffusion' by Ussing, can be analogous to exchange on an ion exchanger and without energy requirement. On the other hand, part may have to be energised if an apparent one-for-one exchange is really the consequence of a separate 'leak' being opposed by an energised movement. This is tested by cutting off the energy, to see if the levels change, and to see to what extent maintenance of different static levels is related to metabolism or its inhibition.

The prevalence of cells containing high K and low Na, in media having low K and high Na, poses the question of how the gradients are maintained. The process is essentially an energy consuming exchange of internal Na^+ for external K^+ [69]. This exchange has been shown to be related to the activity of a widely distributed membrane-bound ATPase[70]. The activity of a part, often a major fraction, of the total cell ATPase depends on the presence of Na^+, K^+, and Mg^{2+}. The remainder only requires Mg^{2+} for activation (Table 2.1). The proportions of the two fractions, or their relative activities, depend on the treatments given and it is questioned whether they are modifications of the same basic enzyme[72, 73]. Using red cell ghosts loaded with ATP and cations, Hoffman[74] showed clearly that the ATPase activity requires internal Na^+ and external K^+; the requirement was asymmetric and presumed to be related to the active movement of Na. The Na^+- and K^+-dependent ATPase is competitively inhibited by the same agents as those that hinder Na transport, such as the cardiac glycosides[75]. High concentrations of oligomycin also inhibit.

Experiments made with preparations of natural membranes have indicated that the labelling of a phosphoprotein is dependent upon

Table 2.1 THE CATION DEPENDENCE OF AN ATPase PREPARED FROM BRAIN, AND INHIBITION BY OUABAIN (FROM AHMED AND JUDAH[56])*

Added cations	Rate	Rate after ouabain addition
Nil	1·15	0·16
Mg	4·4	0·16
Mg, Na	4·4	0·20
Mg, K	4·7	0·66
Mg, K, Na	10·6	6·11
K, Na	1·7	0·81

* Units: μmol split per mg per h.

the presence of Mg^{2+} and Na^+, with K^+ acting as a competitor against the Na^+ (Table 2.2). Ahmed, Judah and Scholefield[76] showed that the activation involves two Na^+ atoms with which K^+ compete, and the K^+ activation involves one K^+, with Na^+ as competitor. The rate of removal of phosphate from the labelled protein of the membrane depended upon the presence of K^+ (Table 2.3) and Na^+ was inhibitory. Ouabain interfered with the K^+-dependent dephosphorylation of the protein, while oligomycin interfered with the Na^+-dependent phosphorylation.

Table 2.2 CATION AND OUABAIN EFFECTS ON THE LABELLING OF ACID STABLE PHOSPHOPROTEIN FROM —ATP^{32}*

Cations present beside Mg	No ouabain	Relative ^{32}P in protein preincubated with ouabain 1 mM
Nil	327	200
120 mM Na	2366	256
30 mM K	317	209
120 mM Na + 30 mM K	646	169

* Mg at 0·25 mM. Time 15 s. Temp. 0 °C

Table 2.3 POTASSIUM REQUIREMENT FOR REMOVAL OF ^{32}P FROM PHOSPHOPROTEIN AND INHIBITION BY OUABAIN*

Addition to wash solution	Relative ^{32}P before final washing	Relative ^{32}P after final washing
Nil	2·8	2·3
1 mM KCl	2·5	0·26
1 mM KCl + ouabain 0·2 mM	2·1	1·9

* Labelling for 15 s; subsequent washing for 5 s

In a comparison between the electrophoretic patterns of the active protein obtained from a series of tissues (14 in number), Post[77] found a common component that became phosphorylated. It was inferred that energising Na^+ movement always involved a specific protein that undergoes a cycle of phosphorylation.

It seems likely that the phosphorylated form, possibly acyl phosphate[78], that takes the Na^+ outward is dephosphorylated on contact with an external K^+ ion. The important feature of the system is that the more energised movement there is the fewer unphosphorylated carrier molecules there will be to give rise to leakage[79] (Figure 2.8).

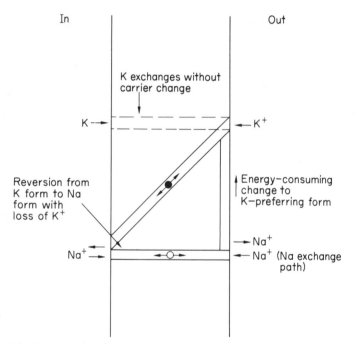

Figure 2.8. Diagram of Na^+-transferring system in which the carrier either can effect exchanges without undergoing a transformation or can be changed with energy consumption to a form that moves K^+ ions inwards (After Harris[79]; Courtesy J. Physiol.)

The mechanism by which Na^+ is moved across epithelial membranes is still obscure (see Chapter 6). It evidently involves the same enzyme because sensitivity to inhibitors and requirement for K^+ are shared with the simpler system.

ENERGISING ION MOVEMENT

The ability to move Na^+ ions against an electrochemical gradient is associated with the ATPase just discussed. A list of tissue contents of the enzyme is given by Bonting, Caravaggio, and Hawkins[80]. There have been a series of studies of the stoichiometry of Na^+ movement relating the ion transfer either to the ATP split or to the O_2 consumption. Under optimal conditions, about three Na^+ ions are removed out of red cells or muscle per ATP split. When respiration has not been inhibited this ADP generation leads to a Na^+-dependent part of the total O_2 consumption (see Table 2.4). The value is usually less than corresponds to a yield of 3 ATP per O.

Table 2.4 YIELDS IN ENERGISED TRANSPORT PROCESSES*

Tissue	Ions moved per oxygen atoms respired or ATP split	Reference
Muscle	3 Na^+ per ATP	81
	4 Na^+ per extra O	(see Ch. 6)
Frog skin	2 to 6 Na^+ per ATP	82
	2 to 4 Na^+ per extra O	83
Gastric mucosa	1·5 H^+ per ATP	84
	6 H^+ per extra O	85
Mitochondria	12 K^+ per extra O	86
	4 K^+ per ATP	86

* The oxygen used refers to the excess over the resting rate

Secretion of Na^+ by frog skin occurs with 3–8 Na^+ moved per O_2[82, 83]. In the stomach, the number of protons produced for each ATP split can be found. The production of protons in the gastric mucosa requires more energy per mole than does Na^+ extrusion, and it may be that for this reason only 1·5 H^+ are secreted per ATP split[84].

CO TRANSPORT OF SUBSTANCES LINKED TO ACTIVE NA–K MOVEMENT

Much of the foregoing discussion has centred on the active interchange of Na^+ and K^+; there is good reason for this, because the uptake of many other substances depends on the continued progress of the Na^+ movement[87, 88]. Amino acid and sugar accumulation by

Table 2.5 METABOLIC RESPONSES TO Na MOVEMENT*

Tissue	Metabolite measured	Rate of use by Na-moving control	Rate of use by inhibited system	Reference
Frog muscle	O_2 consumption	8 µmol/g h	4 µmol/g h	(see Ch. 6)
Rat diaphragm	CO_2 from glucose	1·1 µmol/g h	0·75 µmol/g h	89
Brain slices	O_2 consumption	8·8 ml/g h	6·3 ml/g h	90
	CO_2 from citrate	3·2 µmol/g h	0·7 µmol/g h	90
Kidney	O_2 consumption	24 ml/g dw h	11 ml/g dw h	91
	ATP consumption	9 µmol/mg h	6 µmol/mg h	91
Intestinal sac	O_2 consumption	230 units	150 units	92

* Control values were measured in presence of Na and K. Comparison is with either a Na-free medium or one to which ouabain has been added to inhibit Na movement.

kidney and intestinal cells afford examples. Specific poisons of active Na^+ movement, such as ouabain, or lack of K, inhibit the movement of the other substances[89]. Tables 2.5 and 2.6 summarise examples and more will be found in Riklis and Quastel[97].

There is a converse interaction between sugar movement and movement of NaCl and water through the epithelium of the gut (Table 2.7). During absorption of glucose, galactose, and even the unmetabolised 3-methyl glucose, there is a significantly higher NaCl transfer than from solutions containing compounds not carried specifically, e.g., fructose or mannitol. Chloride is absorbed in excess of the Na^+ and electroneutrality is maintained by a contrary movement of bicarbonate into the lumen. Replacement of Na^+ by K^+ led to inhibition of active movement of Na^+ and sugars.

Another index of interaction between sugar and Na^+ movement that has been applied is the potential difference across the membrane. It rises in response both to Na^+ and to sugar and is depressed by substitution of other ions for the Na^+ [99].

Table 2.6 EFFECT OF THE ACTIVITY OF THE Na TRANSPORT SYSTEM ON THE ACCUMULATION OF SUBSTANCES BY BRAIN SLICES*

Substance measured	Ratio with Na^+ movement in progress	Ratio with Na^+ movement inhibited	Reference
Creatine	5·7	1·8	93
Glycine	8·6	4·5	93
Thiamine	2·3	0·8	94
Acetylcholine	6·4	4·2	95
Glutamic acid	5·5	0·8	96

* Values are of the ratio of concentration in tissue fluid to that in the medium.

Competition for transport can be seen between amino acids and sugars. Histidine and methionine reduce the rate of transport of glucose and substituted glucoses from mucosal to serosal side of the gut, and also lessen the final trans-wall ratio attained, yet no effect is seen on passively absorbed sugars. Only those amino acids that are actively absorbed, as judged from final concentration ratios exceeding unity, have this action on the transport of glucose[100, 101].

Table 2.7 DEPENDENCE OF SUGAR CONTENT OF BRUSH BORDER CELLS OF GUT EPITHELIUM ON IONS IN THE INCUBATION MEDIUM[98]*

1. Incubation with 10 mM sucrose
   (30 min at 37 °C)

Saline	Glucose found in	
	Tissue water	*Medium*
Na salt	28 mM	10·7 mM
K salt	6·1	1·35

2. Incubation with 10 mM glucose-1-phosphate
   (30 min at 37 °C)

Saline	Glucose found in	
	Tissue water	*Medium*
Na salt	26·3 mM	1·94 mM
K salt	1·57	2·74

* Either a Na salt saline mixture or a K salt saline mixture was used.

Crane[98] has proposed the common use of a carrier by both Na and the Na-dependent sugars for transport into and through the cells of the gut (see Table 2.7). One elaboration of the theory, based on the work of Oxender and Christensen[102], is that the down-gradient leak of Na brings about the uptake of one class of amino acids; the class includes alanine, glycine, serine, and threonine. Amino acids of another class, which includes leucine, valine, phenylalanine, and tryptophane, were supposed not to share the same carrier. However, exchanges between amino acids of the two classes occur, so that a Na-independent acid might appear to be Na-dependent because it was exchanging for an acid of the latter class. In support of the concept that the Na^+ gradient is indeed involved in amino acid uptake, Eddy[103, 104] showed that the in-

65

side : outside ratio for glycine was Na-dependent. Similar results are obtained with pigeon erythrocytes[105]. An articially induced outward Na$^+$ gradient has an inhibitory effect on amino acid entry. Normally, the stoichiometry of amino acid entry seems to involve entry of 0·6 molecules per Na entering and at the same time 0·6 ions of K leave.

The interactions between sugars and Na$^+$ can also be exposed by inhibiting one process, when usually the other is found also to be inhibited[97]. Phlorizin, an inhibitor of sugar transport, has an action on Na movement in the presence of Na-dependent sugars but not in their absence.

REVERSAL OF ACTIVE ION MOVEMENT

The movement of Na$^+$ from a cell, usually in exchange for a mixture of K$^+$- and H$^+$, involves ATP hydrolysis. There have been many attempts to reverse the process by discharging the salt gradient in presence of ADP and phosphate with the object of making ATP. Red cell ghosts can conveniently be loaded with a mixture of K, ADP, and phosphate, then exposed to a Na medium. The K emerges and by use of ^{32}P labelled phosphate it can be shown that the isotope is incorporated into ATP[106]. About two to three K ions are lost per phosphate esterified. The amounts of phosphate incorporated are small and do not for certain represent net increments of ATP in the system.

The much higher ATPase activity of mitochondria renders them more suitable for short-term experiments. It is relatively easy to generate ATP in assayable quantities by discharging their K$^+$ into a K$^+$-free medium using valinomycin to increase their K permeability[107]. Rossi and Azzone[108] made the important observation that external ADP increases the rate of K discharge in such circumstances and is phosphorylated. The most likely reason is that the ADP-in-for-ATP-out exchange occurs with the emergent K ions maintaining electrical balance. The stoichiometry was 4 K lost per ATP synthesised. In these experiments, the normal oxidative metabolism was inhibited by respiratory poisons.

Remembering that a high K permeability has to be provided by adding valinomycin to the system, and that Ca moves readily in and out of animal mitochondria without an ionophore, it seems possible that Ca may be involved in the process of phosphorylation. A counter-observation to this suggestion is however that yeast and plant mitochondria are not Ca-sensitive, so a different mechanism would appear necessary.

CONCLUSIONS

The kinetics of membrane limited processes range from the responses taking milliseconds of the components of the mitochondrial respiratory chain through substrate movements taking seconds to cellular equilibrations occurring over hours. Most changes follow the first-order law, provided that account is taken of the delaying consequences of diffusion and passage through intermediate compartments. Rates of flux found from tracer experiments unsupported by analysis may be quite misleading when mixing occurs in intermediate compartments.

Many cellular equilibration processes can be described by assuming two opposed saturable fluxes whose magnitudes depend on the degree of loading of the putative carriers that cycle across the membrane. In general, the carriers behave as if unequal mobilities hold for the loaded and unloaded conditions; this leads to a non-uniform total carrier distribution when the transported substance is at unequal concentrations on the two sides of the membrane.

REFERENCES

1. BERNSTEIN, J., 'Untersuchungen zur Thermodynamik der bioelektrisch Ströme', Arch. für Physiol., 92, 521 (1902)
2. TEORELL, T., 'Transport processes and electrical phenomena in ionic membranes', Prog. Biophys., 3, 305 (1953)
3. TEORELL, T., 'Zur quantitativen Behandlung der Membranpermeabilität', Z. Elektrochem., 55, 460 (1958)
4. JACOBS, M. H., 'Diffusion processes', Ergebn. Biol., 12, 1 (1935)
5. HILL, A. V., 'Diffusion of oxygen and lactic acid through tissues', Proc. R. Soc. Lond. (B), 104, 39 (1938)
6. ROUGHTON, F. J., 'Diffusion and chemical reaction velocity in cylindrical and spherical systems of physiological interest', Proc. Roy. Soc. Lond. (B), 140, 203 (1952)
7. CRANK, J., The Mathematics of Diffusion, Clarendon, Oxford (1956)
8. LAIDLER, K. H., and SCHULER, K. E., 'The kinetics of membrane processes', J. chem. Phys., 17, 851, 856 (1949)
9. CHINARD, F. P., 'Derivation of an expression for the rate of formation of glomerular fluid (G.F.R.). Applicability of certain physical and physicochemical concepts', Am. J. Physiol., 171, 578 (1952)
10. BERKELEY, EARL OF, HARTLEY, E. G., and BURTON, C. V., 'Osmotic pressure derived from vapour pressure measurements', Phil. Trans. (A), 218, 295 (1919)
11. KUHN, W., 'Grenze der Durchlässigkeit von Filtrier- und Löslichkeitsmembranen', Z. Elektrochem., 55, 207 (1951)
12. PAPPENHEIMER, J. R., 'Passage of molecules through capillary walls, Physiol. Rev., 33, 387 (1953)
13. STARLING, E. H., On the Fluids of the Body, Constable, London (1909)
14. LANDIS, E. M., 'Microinjection studies of capillary permeability', Am. J. Physiol., 82, 217 (1927)
15. LANDIS, E. M., 'Microinjection studies of capillary permeability', Am. J. Physiol., 83, 528 (1928)

16. LANDIS, E. M., 'Capillary pressure and capillary permeability', *Physiol. Rev.*, **14**, 404 (1934)
17. PAPPENHEIMER, J. R., and SOTO-RIVERA, A., 'Effective osmotic pressure of the plasma proteins', *Am. J. Physiol.*, **152**, 471 (1948)
18. LANDIS, E. M., *Ann. N.Y. Acad. Sci.*, **46**, 713 (1946)
19. PAPPENHEIMER, J. R., RENKIN, E. M., and BORRERO, L. M., 'Filtration, diffusion and molecular sieving through peripheral capillary membranes. A contribution to the pore theory of capillary permeability', *Am. J. Physiol.*, **167**, 13 (1951)
20. ZWEIFACH, B., 'The structural basis of permeability and other functions of blood capillaries', *Cold Spring Harb. Symp. Quant. Biol.*, **8**, 216 (1940)
21. CHAMBERS, R., and ZWEIFACH, B. W., 'Capillary Endothelial cement in relation to permeability', *J. Cell. Comp. Physiol.*, **15**, 255 (1940)
22. DURBIN, R. P., FRANK, H., and SOLOMON, A. K., 'Water flow through frog gastric mucosa', *J. gen. Physiol.*, **39**, 535 (1956)
23. DIAMOND, J. M., *J. gen. Physiol.*, **48**, 15 (1964); DICK, D. A. T., and LOEWENSTEIN, L. M., 'Osmotic equilibria in human erythrocytes studied by immersion refractometry', *Proc. R. Soc. Lond.* (B), **148**, 241 (1958)
24. KITCHING, J. A., 'Contractile Vacuoles', *Biol. Rev.*, **13**, 403 (1938)
25. KITCHING, J. A., *J. exp. Biol.*, **15**, 143 (1938)
26. KITCHING, J. A., 'The physiology of contractile vacuoles. X. Effects of high hydrostatic pressure on the contractile vacuole of a suctorian', *J. exp. Biol.*, **31**, 76 (1954)
27. KITCHING, J. A., 'The physiology of contractive vacuoles. IX. Effects of sudden changes in temperature on the contractile vacuole of a suctorian,' with a discussion of the mechanism of contraction', *J. exp. Biol.*, **31**, 68 (1954)
28. USSING, H. H., 'Transport of ions across cellular membranes', *Physiol. Rev.*, **29**, 127 (1949)
29. HARRIS, E. J., and SJODIN, R. A., 'Kinetics of exchange and net movement of frog muscle potassium', *J. Physiol.*, **155**, 221 (1961)
30. DIXON, M., and WEBB, E. C., in *Enzymes* 2nd edn, Longman, London (1964)
31. MILLER, D. M., 'The kinetics of selective biological transport. I. Determination of transport constants for sugar movements in human erythrocytes', *Biophys. J.*, **5**, 407 (1965)
32. MILLER, D. M., 'The kinetics of selective biological transport. II. Equations for induced uphill transport of sugars in human erythrocytes', *Biophys. J.*, **5**, 417 (1965)
33. WILBRANDT, W., and ROSENBERG, T., 'The concept of carrier transport and its corollaries in pharmacology', *Pharmac. Rev.*, **13**, 109 (1961)
34. BRITTON, H. G., 'Permeability of the human red cell to labelled glucose', *J. Physiol.*, **170**, 1 (1964)
35. REGEN, D. M., and MORGAN, H. E., 'Studies of the glucose transport system in the rabbit erythrocyte', *Biochim. biophys. Acta*, **79**, 151 (1964)
36. LEVINE, M., OXENDER, D. L., and STEIN, W. D., 'The substrate-facilitated transport of the glucose carrier across the human erythrocyte membrane', *Biochim. biophys. Acta*, **109**, 151 (1965)
37. MAWE, R. C., and HEMPLING, H. G., 'The exchange of C^{14} glucose across the membrane of the human erythrocyte', *J. cell. comp. Physiol.*, **66**, 95 (1965)
38. KOTYK, A., 'Mobility of the free and of the loaded monosaccharide carrier in saccharomyces cerevisiae', *Biochim. biophys. Acta*, **135**, 112 (1967)
39. ROSENBERG, T., and WILBRANDT, W., 'Uphill transport induced by counterflow', *J. gen. Physiol.*, **41**, 289 (1957)
40. LEFEVRE, P. G., and MCGINNISS, G. F., 'Tracer exchange vs. net uptake of glucose through human red cell surface. New evidence for carrier mediated diffusion', *J. gen. Physiol.*, **44**, 87 (1960)

41. HARRIS, E. J., and PRESSMAN, B. C., *Nature, Lond.*, **216**, 918 (1967)
42. BÉGIN, N., and SCHOLEFIELD, P. G., 'The uptake of aminoacids by mouse pancreas *in vitro*. III. The kinetic characteristics of the transport of L-proline', *Biochim. biophys. Acta*, **104**, 566 (1965)
43. JACOBS, M. H., and STEWART, D. R., 'Osmotic properties of the erythrocyte. XII. Ionic and osmotic equilibria with a complex external solution', *J. cell. comp. Physiol.*, **30**, 79 (1947)
44. HELFFERICH, F., *Ion Exchange*, McGraw-Hill, New York (1962)
45. WHALEN, W. J., 'Intracellular PO_2: a limiting factor in cell respiration', *Am. J. Physiol.*, **211**, 862 (1966)
46. WHALEN, W. J., in *Glass Microelectrodes* (eds M. LAVALLEE, O. F. SHANNE, and N. C. HERBERT), Wiley, New York, 396 (1969)
47. HARRIS, E. J., and BURN, G. P., 'Permeation and diffusion of Na ions in frog muscle', *Trans. Faraday Soc.*, **45**, 508 (1949)
48. KEYNES, R. D., 'Ionic fluxes in frog muscle', *Proc. R. Soc. Lond.* (B), **142**, 359 (1954)
49. HARRIS, E. J., 'Permeation and diffusion of K ions in frog muscle', *J. gen. Physiol.*, **41**, 169 (1957)
50. HARRIS, E. J., 'The chloride permeability of frog sartorius', *J. Physiol.*, **176**, 123 (1966)
51. CREESE, R., 'Sodium fluxes in diaphragm muscle and the effects of insulin and serum proteins', *J. Physiol.*, **197**, 255 (1968)
52. OCHS, S., and JOHNSON, J., *J. Neurochem.*, **16**, 845 (1969)
53. HARRIS, E. J., *Trans. Faraday Soc.*, **46**, 872 (1950)
54. DAVIES, J. T., and WIGGILL, J. B., *Proc. Roy. Soc. (A)*, **255**, 277 (1960)
55. TING, H. P., BERTRAND, G. L., and SEARS, D. F., 'Diffusion of salts across a butanol–water interface', *Biophys. J.*, **6**, 813 (1966)
56. DAVSON, H., 'Ionic permeability. The comparative effects of environmental changes on the permeability of the cat erythrocyte membrane to sodium and potassium', *J. Cell. comp. Physiol.*, **15**, 317 (1940)
57. DAVSON, H., 'The influence of the lyotropic series of anions on cation permeability', *Biochem. J.*, **34**, 917 (1940)
58. ØRSKOV, S. L., 'Untersuchungen über den Einfluss von CO_2 und Pb auf die Permeabilität der Blut Körperchen für K und Rb', *Biochem. Z.*, **279**, 250 (1935)
59. WIDDAS, W. F., *J. Physiol.*, **125**, 163 (1954)
60. LEFEVRE, P. G., 'Rate and affinity in human red blood cell sugar transport', *Am. J. Physiol.*, **203**, 286 (1962)
61. LEFEVRE, P. G., 'Absence of rapid exchange component in a low-affinity carrier transport', *J. gen. Physiol.*, **46**, 721 (1963)
62. LEFEVRE, P. G., 'The "Dimeriser" hypothesis for sugar permeation through red cell membrane: reinvestigation of original evidence', *Biochim. Biophys. Acta*, **120**, 395 (1966)
63. CHANCE, B., EISENHARDT, R. H., GIBSON, Q. H., and LONBERG-HOLM, K. K., (eds) *Rapid Mixing and Sampling Techniques in Biochemistry*, Academic, London (1964)
64. DICK, D. A. T., *Exp. Cell Res.*, **14**, 608 (1958)
66. CHAPPELL, J. B., and CROFTS, A. R., 'Ion transport and reversible volume changes of isolated mitochondria', in *Regulation of metabolic processes in mitochondria* (B.B.A. library, vol. 7, eds J. M. TAGER *et al.*), Elsevier, Amsterdam, 293 (1966)
67. JACOBS, M. H., and STEWART, D. R. 'The role of carbonic anhydrase in certain ionic exchanges involving the erythrocyte', *J. gen. Physiol.*, **25**, 539 (1941–42)
67. ØRSKOV, S. L., 'Untersuchungen über den Einfluss der Kohlensäure auf die Permeabilität der Ammoniumsalze', *Biochem. Z.*, **269**, 349 (1934)
68. PAGANELLI, C. V., and SOLOMON, A. K., 'The rate of exchange of tritiated water

across the human red cell membrane', *J. gen. Physiol.*, **41**, 259 (1957)

69. JUDAH, J. D., and AHMED, K., 'The biochemistry of sodium transport', *Biol. Rev.*, **39**, 160 (1964)

70. GIBBS, R., RODDY, P. M., and TITUS, E., 'Preparation assay, and properties of an Na^+ and K^+-requiring adenosine triphosphatase from beef brain', *J. biol. Chem.*, **240**, 2181 (1965)

71. AHMED, K., and JUDAH, J. D., 'Identification of active phosphoprotein in a cation-activated adenosine triphosphatase', *Biochim. biophys. Acta*, **104**, 112 (1965)

72. NAKEO, T., TASHIMA, Y., NAGANO, K., and NAKAO, M., 'Highly specific sodium–potassium-activated adenosine triphosphatase from various tissues of rabbit', *Biochem. biophys. Res. Commun.*, **19**, 755 (1965)

73. EMMELOT, P., and BOS, C. J., 'Studies on plasma membranes. III. Mg^{2+}-ATPase, $(Na^+-K^+-Mg^{2+})$-ATPase and $5'$-nucleotidase activity of plasma membranes isolated from rat liver', *Biochim. biophys. Acta*, **120**, 369 (1966)

74. HOFFMAN, J. F., 'Cation transport and structure of the red cell membrane', *Circulation*, **26**, 1201 (1962)

75. CHARNOCK, J. S., and POST, R. L., 'Evidence of the mechanism of ouabain inhibition of cation activated adenosine triphosphatase', *Nature, Lond.*, **199**, 910 (1963)

76. AHMED, K., JUDAH, J. D., and SCHOLEFIELD, P. G., 'Interaction of sodium and potassium with a cation-dependent adenosine triphosphatase system from rat brain', *Biochim. biophys. Acta*, **120**, 351 (1966)

77. POST, R. L., in *Regulatory Functions of Biological Membranes,* (ed. J. JÄRNEFELT), Elsevier, B.B.A. library, vol. 11, Amsterdam

78. NAGANO, K., KANAZAWA, T., MIZUNO, N., TASHIMA, Y., NAKAO, T., and NAKAO, M., 'Some acyl-phosphate-like properties of P^{32} labeled sodium–potassium-activated adenosine triphosphatase', *Biochem. and biophys. Res. Commun.*, **19**, 759 (1965)

79. HARRIS, E. J., 'The dependence of efflux of sodium from frog muscle on internal sodium and external potassium', *J. Physiol.*, **177**, 355 (1965)

80. BONTING, S. L., CARAVAGGIO, L. L., and HAWKINS, N. M., 'Studies on sodium–potassium-activated adenosinetriphosphatase. IV. Correlation with cation transport sensitive to cardiac glycosides', *Arch. biochem. Biophys.*, **98**, 413 (1962)

81. HARRIS, E. J., 'The stoicheiometry of sodium ion movement from frog muscle', *J. Physiol.*, **193**, 455 (1967)

82. LEAF, A., and RENSHAW, A., 'Ion transport and respiration of isolated frog skin', *Biochem. J.*, **65**, 82

83. HUF, E. G., DOSS, N. S., and WILLS, J. P., 'Effects of metabolic inhibitors and drugs on ion transport and oxygen consumption in isolated frog skin', *J. gen. Physiol.*, **41**, 397 (1957)

84. FORTE, J. G., ADAMS, P. H., and DAVIES, R. E., 'Acid secretion and phosphate metabolism in bullfrog gastric mucosa', *Biochim. biophys. Acta,*, **104**, 25 (1965)

85. CRANE, E. E., and DAVIES, R. E., 'Chemical and electrical energy relations for the stomach', *Biochem. J.*, **49**, 169 (1951)

86. HARRIS, E. J., COCKRELL, R., and PRESSMAN, B. C., 'Induced and spontaneous movements of potassium ions. Mitochondria', *Biochem. J.*, **99**, 200 (1966)

87. CRANE, R. K., 'Uphill outflow of sugar from intestinal epithelial cells induced by reversal of the Na^+ gradient: its significance for the mechanism of Na^+ dependent active transport', *Biochem. biophys. Res. Commun.*, **17**, 481 (1964)

88. INUI, Y., and CHRISTENSEN, H. N., 'Discrimination of single transport systems. The Na^+-sensitive transport of neutral aminoacids in the Ehrlich cell', *J. gen. Physiol.*, **50**, 203 (1966)

89. CLAUSEN, T., 'The relationship between the transport of glucose and cations across cell membranes in isolated tissues. II. Effects of K^+-free medium, ouabain and insulin upon the fate of glucose in rat diaphragm', *Biochim. biophys. Acta,* **120**, 361 (1966)

90. GONDA, O., and QUASTEL, J. H., 'Effects of ouabain on cerebral metabolism and transport mechanisms *in vitro*', *Biochem. J.,* **84**, 394 (1962)

91. BLOND, D. M., and WHITTAM, R., 'The regulation of kidney respiration by sodium and potassium ions', *Biochem. J.,* **92**, 158 (1964)

92. PONZ, F., 'The active transport of sugars through the intestine and its regulation with the Na ion', *Arquivos Portugueses de Bioquimica,* 7, 93 (1963/64)

93. QUASTEL, J. H., 'Molecular transport at cell membranes', *Proc. R. Soc. Lond.* (B), **163**, 169 (1965)

94. SHARMA, S. K., and QUASTEL, J. H., 'Transport and metabolism of thianine in rat brain cortex *in vitro*', *Biochem. J.,* **94**, 790 (1965)

95. LING, C. C., and QUASTEL, J. H., *Biochemical Pharm.,* **18**, 1169 (1969)

96. TAKAGAKI, G., HIRANO, S., and NAGATO, Y., 'Some observations on the effect of D-glutamate on the glucose metabolism and the accumulation of potassium ions in brain cortex slices', *J. Neurochem.,* **4**, 124 (1959)

97. RIKLIS, E., and QUASTEL, J. H., 'The effects of inhibitors and cations on sugar transport by guinea pig intestine', *Can. J. Biochem. Physiol.,* **36**, 347 & 363 (1965)

98. CRANE, R. K., 'Hypothesis for mechanism of intestinal active transport of sugars', *Fedn Proc. Fedn Am. Socs exp. Biol.,* **21**, 891 (1962)

99. LYON, I., and CRANE, R. K., 'Studies on *in vitro* transmural potentials in relation to intestinal absorption. III. Inhibition of Na^+-dependent transmural potential of rat intestine', *Biochim. biophys. Acta,* **135**, 61 (1967)

100. HINDMARSH, J. T., KILBY, D., and WISEMAN, G., 'Effect of aminoacids on sugar absorption', *J. Physiol.,* **186**, 166 (1966)

101. BINGHAM, J. K., NEWEY, H., SMYTH, D. H., 'Interactions of sugars and aminoacids in intestinal transfer', *Biochim. biophys. Acta,* **130**, 281 (1966)

102. OXENDER, D. L., and CHRISTENSEN, H. N., 'Distinct mediating systems for the transport of neutral aminoacids by the Ehrlich cell', *J. Biol. Chem.,* **238**, 3686 (1963)

103. EDDY, A. A., 'A net gain of sodium ion and a net loss of potassium ions accompanying the uptake of glycine by mouse ascites-tumour cells in presence of sodium cyanide', *Biochem. J.,* **108**, 195 (1968)

104. EDDY, A. A., HOGG, C., and RAID, M., 'Ion gradient and the accumulation of various aminoacids by mouse ascites tumour cells depleted of adenosine triphosphate', *Biochem. J.,* **112**, 11P (1969)

105. VIDAVER, G. A., 'Some tests of the hypothesis that the sodium ion gradient furnishes the energy of glycine-active transport by pigeon red cells', *Biochemistry,* **3**, 803 (1964)

106. GLYNN, I. M., LEW, V. L., 'Synthesis of adenosine triphosphate at the expense of downhill cation movements in intact human red cells', *J. Physiol.,* **207**, 393 (1970)

107. COCKRELL, R. C., HARRIS, E. J., and PRESSMAN, B. C., 'Synthesis of ATP driven by a potassium gradient in mitochondria', *Nature, Lond.,* **215**, 1487 (1967)

108. ROSSI, E., and AZZONE, G. F., 'The mechanism of ion transport in mitochondria, Pt. III', *Eur. J. Biochem.,* **12**, 319 (1970)

3

DATA CONCERNING SOLUTIONS AND BIOLOGICAL FLUIDS. EQUATIONS RELATING POTENTIAL DIFFERENCES TO ION GRADIENTS

E. J. HARRIS

Department of Biophysics, University College, London

INTRODUCTION

Much of the information about cellular permeability depends upon the conditions under which the measurements have been made. The total osmolarity is an important factor and this can usually be found for the natural medium, and matched by making use of the freezing-point depression as an index of the activity of water in the solution (Table 3.1). It is easy to show that the depression of freezing point below the freezing point of pure solvent is related approximately to the reduction of vapour pressure by the equation $\Delta p/p = \Delta T L/RT^2$, and, combining this with the relation between $\Delta p/p$ and π, the osmotic pressure can be calculated from ΔT (L is the latent heat of fusion).

The cryoscopic method has been applied to small quantities of fluid and also to the fluids within cells. The latter application has not been without error because it is easy to supercool an unstirred fluid, in which case the depression appears to be greater than the correct amount. Perhaps the most reliable method when the fluid is accessible is that of seeding the cooled solution with a crystal of frozen solvent (or solution) or measuring the temperature at which the freezing process proceeds by means of a thermocouple. This technique has been described by Davson and Purvis[2] and applied by Conway and McCormack[3] to various body fluids. Brodsky et al.[4] have investigated the freezing points of tissue homogenates.

Most biological extracellular media are mainly Na^+ salts (though insect fluid offers an exception). When one seeks to use a medium

73

Table 3.1. FREEZING-POINT DEPRESSIONS OF SOME BODY FLUIDS

Group	Animal	Freezing-point depression of blood serum (°C)
Insect	*Carabus intricatus*	0·943
	Mantis religiosa	0·885
	Spider	0·894
	Scorpion	1·125
Snail	*Helix pomatia*	0·421
Elasmobranch fish	*Raja diaphenes* (marine)	1·924 (contains 21·4 g urea per kg)
	Pristis microdon (fresh water)	1·02 (contains 7·8 g urea per kg)
Teleost fish	*Conger vulgaris* (marine)	0·77
	Carp (fresh water)	0·52
	Anguilla vulgaris	
	(in fresh water)	0·61–0·62
	(in sea water)	0·69–0·73
Amphibia	Rana	0·4–0·7
	Bufo	0·76
Crustacea	Eriocheir	
	(in fresh water)	1·18
	(in sea water)	1·85
	Astacus fluvatialis	0·80
Mammals	Dog	0·576

After Prosser[1]

with the correct water activity, but with low or no sodium, difficulties are met. To replace salts with sugar involves a drastic reduction of ionic strength and produces Donnan potentials between the charged materials of the membrane and the solution. Potassium ions enter many cells when Cl is present so that swelling occurs if NaCl is replaced by KCl.

Since the activity of ions in the mixed media in common use need to be known, there follows a collection of data to assist in making estimates.

THE OSMOTIC EFFECT OF IONISED SUBSTANCES

All biological fluids consist of aqueous solutions containing a proportion of ionised substances. The osmotic effect of an ionised substance is greater than that of a non-ionised one because the number of particles formed per dissolved molecule exceeds unity. A very dilute solution of a uni-univalent salt like NaCl behaves as if it contained nearly two particles per molecule. As the solution reaches appreciable concentration, the electrostatic attraction between ions of unlike sign reduces the kinetic energy they possess and therefore reduces the osmotic effect, rendering the vapour pressure lowering less than that calculated for the presence of two ions per molecule. To express the osmotic effect of a salt in a given solution it is then necessary to know to what extent the ions still behave as unhindered particles. This can be formulated in two ways. Either we can use the 'osmotic coefficient', G, which is the factor by which the observed effect exceeds that calculated for no, or complete, dissociation* (according to the author's choice), or we can introduce what is called an 'activity coefficient', f, into the equation along with each of the various ion concentration terms, choosing the value to give agreement with experiment. In this latter case, complete dissociation into ions is assumed.

Table 3.2 VALUES OF G, THE APPARENT NUMBER OF OSMOTICALLY ACTIVE PARTICLES FURNISHED BY A SALT IN SOLUTION. TEMPERATURE 25 °C, UNLESS INDICATED OTHERWISE

Salt	Molarity		
	0·1	0·2	0·5
HCl	1·886	1·890	1·948
NaCl	1·864	1·850	1·842
NaBr	1·868	1·856	1·866
$NaNO_3$	1·842	1·804	1·746
NaH_2PO_4	1·822	1·768	1·664
Na_2HPO_4 (0°C)	2·336	—	—
Na_3PO_4 (0 °C)	3·062	—	—
KCl	1·854	1·826	1·798
KNO_3	1·812	1·746	1·634
K_2SO_4 (0 °C)	2·325	—	—
KH_2PO_4	1·802	1·736	1·610
$CaCl_2$ (0 °C)	2·601	—	—
$MgCl_2$ (0 °C)	2·658	—	—
$MgSO_4$	1·212	1·124	1·044

The 25 °C figures are taken from Conway[5]. The 0 °C figures are from a collection of values given by Heilbrunn[6].

* The theoretical equation relating the pressure equivalent of the osmotic effect is $\pi\bar{v} = -GRT \ln \mathcal{N}_1$, where $\mathcal{N}_1$ is the molar fraction of solvent, that is, $\mathcal{N}_1 = \mathcal{N}/(\mathcal{N} + \Sigma n_i)$ with $\mathcal{N} = $ mol solvent associated with Σn_i mol ions. Then, $\pi\bar{v} = GRT \ln(1 + \Sigma n_i/\mathcal{N})$; the logarithm approximates to $\Sigma n_i/\mathcal{N}$; so a practical equation is $\pi\bar{v} = \phi RT \Sigma n_i/\mathcal{N}$, where ϕ is slightly greater than G.

Some values of G are tabulated in Table 3.2; these are the factors by which osmotic and similar effects exceed those calculated for *no* dissociation. For salts forming two ions, the values are somewhat less than 2. For polyvalent salts, the reduction below the number of particles expected for full ionisation is greater than for the uni-univalent salts because the interaction between oppositely charged ions varies as the square of the charge. The interaction also depends upon the concentration and composition of the solution; therefore, the tabulated G values cannot be used in mixtures of salts. The relation between the practical osmotic coefficient ϕ and the activity coefficient f was deduced by Bjerrum[7] (see Harned and Owen reference 8, pp. 12–13).

The osmotic coefficient of proteins varies with the concentration of both protein and salts. The dependence on protein concentration can be put in terms of the virial coefficients B_1, B_2, which appear in the equation:

$$\pi = m \frac{RT}{V_0} (1 + B_1 m + B_2 m^2 + \ldots) = \phi m \frac{RT}{V_0} \qquad (3.1)$$

Here, the concentration (m) is expressed as molality, V_0, and is the volume of 1 kg of protein-free solution.

Albumin has values of B between 85 and 320: see reference 9; the same authors have also provided values for globulin–albumin mixtures. The value of ϕ (above) for pseudoglobulin rises from 0.82 when 3 mM salt is present together with 50 g globulin (kg water)$^{-1}$ to 1.10 with 150 mM salt and 50 g globulin (kg water)$^{-1}$ and to 1.20 with 150 mM salt and 100 g globulin (kg water)$^{-1}$.

Haemoglobin has an osmotic coefficient ϕ which can be expressed, according to Dick and Lowenstein[10] from data of Adair[11], as

$$\phi = 1 + Ax + Bx^2$$

where $A = 1.59$, $B = 4.49$, and x is the concentration in g ml^{-1} of solvent water.

To deal with mixtures, it is convenient to make use of the appropriate activity coefficients for the salts or ions present. It is convenient to assign to each ion an activity, but as a solution inevitably contains at least two species of ions it proves in practice difficult to evaluate the separate contributions. For that reason, most tables specify mean activity coefficients for salts, although some values of separate ion activity coefficients (denoted f_i) have been estimated. Before giving values we shall deal shortly with the theoretical formulae, to show how f depends upon the composition of the solution. The activity, 'a', of a substance is given by the product of its activity coefficient, f, and its concentration. The activity is the effective concentration on the basis of ideal behaviour. This requires a linear correspondence between vapour pressure and the amount present. Thus the vapour pressure is proportional to a.

The theory of the mutual influence of the ions was worked out by Debye and Hückel[12]. The interaction between opposite charges depends upon a quantity called the *ional strength*,

$$\Gamma = \sum_1^s c_i z_i^2$$

for s different ions present where c_i is the molar concentration of ion i of valency z_i. Γ is related to another quantity called the *ionic strength*

$$\mu = \tfrac{1}{2} \sum_1^s m_i z_i^2 \qquad (3.2)$$

(m_i = molal concentration of i) in the same way as molar strengths are related to molal strengths, with the additional factor of $\tfrac{1}{2}$. In the following, we shall use Γ and, when univalent ions only are in question, c_i for the molar concentrations.

The theoretical equation relating the activity coefficient, $f_\pm$, of a salt to the ional strength is:

$$-\log f_\pm = \left(\frac{1}{v} \sum_1^p v_i z_i^2 B\Gamma^{\frac{1}{2}}\right) \Big/ (1 + \mathring{a} A \Gamma^{\frac{1}{2}}) \qquad (3.3)$$

where $A = (35\cdot57)/(DT)^{\frac{1}{2}} \simeq 0\cdot23$ at 20 °C; $\mathring{a}$ is the mean distance of approach of ions in ångströms; $B = 1\cdot28 \times 10^6 (DT)^{-\frac{3}{2}} = 0\cdot355$ at 20 °C; the salt splits into p kinds of ions, v_i of ion i, and so on; v is the total number of ions produced by complete dissociation; D is the dielectric constant of water (taken as 80 for evaluation).

The activity coefficients of single ions can be calculated from the equation:

$$-\log f_i = \frac{z_i^2 B\Gamma^{\frac{1}{2}}}{1 + \mathring{d}_i A \Gamma^{\frac{1}{2}}} \qquad (3.4)$$

Here, $\mathring{d}_i$ is the effective 'diameter' of the ion, i, in solution in Å. It is difficult to evaluate $\mathring{d}_i$ or $\mathring{a}$ in equations (3.3) and (3.4). For uni-univalent salts, $\mathring{d}$ is between 3 and 5 Å, while for uni-divalent salts the range is 4·5–6 Å (reference 8, p. 381; reference 5, p. 62). These values are usually deduced from diffusivity or conductivity data.

The equations (2.7) and (2.8) do not take into account the effects of hydration of the ions or the reduction of the dielectric constant of the water by the ions. As a result of these factors, the activity coefficient–Γ curve passes through a minimum at between 0·5 and 1 M and calculated values for over 0·5 M are not valid, and even those for strengths between 0·5 and 0·1 are inaccurate. To take the other factors into account, Hückel[13] added to equation (2.7) a further term proportional to concentration. The equation becomes:

$$\log f_{\pm} = -\frac{1}{v} \sum_{1}^{p} \frac{v_i z_i^2 B \Gamma^{\frac{1}{2}}}{1 + \mathring{a} A \Gamma^{\frac{1}{2}}} + B' C_s \qquad (3.5)$$

A and B are the same as in equation (2.7), C_s is the total salt concentration; values of $\mathring{a}$ and B' for three substances are given in Table 3.3 (see reference 8, p. 381, for further figures). Equation (3.5) agrees to 1 part in 500 with experimental values of f up to 1 M for most uni-univalent compounds.

Table 3.3 CONSTANTS FOR EQUATION (3.5)

Salt	$\mathring{a}$	B'
HCl	4·3	0·133
NaCl	4·0	0·0521
KCl	3·8	0·0202

Guggenheim[14, 15] has given a useful semi-empirical equation. This relies upon the fact that $\mathring{a}A$ for many ions is approximately $1/\sqrt{2}$ and that for uni-univalent electrolytes

$$\Gamma = c_+ + c_- = 2C_s$$

When these values are substituted in equation (3.5), but with a new constant λ instead of B', one has, for uni-univalent salts,

$$\log f_i = \log f_{\pm} = \frac{-\sqrt{(2)} B C_s^{\frac{1}{2}}}{1 + C_s^{\frac{1}{2}}} + \lambda C_s$$

(Note: $\sqrt{(2)}B \simeq 0·5$.)

This equation agrees with experiment up to 0·1 M to 1 in 200. Some values of λ are listed in Table 3.4, more values are given in reference 8, p. 565.

A similar equation, satisfactory for NaCl and KCl at concentrations up to 0·5 M, is obtainable from equation (2.7) by putting

Table 3.4 CONSTANTS FOR EQUATION (3.6) AND VALUES OF f FOR VARIOUS VALUES OF C_s WHEN $\lambda = 0$. $\log f \pm = \log f(\lambda = 0) + \lambda C_s$

Salt	λ	C_s (M)	$f(\lambda = 0)$
HCl	0·240	0·01	0·901
NaCl	0·130	0·02	0·867
KCl	0·072	0·05	0·810
NaNO$_3$	0·000	0·10	0·758
KNO$_3$	−0·206		

Figures from Harned and Owen[8].

$\mathring{a}A = 1/\sqrt{2}$ and $B = 0.306$. These substitutions give

$$\log f_i = \log f_\pm = \frac{-0.43 C_s^{\frac{1}{2}}}{1 + C_s^{\frac{1}{2}}} \tag{3.7}$$

in natural logarithms the simple form

$$\ln f_\pm = \frac{-C_s^{\frac{1}{2}}}{1 + C_s^{\frac{1}{2}}} \tag{3.8}$$

is obtained.

This last equation is particularly useful because it lends itself to mathematical treatment. The agreement with experiment is to 3 parts in 100 at 0·5 M, as may be seen in Table 3.6.

The tables of experimentally determined values of activity coefficients (Tables 3.5 and 3.6) have been prepared from tables in the two works cited, which should be consulted for references to the original sources and for more complete information.

Table 3.5 ACTIVITY COEFFICIENTS FOR SOME SUBSTANCES AT VARIOUS CONCENTRATIONS AND 25 °C

Salt	Molarity		
	0·1 M	0·2 M	0·5 M
NaCl	0·778	0·735	0·681
KCl	0·770	0·718	0·649
CaCl$_2$	0·518	0·472	0·448
MgCl$_2$	0·529	0·489	0·481
MgSO$_4$	0·150	0·108	0·068
H$_2$SO$_4$	0·265	0·209	0·154
HCl	0·796	0·767	0·756
Na acetate	0·791	0·757	0·735

Figures from Conway[5] or Harned and Owen[8].

Table 3.6 MEASURED AND CALCULATED ACTIVITY COEFFICIENTS (EQUATION 3.8) FOR NaCl AND KCl AT 25 °C

Molal concentration	$f_{\pm}$ calculated	$f_{\pm}$ measured	
		KCl	NaCl
0·01	0·91	0·903	0·903
0·05	0·83	0·820	0·821
0·10	0·785	0·770	0·778
0·20	0·732	0·718	0·735
0·50	0·66	0·649	0·681

THE DONNAN DISTRIBUTION

When charged particles are retained in one of two connected compartments, a special case of a diffusion potential is present. The mobile ions tend to reach equality of electrochemical activity in the two compartments while, bearing in mind the chemically negligible displacement needed to set up a potential difference, we can take the total number of plus and minus charges in each respective compartment as being equal. This, so far as the ions are concerned, gives two sets of equations which allow calculation of the distribution (an example follows).

It is, however, important to stress that there cannot be an equilibrium if the immobile ions are all on one side, unless pressure is used to equalise the water activities. In practical cases, this pressure may be produced by internal strain in the structure of an ion exchanger, or in the cellulose wall of a plant cell. An equilibrium condition can also be attained if a non-penetrating substance, not necessarily ionised, is added to the other side. The fact that, failing one of these means of equalising the water activities, an immobile substance will continually cause water to diffuse into its compartment, illustrates the importance of the colloid's contribution to the reduced water activity. An activity difference due to salts will be equalised on the two sides by diffusion of the salts; only the colloid contribution persists. In other words, the activity difference corresponds to the 'colloid osmotic pressure'. In the following simple example, it will be seen that the unbalance of water activity caused by immobile ions exceeds that corresponding to the concentration of ions present. We shall calculate the osmotic unbalance between the two sides of a simple system containing on side (a) pure NaCl and on side (b) NaCl solution and Na proteinate with a multi-valent anion (P^{-n}).

The conditions of electroneutrality on sides (a) and (b) require:

$$(1)\ \mathrm{Na_a} = \mathrm{Cl_a};\qquad (2)\ \mathrm{Na_b} = \mathrm{Cl_b} + nP^{-n}$$

The respective equalities of electrochemical activities of Na and of Cl requires, from (2),

$$\mathrm{Na_b} = \mathrm{Na_a}\exp(EF/RT);\qquad \mathrm{Cl_b} = \mathrm{Cl_a}\exp(-EF/RT)$$

so

$$\mathrm{Na_a}/\mathrm{Na_b} = \mathrm{Cl_b}/\mathrm{Cl_a};\qquad \mathrm{Na_a^2} = \mathrm{Na_b}(\mathrm{Na_b} - nP^{-n})$$

hence

$$\mathrm{Na_b} = \tfrac{1}{2}\{nP^{-n} + [(nP^{-n})^2 + 4\mathrm{Na_a^2}]^{\frac{1}{2}}\}$$

The total concentration of particles on side (b) is then

$$\mathrm{Na_b} + \mathrm{Cl_b} + P^{-n} = [(nP^{-n})^2 + 4\mathrm{Na_a^2}]^{\frac{1}{2}} + P^{-n}$$

while on side (a) it is $2\mathrm{Na_a}$.

The difference in number between the sides is positive when the concentration of P exceeds zero; also, so long as $n > 0$ the difference is greater than $[P^{-n}]$, i.e. the 'colloid osmotic pressure' of an ionised colloid will exceed that calculated for concentration $[P^{-n}]$ of colloid ions. Examples of model systems in which predicted concentration ratios of penetrating ions were reached have been given by Sollner[16]; water movement was prevented in these models by adding a non-electrolyte to balance the colloid osmotic pressure. The ionic concentrations calculated from the above equations for several different initial values of salt concentration are given in Table 3.7. The result can be generalised to mixtures of uni-univalent

Table 3.7 INITIAL AND FINAL STATES OF SYSTEMS CONTAINING A NON-PENETRATING ION X^-, HAVING TWO EQUAL COMPARTMENTS SEPARATED BY A MEMBRANE PERMEABLE TO THE OTHER IONS. CONCENTRATIONS IN ARBITRARY UNITS

Initial states		Final states		
		Side 1		Side 2 (X = 0)
Side 1 Na$^+$ and X$^-$	Side 2 Na$^+$ and Cl$^-$	Na$^+$	Cl$^-$	Na$^+$ and Cl$^-$
100	100	133·3	33·3	66·6
100	30	105·6	5·6	24·4
100	1	100·01	0·01	0·99
10	100	57·6	47·6	52·4

salts (one of which is denoted by BA, another by $B'A'$) since at equilibrium in the system $B_a/B_b = B'_a/B'_a = r$, and so on. Also, $\Sigma B_a/\Sigma B_b = r$. Similarly, for the anions all ratios: $A_b/A_a = r$.

The equations for the concentrations on side (b) (which holds the large ion) in terms of the concentration P^{-n}, the charge n of the large anion, and concentrations on side (a) are:

$$\Sigma B_b = \tfrac{1}{2}\{nP^{-n} + [(nP^{-n})^2 + 4(\Sigma B_a)^2]^{\frac{1}{2}}\}$$

$$\Sigma A_b = \tfrac{1}{2}\{-nP^{-n} + [(nP^{-n})^2 + 4(\Sigma A_a)^2]^{\frac{1}{2}}\}$$

These values can be used in the electrochemical activity equations. For any cation, since $B_b = B_a \exp EF/RT$:

$$E_b - E_a = \frac{RT}{F}\ln\frac{B_b}{B_a} = \frac{RT}{F}\ln\frac{\Sigma B_b}{\Sigma B_a}$$

$$= \frac{RT}{F}\ln\frac{\{nP^{-n} + [(nP^{-n})^2 + 4(\Sigma B_a)^2]^{\frac{1}{2}}\}}{2\Sigma B_a}$$

This is the Donnan potential difference, corresponding to a fixed anion. If, instead, a fixed cation had been present, it would be only necessary to change the sign of n. When a penetrating divalent cation is added to a Donnan system, the equations for equal electrochemical activity on sides (a) and (b) is $C_b = C_a \exp (EF/2RT)$. Using this, together with the relations for univalent ions, leads to $(C_a/C_b)^{\frac{1}{2}} = B_a/B_b$, etc. A cubic equation in r can be deduced; it is

$$2C_a r^3 + r^2 \Sigma B_a - nP^{-n} - \Sigma A_a = 0$$

The value of r for given ΣB_a, P^{-n} and ΣA_a falls as C_a increases.

In any system in which a membrane having ion exchange properties, that is, possessing fixed charges, separates dissimilar solutions, two Donnan potentials can arise, one at each membrane boundary. This has been stressed by Wilbrandt[17], Teorell[18, 19], and Meyer and Bernfeld[20], but the point is often ignored. Only in the exceptional conditions of all the salts being of equal valency, and the total salt concentration on each side being the same, will the two boundary potential differences cancel out. Even this will not be true if the fixed charge concentration differs on the two sides, as is likely to be the case of the pH values on the two sides are unequal. The Donnan potential will have to be considered in a system comprising an ion exchanger and a solution; this might correspond to the protein of a cell and the external medium. The difficulty is that there is insufficient information about the fixed charge density of cellular materials. It is probable that the Donnan effect is the most important single factor determining total ionic contents of cells, but some further effect has to be invoked to explain the selection of particular ions.

When a pair of membranes bearing fixed charges of opposite signs are in juxtaposition they provide anomalous electrical properties if an ionised solute is present. This is because there is a time factor for migration of the ions to the fixed charges and the conductivity has a time dependence. The system behaves as a lossy capacitance, but the time factor can be lead to a resistance that falls with time, as in an inductive circuit (for details see references 21 and 22).

The breakdown characteristics of model and biological fixed-charged membranes have been analysed by Coster[23], who compared the membrane properties to those pn junction diodes.

CONCENTRATION CELLS

When two solutions of unequal concentration but of the same ionised substance are connected by a salt bridge (usually concentrated KCl), the arrangement is called a 'concentration cell without transfer'. The salt bridge temporarily prevents interdiffusion of the two solutions, though the arrangement does eventually run down. So long as a given ion is at unequal electrochemical potential in the two compartments, the difference can be measured electrically using electrodes reversible to the ion. If this is done with cation reversible electrodes,

$$\text{potential difference} = E_1 - E_2 = RT/zF \ln c_1 f_1 / c_2 f_2$$

where z is the charge on the cation, c is the concentration, and f is the Debye–Hückel activity factor. The p.d. measured with an anion reversible electrode would however be of opposite sign. In this respect, the concentration cell differs from the Donnan system, in which the anion concentration ratio is the inverse of the cation concentration ratio, so that the sign of the p.d. is the same whichever kind of electrode is used.

THE DIFFUSION POTENTIAL

When dissimilar solutions are put in contact, the mobile ions tend to move towards a uniform electrochemical activity. The system when established is of transient character unless streams of material are added at one side and removed at the other. A model of these lines was used by Teorell[24] to demonstrate that a diffusion potential established by HCl being poured into one compartment did lead to the establishment of unequal concentration ratios of other ions.

The potential established when a uni-univalent salt DA is added to one side of a system containing a mixture of salts causes the univalent cation concentration ratio between the two sides to reach a figure given by:

$$\ln \frac{M_i^+}{M_o^+} = \frac{u_D - v_A}{u_D + v_A} \ln \frac{D_i + \Sigma M_i^+}{D_o + \Sigma M_o^+}$$

where u_D is the mobility of the added cation, v_A is the mobility of added anion, D_i, D_o are the concentrations of D, and M_i^+ and M_o^+ are the concentrations of a cation M. These are summed on the respective sides. A similar equation is obtained for anions.

A general treatment of the diffusion potential set up between mixtures without restriction of valency is so complex that a summary only has been included; this will be found below. Here some of the simpler cases will be outlined.

Equation (2.3) (p. 36) is the basic equation for movement of a given ion (called for generality 'i' and distinguished by a subscript). When expanded, and using the Einstein relation

$$D = uRT/E$$

the equation can be written

$$\frac{dn}{dt} = - \frac{u_i RT}{F} \left(\frac{dc_i}{dx} + c_i \frac{d \ln f_i}{dx} + \frac{z_i F c_i}{RT} \frac{dE}{dx} \right) \tag{3.9}$$

where c_i, f_i, and u_i are respectively concentration, chemical activity factor, and mobility. This is the Nernst–Planck equation with activity factor included. The current j_i carried by the net diffusion is obtained by multiplication by $z_i F$, when n is in moles.

$$j_i = - z_i u_i RT \left(\frac{dc_i}{dx} + c_i \frac{d \ln f}{dx} + \frac{z_i F c_i}{RT} \frac{dE}{dx} \right) \tag{3.10}$$

When only a single mobile ion is present and there is no external source of current $j_i = 0$ and (putting $\mathscr{A} \equiv cf$):

$$dE(dx = - (RT/zF) \, d \ln (\mathscr{A}/dx)$$

When integrated across an activity difference $\mathscr{A}_2$ to $\mathscr{A}_1$:

$$E_2 - E_1 = (RT/zF) \ln (\mathscr{A}_2/\mathscr{A}_1)$$

In the case when cations and anions have equal mobilities, the cation current equals the anion current. Equating j_+ and j_- and rearranging gives:

$$\frac{dE}{dx} = \frac{RT}{F} \frac{z_+(d\mathscr{A}_+/dx) + z_-(d\mathscr{A}_-/dx)}{z_+^2 c_+ + z_-^2 c_-} \tag{3.11}$$

In a system which does not have a gradient of fixed charges, $z_+c_+ + z_-c_- = $ constant (the fixed charge density) irrespective of position so that:

$$z_+(dc_+/dx) + z_-(dc_-/dx) = 0$$

Apart from possible differences of gradient of the respective f's we then have $dE/dx = 0$. This is an important result, since it means that we can make a junction between two unlike solutions with small potential difference by using in the junction a high concentration of a salt such as KCl, whose cation and anion have nearly equal mobility. If the KCl concentration is high compared with that of the other solutions, it tends to swamp the p.d. arising from inequality of the mobilities of the other ions.

In order to simplify the diffusion equation when the system contains several mobile ions, it is usual to assume that there is zero net charge in every small element of the space (the principle of *microscopic electroneutrality*). This assumption will not be justified if material of strong adsorbing property is present, for then there is a sharp discontinuity of charge density in the vicinity of the surface. One of the three following possible assumptions is also made:

1. that the gradient dE/dx is variable and determined by the progress of diffusion: this is the system dealt with by Planck[25, 26], Behn[27], and Schlögl[28];
2. that the gradient dE/dx is constant, which is one of the two cases examined by Goldman[29], or
3. that the concentration gradients of the respective ions are all constant, which is the kind of boundary postulated by Henderson[30, 31].

Goldman compared the merits of the results derived from assumptions (1) and (2) and concluded that (2) provides a qualitative but not quantitative agreement with the behaviour of biological membranes. It is pointed out that materials of high dielectric constant, as are aqueous media, tend to provide uniform electrical fields because the dipole groups become orientated in the field; effects of ions present might then be small.

DERIVATION OF DIFFUSION POTENTIAL ACROSS A HOMOGENEOUS SYSTEM WITH FIXED CHARGES (FOR RESULTS SEE EQUATIONS (3.13), (3.17), AND (3.18))

Let there be a uniform concentration $\overline{N}$ of fixed charges. In unit volume of the system, for electroneutrality: $\overline{N} + \Sigma z_i c_i = 0$ where z_i and c_i are respectively the algebraic valency and the concentration of ion type i, and the sum is taken over all ions present. Let the mobility of i be $z_i u_i$; z has to be restricted to a number $+$ or $- m$. If $\Sigma c_{i+} + \Sigma c_{i-} = N$, which is the total concentration of mobile ions, and $F/RT = \beta$, equation (3.10) (with f = unity) can be written as

$$j/RT = - z_i u_i (dc_i/dx + z_i c_i \beta \, dV/dx) \tag{3.12}$$

In an integral form, it is

$$j_i = \frac{(gt - 1)z_i u_i RT(\Delta N + \beta \overline{N} \Delta V)[c_{i_2} \exp(- z_i \beta \Delta V) - c_{i_1}]}{a\{N_2 \exp(- z_i \beta \Delta V) - N_1 - (\overline{N}/z_i)[\exp(- z_i \beta \Delta V) - 1]\}} \tag{3.13}$$

where $m/z = t$, so $t = + 1$ or $- 1$, corresponding to the sign of the ion, and

$$g = \frac{mF(E_o - E_a)}{RT \ln \dfrac{mN_2 - g\overline{N}}{mN_1 - g\overline{N}}} = \frac{mF \Delta V}{RT \ln \dfrac{mN_2 - g\overline{N}}{mN_1 - g\overline{N}}} \tag{3.14}$$

This equation is equivalent to the one derived by Behn[27] and discussed by Teorell[19]. When comparing the equations, as well as to facilitate the derivation of the conductance, it is useful to make the substitution:

$$\ln \xi = - \frac{F \Delta V}{RT}$$

so

$$g = - m \ln \xi \bigg/ \left(\ln \frac{mN_2 - g\overline{N}}{mN_1 - g\overline{N}} \right) \tag{3.15}$$

When fixed charges are absent $\overline{N} = 0$. The derivation can then be carried through in the same manner, but with the activity as variable instead of concentration, provided only uni-univalent salts are present. This form is better suited for use when considerable activity differences exist across the system, but its scope is limited by the valency restriction. This is imposed because one has to relate the activity coefficient f to the total salt concentration c_s by use of a manageable equation. Equation (3.8) is suitable. This equation is equivalent to:

$$-\ln f_i = \frac{c_s^{\frac{1}{2}}}{1 + c_s^{\frac{1}{2}}} \tag{3.16}$$

where $c_s = \frac{1}{2}(\Sigma c_{i+} + \Sigma c_{i-}) = \frac{1}{2}N$.

Writing the total salt activity $A_s = f_s c_s$; $z_i = +1$ or -1; $a_i =$ activity of ion i; the result equivalent to equation (4.10) for $\bar{N} = 0$ is

$$j_i = \frac{(g - z_i)u_i RT[a_{i_2} \exp(-z_i\beta\Delta V) - a_{i_1}](c_{s_2} - c_{s_1} - \psi)}{a[A_{s_2} \exp(-z_i\beta\Delta V) - A_{s_1}]} \tag{3.17}$$

$$g = \frac{F(E_o - E_a)}{RT \ln A_{s_2}/A_{s_1}} \tag{3.18}$$

$$\psi = \left(c_{s_2}^{\frac{1}{2}} - \frac{1}{1 + c_{s_2}^{\frac{1}{2}}}\right) - \left(c_{s_1}^{\frac{1}{2}} - \frac{1}{1 + c_{s_1}^{\frac{1}{2}}}\right) + 2\ln\frac{1 + c_{s_1}^{\frac{1}{2}}}{1 + c_{s_2}^{\frac{1}{2}}}$$

Using equation (3.15), (3.17) can be written as

$$j_i = \frac{z_i u_i RT \ln(A_{s_2}\zeta^{z_i}/A_{s_1})[a_{i_2} \exp(-z_i\beta\Delta V) - a_{i_1}](c_{s_2} - c_{s_1} - \psi)}{(a \ln A_{s_2}/A_{s_1})[A_{s_1} \exp(-z_i\beta\Delta V) - A_{s_1}]} \tag{3.19}$$

(reference 32).

Pleijel[33] gives solutions for zero current and no fixed charges in systems with mixtures of different valencies.

SIMPLIFIED FORMULAE FOR A SINGLE 1-1-VALENT ELECTROLYTE

When only one salt is present in a system, but at unequal concentrations at the two boundaries, the ion activities are equal to the mean salt activity and some terms in equation (3.17) cancel. For this special case the alternative forms of the equation for current carried by ion i are:

$$j_i = u_i \frac{RT}{a}(g - z_i)(C_2 - C_1 - \psi) \tag{3.20}$$

$$j_i = -z_i u_i \frac{RT}{a}(C_2 - C_1 - \psi)\frac{\ln A_2 \zeta^{z_i}/A_1}{\ln A_2/A_1} \tag{3.21}$$

where $C =$ salt concentration and $A =$ salt activity, and the total current carried by the salt is $j_+ + j_- = J$. By addition

$$J = (C_2 - C_1 - \psi) \frac{RT}{a} [z_+ u_+ (g - 1) - z_- u_- (g + 1)] \quad (3.22)$$

When no external current flows $J = 0$; rearrangement of equation (3.22), with (3.18), gives

$$\Delta V = \frac{RT}{F} \frac{u - v}{u + v} \ln A_2/A_1 \quad (3.23)$$

where u is the mobility of the cation, v is the mobility of the anion. The equation (3.22) also allows the conductance G to be found by differentiation with respect to E,

$$G = \frac{dJ}{dE} = \frac{F}{a} (C_2 - C_1 - \psi) \frac{(u + v)}{\ln A_2/A_1}$$

This conductance is made up of two terms, one for each ion. One can combine the formula with the relation between net flux and current $J_i = F\phi_{i\,\text{net}} = F(\phi_{i_{1-2}} + \phi_{i_{2-1}})$ (Note: $\phi_{i_{2-1}}$ has sign opposite that of $\phi_{i_{1-2}}$).

The relations between fluxes, concentrations, and electrical potential difference discussed for a thin membrane, cf. p. 36, can be expressed as:

$$\frac{\phi_{i_{2-1}}}{\phi_{i_{1-2}}} = - \frac{a_2 \zeta^{z_i}}{a_1}$$

Using (3.15) to put ΔV in convenient form, we can obtain

$$G_i = \frac{\phi_{i_{1-2}} F^2}{RT} \frac{[(a_{i_2} \zeta^{z_i}/a_{i_1}) - 1]}{\ln (A_2 \zeta^{z_i}/A_1)} \quad (3.24)$$

for the partial conductivity of ion i. This, of course, has been derived for an ion subject only to physical forces.

THE CONSTANT-FIELD ASSUMPTION

The equations so far given were obtained for a diffusion boundary. When the material contains a high density of dipoles any non-linearity of the electrical field tends to be reduced by a changed orientation of the dipoles, so that the field tends to be made into a constant given by the quotient $-\Delta V/a$. This makes the mathematical treatment of the problem much simpler. The constant-field treatment has been applied to describe the behaviour of several biological membranes.

When the field $dE/dx = \Delta V/a$; where ΔV is the p.d. and a the thickness of the layer, equation (3.10) becomes

$$j_i = - z_i u_i (RT \, dc_i/dx + Fz_i c_i \Delta V/a)$$

With $c_i = c_{i_2}$ and c_{i_1} at the limits, the integral for the current carried by ion i is

$$j_i = - \frac{z_i^2 u_i F \Delta V}{a} \frac{c_{i_1} - c_{i_2} \exp(-z_i F \Delta V/RT)}{1 - \exp(-z_i F \Delta V/RT)} \tag{3.25}$$

This holds for mixtures of different ions and also in presence of fixed charges.

When all the ions do have equal numerical valency, given by $+z$ or $-z$, the contributions they make to the current can be added directly. Introducing the symbols:

$$\Lambda_+ \equiv \underset{\text{side 2 cations}}{\Sigma u_i c_{i+}} (2) + \underset{\text{side 1 anions}}{\Sigma v_i c_{i-}} (1)$$

and

$$\Lambda_- \equiv \underset{\text{side 1 cations}}{\Sigma u_i c_{i+}} (1) + \underset{\text{side 2 anions}}{\Sigma v_i c_{i-}} (2)$$

then the total current

$$J = \frac{z^2 F \Delta V}{a} \left[\frac{\Lambda_+ - \Lambda_- \exp(-zF \Delta V/RT)}{1 - \exp(-zF \Delta V/RT)} \right] \tag{3.26}$$

When $J = 0$ the p.d. ΔV is

$$\Delta V = \frac{RT}{zF} \ln \frac{\Lambda_-}{\Lambda_+}$$

The conductivity $G = dJ/dV$ is

$$G = \frac{z^3 F^2 \Delta V}{aRT} \frac{(\Lambda_- - \Lambda_+) \exp(-zF \Delta V/RT)}{[\exp(-zF \Delta/RT) - 1]^2}$$

$$+ \frac{z^2 F}{a} \left[\frac{\Lambda_- \exp(-zF \Delta V/RT) - \Lambda_+}{\exp(-zF \Delta V/RT) - 1} \right]$$

When $J = 0$ the value of G simplifies to

$$G_0 = \frac{\Lambda_- \Lambda_+}{\Lambda_- - \Lambda_+} \frac{z^3 F^2 \Delta V}{aRT} \tag{3.27}$$

These formulae will be encountered later when the permeability p.d. relations of nerve and muscle are discussed.

FORMULAE OBTAINED DIRECTLY FOR SYSTEMS WITH NO CURRENT FLOW

The equation connecting the potential difference across the junction between unequal concentrations of single salt but with diffusion of both ions going on is obtained by summing the contributions of + and − ion movements to the current. As there is no external current, the sum is zero. For neutrality

$$z_+c_+ + z_-c_- = 0$$

The + and − ion currents are, respectively,

$$j_+ = -z_+u(RT\,dc_+/dx + Fz_+c_+\,dE/dx)$$

$$j_- = -z_-v(RT\,dc_-/dx + Fz_-c_-\,dE/dx)$$

as equation (3.8). The sum = 0, so by rearrangement:

$$dE/dx = \frac{(u-v)}{(z_+u-z_-v)}\frac{RT}{F}\frac{d\ln c_+}{dx} \tag{3.28}$$

Integrating,

$$E_2 - E_1 = \frac{(u-v)}{(z_+u-z_-v)}\frac{RT}{F}\ln\frac{c_2}{c_1} \tag{3.29}$$

This can be repeated in a similar way for + and − ions moving in a material having algebraic concentration $\bar{N}$ of fixed charges. The concentrations of ions are now related by

$$\bar{N} + z_+c_+ + z_-c_- = 0$$

The current due to + ions is

$$j_+ = -(z_+RT\,dc/dx + z_+{}^2c_+Fu\,dE/dx)$$

and that due to the negative ions is, with elimination of one z_-c_-

$$j_- = -[z_-vRT\,dc_-/dx - z_-(z_+c_+ \bar{N})Fv\,dE/dx]$$

For the sum of the two currents to be zero, and remembering that

$$z_+\,dc_+/dx + z_-\,dc_-/dx = 0$$

we get

$$dE/dx = \frac{RT}{F}\frac{z_+(u-v)}{[z_+^2c_+u - (z_+c_+ + \bar{N})z_-v]}\frac{dc_+}{dx} \tag{3.30}$$

By integration between concentrations c_2 and c_1,

$$E_2 - E_1 = \frac{RT}{F}\frac{(u_+ - v)}{(z_+u - z_-v)} \ln \frac{z_+^2 uc_{+(2)} + z_-^2 vc_{-(2)}}{z_+^2 uc_{+(1)} + z_-^2 vc_{-(1)}} \quad (3.31)$$

A solution to the diffusion equation for a mixture of salts in which all anions have one valency q, and all cations valency p, was found by K. R. Johnsen[34].

Let
$$U_2 = u_1'c_1' + u_2''c_2'' + \text{etc.} \qquad \text{cations}$$
$$V_1 = v_1'\bar{c}_1' + v_1''\bar{c}_1'' + \text{etc.} \qquad \text{anions} \Big\} \text{at one side}$$

$$C_1 = c_1' + c_1'' + \text{etc. on side 1}$$

$$C_2 = \dot{c}_2' + c_2'' + \text{etc. on side 2}$$

u and $\bar{v}$ here being the conventional mobilities, and U_2, V_2, C_2 the analogues of U_1, V_1, C_1 on the other side. Then the potential difference E appears in the transcendental equation

$$\frac{U_2 \exp(pEF/RT) - U_1}{V_2 - V_1 \exp(qEF/RT)} =$$

$$\left\{\frac{q\ln(C_2/C_1) - (qEF/RT)}{p\ln(C_2/C_1) + (pEF/RT)}\right\}\left\{\frac{[C_2/C_1 \exp(pEF/RT) - 1]}{C_2/C_1 - \exp(qEF/RT)}\right\} \quad (3.32)$$

THE CONSTANT CONCENTRATION GRADIENT CASE

In practice, it is difficult to apply equation (3.32) and more convenient to use the simpler one obtained on the assumption of a constant concentration gradient[30, 31]. For a system having the components described by the following:

One side	Other side
One side	*Other side*
$U_1 = u_1c_1 + u_2c_2 + \text{etc.}$	$U_2 = u_1'c_1' + u_2'c_2' + \text{etc.}$
$V_1 = v_1\bar{c}_1 + v_2\bar{c}_2 + \text{etc.}$	$V_2 = v_1'^{-}{}_1 + v_2'\bar{c}_2 + \text{etc.}$
$U_1' = z_1u_1c_1 + z_2u_2c_2 + \text{etc.}$	$U_2' = z_1'u_1'c_1' + z_2'u_2'c_2' + \text{etc.}$
$V_1' = z_{-1}v_1\bar{c}_1 + z_{-2}v_2\bar{c}_2 + \text{etc.}$	$V_2' = z_{-1}v_1'\bar{c}_1' + z_{-2}'\bar{c}_2' + \text{etc.}$

$\bar{c}$, concentration in equivalents of anion valency z_-, mobility z_-v
c, concentration in equivalents of cation valency, z_+, mobility z_+u
One takes each c_x at a distance x in the boundary to be $= c(1 - x)$

91

and each $c_x' = xc'$. By writing down the equation for the separate ions and adding

$$dE/dx =$$

$$\frac{RT}{F}\left[\left(\Sigma u_1 \frac{dc_1}{dx} - \Sigma v_1 \frac{d\bar{c}_1}{dx}\right) - \left(\Sigma u_1' \frac{dc_1'}{dx} - \Sigma v_1' \frac{d\bar{c}_1'}{dx}\right)\right]$$

$$\overline{[\Sigma z_1 u_1 c_1(x) + \Sigma z_{-1} v_1 \bar{c}_1(x)] + [\Sigma z_1' u_1' c_1'(x) + \Sigma z_{-1}' v_1' \bar{c}_1'(x)]}$$

and
$$E = \frac{RT}{F} \frac{(U_1 - U_2) - (V_1 - V_2)}{(U_1' + V_1') - (U_2' + V_2')} \ln \frac{(U_1' + V_1')}{(U_2' + V_2')} \quad (3.33)$$

REFERENCES

1. PROSSER, C. L., *Comparative Animal Physiology*, Saunders, Philadelphia (1950)
2. DAVSON, H., and PURVIS, C., *J. Physiol.*, **124**, 12P (1954)
3. CONWAY, E. J., and MCCORMACK, J. I., *J. Physiol.*, **120**, 1 (1953)
4. BRODSKY, W. A., *et al.*, *J. gen. Physiol.*, **40**, 183 (1956/7)
5. CONWAY, B. E., *Electrochemical Data*, Elsevier (1952)
6. HEILBRUNN, L. V., *Outline of General Physiology*, 2nd edn, Saunders, Philadelphia (1947)
7. BJERRUM, N., *Z. Elektrochem.*, **24**, 231 (1918)
8. HARNED, H. S., and OWEN, B. B., *Physical Chemistry of Electrolytic Solutions*, 2nd edn, Reinhold, New York (1950)
9. SCATCHARD, G., GEE, A., and WEEKS, J., *J. phys. Chem.*, **58**, 783 (1954)
10. DICK, D. A., and LOWENSTEIN, L. M., *Proc. Roy. Soc. (B)*, **148**, 241 (1958)
11. ADAIR, G. S., *Proc. Roy. Soc. (A)*, **126**, 16 (1929)
12. DEBYE, P., and HÜCKEL, E., *Phys. Z.*, **24**, 185 (1923)
13. HÜCKEL, E., *Phys. Z.*, **26**, 93 (1925)
14. GUGGENHEIM, E. A., *Phil. Mag.*, **19**, 588 (7th series) (1935)
15. GUGGENHEIM, E. A., *Phil. Mag.*, **22**, 322 (7th series) (1936)
16. SOLLNER, K., *Ann. N.Y. Acad. Sci.*, **57**, 177 (1953)
17. WILBRANDT, W., *J. gen. Physiol.*, **18**, 933 (1935)
18. TEORELL, T., *J. biol. Chem.*, **113**, 735 (1936)
19. TEORELL, T., *Progress in Biophysics*, vol. 3, Pergamon, London, 305 (1953)
20. MEYER, K. H., and BERNFELD, P., *J. gen. Physiol.*, **29**, 353 (1946)
21. MAURO, A., *Biophys. J.*, **1**, 353 (1961)
22. MAURO, A., *Biophys. J.*, **2**, 180 (1962)
23. COSTER, H. G., *Biophys. J.*, **5**, 669 (1965)
24. TEORELL, T., *Proc. nat. Acad. Sci., Wash.*, **21**, 152 (1935)
25. PLANCK, M., *Ann. Phys. (Chem.)*, **39**, 161 (1890)
26. PLANCK, M., *Ann. Phys. (Chem.)*, **40**, 561 (1890)
27. BEHN, U., *Ann. Phys. (Chem.)*, **62**, 54 (1897)
28. SCHLÖGL, R., *Z. phys., Chem. N.F.*, **1**, 305 (1954)
29. GOLDMAN, D. E., *J. gen. Physiol.*, **27**, 37 (1943)
30. HENDERSON, P., *Z. phys. Chem.*, **59**, 118 (1907)
31. HENDERSON, P., *Z. phys. Chem.*, **63**, 325 (1908)
32. LINDERHOLM, H., *Acta physiol. Scand.*, **27**, Suppl. 97 (1952)
33. PLEIJEL, H., *Z. phys. Chem.*, **72**, 1 (1910)
34. JOHNSON, K. R., *Ann. Phys., Lpz.*, **14**, 995 (1904)

4

RED BLOOD CELLS

LIANA BOLIS
Department of General Physiology, University of Rome

and

B. D. GOMPERTS
Department of Experimental Pathology, University College Hospital Medical School, London

INTRODUCTION: RED CELL FUNCTION

An awareness of the almost total functional specialisation of red blood cells, namely, the transport of oxygen and carbon dioxide between lungs and tissues, must underlie the consideration of any aspect of their physiology. The red cell has evolved so that its capacity for O_2 is very great and it has the ability to exchange O_2 with the tissues at a very rapid rate. Most properties of the red blood cell seem to have evolved especially to facilitate this role. In particular, this is true of the mature mammalian red cell, which is devoid of nucleus, mitochondria, and all other subcellular structures. At the outset of a discussion of membrane physiology, it is worth noting the deficiency of the extensive ion exchange surfaces that intracellular membranous structures provide, in comparison with almost all other cell types. It is possible that the loss of control functions that ion exchange surfaces can provide may be compensated for in part by the presence of 5 mM 2,3-diphosphoglycerate, a chelator at physiological pH, and an important red cell metabolite. At a lower level of development, or because of evolutionary (Darwinian) advantages, the red cells of most nonmammalian vertebrates retain nuclei and mitochondria, and are not therefore solely dependent on the availability of glucose and the reactions of the anaerobic glycolytic pathway.

HETEROGENEITY OF RED CELL POPULATIONS

In spite of this deficiency of membrane structures, the mammalian red cell has long been a cell of choice for those who wish to study membrane structure and function, but one is bound to ask if this choice is not dictated by the ready availability of red cells and the ease of handling them? Red cells do not constitute a homogeneous cell population. They have a definite life span (about 120 days in humans) and many functions of the ageing process have been observed (for summary, see reference 1). The red cell membrane alters in a number of respects during ageing (see Figure 4.1). Surface charge is reduced in old red cells[2], owing to the loss of sialic acid. The surface area is decreased in old cells (see Table 4.1), but the total lipid per unit area of surface and the proportions of the individual classes of phospholipid are unaltered. The amount of cholesterol per unit area of surface is decreased[3, 4]. In terms of membrane function, the reduced intracellular potassium concentration of old red cells[5, 6] can most probably be related to a loss of sodium- and potassium-dependent ATPase[7].

Table 4.1 DIMENSIONS OF THE HUMAN RED CELL

Diameter (dry films)	$7.5 \pm 0.3 \, \mu m \, (\pm s.d.)$
(wet films)	$8.5 \pm 0.41 \, \mu m$
Greatest thickness	$2.4 \pm 0.13 \, \mu m$
Least thickness	$1.0 \pm 0.08 \, \mu m$
Mean cell average thickness	$2.1 \, \mu m$
Thickness : Diameter ratio	$1:3.9$
Mean surface area	$150 \pm 5.2 \, \mu m^2$ ('young' cells)
	$140 \pm 8.5 \, \mu m^2$ ('old' cells)
Mean volume	$87 \, \mu m^3$

After Wintrobe[23], Ponder[118], and Westerman et al.[4]

In dealing with the pathological conditions arising from inherited abnormalities, genetic mosaicism producing two distinct cell populations often occurs. More than one cell population exists in the normal newborn infant, and in subjects with thalassaemia and sickle cell disease; also in heterozygotes for glucose-6-phosphate dehydrogenase deficiency[8] and in one subject who suffered from the disease, paroxysmal nocturnal haemoglobinuria[9] in which there is a deficiency of membrane acetyl choline esterase[10]. In terms of transport properties, red cells are rather inactive when compared with cells from neuromuscular sources. One particular advantage red cells possess over other cell types is the possibility of preparing 'ghosts' with defined interior contents, and this in itself is sufficient

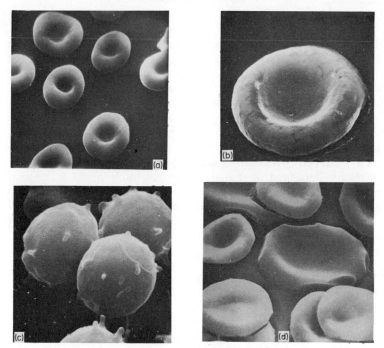

Figure 4.1. Stereoscan electronmicroscope photographs of normal and abnormal red cells. (a) normal red cells ($\times 4000 \times \frac{1}{2}$); (b) an ageing red cell. showing patchy loss of membrane material ($\times 16\,000 \times \frac{1}{2}$); (c) Hereditary spherocytosis microspherocytes, showing surface projections ($\times 12\,500 \times \frac{1}{2}$); (d) Paroxysmal nocturnal haemoglobinuria red cell showing collapse distortion produced by adjacent cells and small surface defects ($\times 5700 \times \frac{1}{2}$). Previously unpublished photographs taken by Dr J. A. Clarke (St Bartholomew's Medical College, London) and Dr A. Salsbury (Brompton Hospital, London)

to justify the use of red blood cells in a wide range of experiments concerned with membrane transport.

SECONDARY FUNCTIONS OF THE RED CELL

The red cell also has secondary functions.

1. The highly concentrated haemoglobin solution enables the cell to act as the main buffering component of the blood, by providing over 80% of the total Brønsted basic groups capable of accepting protons from carbonic acid in the physiological pH range[11].

2. There is some evidence that red cells (of mice) have a specific function in the transport of purines[12].

RED CELL DIMENSIONS

The volume of the human red cell is about half that of a sphere having the same surface area. The largest red cells are those of the urodele amphibians (56–65 μm long by 37–38 μm wide). These are of special interest to the investigator who wishes to measure transmembrane potentials with microelectrodes.

Normal human red blood cells contain 340 g of haemoglobin per litre of packed cells, i.e. they are about 5 mM in haemoglobin, and this is close to the concentration at which crystallisation takes place[14]. The cells, with their normal gel-like contents of haemoglobin, are readily deformed so that they can pass through the fine blood vessels having diameters considerably less than the natural dimensions of the cell. Rand[13] has described experiments in which red cells were drawn into a micropipette of 3 μm diameter without damage to the membrane.

THE DEVELOPMENT OF IDEAS OF MEMBRANE STRUCTURE

A considerable fraction of the red cell membrane can be extracted into organic solvents. The lipid material isolated in this way can be spread on the surface of an aqueous solution, and by doing this in a Langmuir trough the spreading qualities of the lipid mixture can be determined. It was from experiments of this kind that it was first suggested that the lipid molecules of the red cell membrane are organised as a bilayer over the entire surface of the cell[15]. This idea was incompatible with the observation that the surface tension of lipid monolayers at air–water interfaces is more than an order of magnitude greater than the surface tension of mammalian cell membranes. To account for this discrepancy, models were developed in which proteins occupy space on the membrane surface, either by covering the entire bilayer[16] or by interdigitation with lipid molecules[17]. The more recent development of lipid bilayer membranes separating two aqueous phases[18] has allowed the physical properties of a more realistic model membrane to be examined. In all respects, the intrinsic properties of lipid bilayers have been found to match exactly the corresponding passive or linear properties of natural cell membranes[19a, b], and the surface tension of a phosphatidyl choline bilayer is right in the range for cell membranes[20]. Thus the original discrepancy that led away from the idea of the cell membrane being based upon a simple lipid bilayer had disappeared, but in the meantime much evidence has accumulated to support the concept of strong protein–lipid interactions in cell membranes. Recent determinations of the lipid content and spreading characteristics have led to the suggestion that there is insufficient lipid present to cover the red cell membrane as a bilayer[21a] but there are many assumptions involved in this deduction, and the bilayer concept still receives strong support[21b].

The similarity of electronmicroscope pictures of many membrane types suggests that there is a universal structural unit in membranes. The 'unit membrane' concept, which proposes that membranes consist of a lipid double layer spread with protein attached to the outer polar 'head' groups of the lipids, is derived mainly from the study of myelin membranes[22a, b]. Myelin may be a special case, however, whose function is determined by its very inert character, and some critics now believe that the similarity of the electron photomicrographs of myelin and various other membrane types is artefactual[21a]. Consideration of spectroscopic data and of the forces involved in membrane stability leads to the idea that it is the hydrophobic side chains of the proteins which interact with the lipid, via the lipid hydrocarbon regions. Recombination studies using isolated ghost proteins with negatively charged lipids favour some measure of electrostatic interaction[24]. Many model structures embracing these opposing ideas have been proposed[25, 26a]; but the lipid bilayer as a central structural fact once again forms the basis of most current discussions of membrane structure, and an impressive array of observations and data has all but proved the point[19b, 26b].

FORCES STABILISING MEMBRANE STRUCTURE

The most important forces acting within membranes are (1) Coulombic, (2) van der Waals and (3) hydrophobic. The first are due to polar groups such as $-NH_3^+$ and $-NR_3^+$, the first of which will have a characteristic ionisation constant depending on its environment in the membrane; the second will be positively charged under all conditions. Typical negatively charged groups forming part of the membrane structure are $-COO^-$ and phosphate. The second ionisation constant of orthophosphoric acid ($pK_2 = 7.2$ at $25\,°C$ and zero ionic strength[27]) lies in the physiological range of pH, and in considering Coulombic forces this must be borne in mind. The ionisation of individual phosphate groups will, of course, depend on the nature of their immediate microenvironment. Van der Waals forces arise from the sum of interactions between (1) ion-induced dipoles, (2) dipole-induced dipoles, and (3) temporary distortions of time-average symmetric fields about uncharged molecules (these

97

are known also as 'London dispersion forces'). They diminish according to the seventh power of the distance and are therefore termed 'very-short-range' forces. Hydrophobic forces are essentially due to the reluctance of nonpolar assemblies to enter water because of its very high dielectric constant. When a hydrocarbon is dissolved in water, the entropy of the solution is negative compared with that of solutions of ionisable or polar molecules of similar dimensions. This means that water molecules are ordered in the vicinity of hydrocarbon chains. This ordering, leading to a negative entropy contribution, is minimised if hydrocarbon chains interact with each other. As a result, nonpolar portions of lipids and proteins tend to associate, and this is just the situation that arises in the liposome and cell membrane.

RED CELL MEMBRANE LIPIDS

The membrane of the red cell accounts for 3.4% of the dry weight of the cell[28]. Microscopy and electronmicroscopy[29] and studies of nucleoside metabolism[30], Na^+ and K^+ transport and the ATPase[31], transport of glycerol[32], and glucose[33] have shown that red cell ghosts, obtained by haemolysis of red cells in hypotonic media, closely resemble the membrane of intact cells, but some electron-microscopic (see Figure 4.2)[34] and electrophoretic[35] differences have been described. Much of the present knowledge of red cell membranes has been derived from the experimental study of ghosts.

Ghosts have been prepared from which it was possible to recover the entire red cell lipids[36]. Detailed analyses of ghost lipids have been reported from time to time[37, 38], and alterations in these estimates reflect the continual improvements in the technology used in their fractionation, preservation, identification, and assay. Some recent compilations[38] of the distribution of the groups of lipid molecules and of the phospholipids in a number of species are shown in Tables 4.2, 4.3, and 4.4.

Recent work shows that cholesterol is the only neutral lipid in the membranes of red cells of many species and comparative studies show a rather consistent 1 : 1 ratio of cholesterol to total phospholipid[39]. Alteration of the lipids of intact cells by the application of pancreatic lipase affects their permeability to hydrophilic non-electrolytes profoundly[40]: trypsin and pepsin are without effect in this respect. Lipid variation arises naturally, from dietary variation, and in pathological states arising from hepatic disease[41]. In general, no appreciable differences in the quantitative proportions of lipid and phospholipid classes are observable, but the fatty acid composition can be altered to a significant degree[37]. Thus, rats on a fat-free diet have decreased red cell membrane linoleic and arachidonic acids, and decreased 'cephalin' (generally taken to mean phosphatidyl serine and phosphatidyl ethanolamine). The osmotic fragility is increased, and there is electronmicroscopic evidence of structural alteration[42].

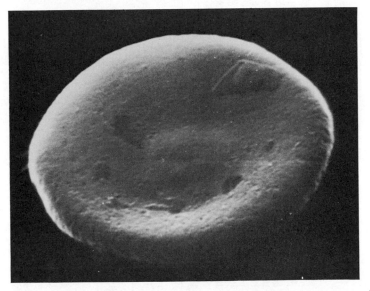

Figure 4.2. Scanning electronmicrograph of a typical reconstituted human red cell ghost. The ghosts vary in shape and are pitted all over and have several relatively large indentations. The ghosts are quite unlike the smooth cell membrane seen in intact red cells by the scanning technique. The appearance of the ghosts varies, depending on the method of preparation ($\times 13200 \times \frac{2}{3}$) (From O'Donnell and Ellory[34]; courtesy Experientia)

The lipids of the red cell membrane exchange continuously with those of the plasma. Cholesterol has a half-time of exchange of one hour[43] and cholesterol exchange has been demonstrated *in vivo*[44]. Phospholipids exchange at different rates, in the order lysolecithin > lecithin > sphingomyelin[45]. 26 % of membrane-bound α-tocopherol exchanges hourly[46].

The lipids differ in their ease of extraction from the membrane with solvents, and a scheme for extraction of strongly and loosely bound components is indicated in Figure 4.3. There is disagreement on whether individual lipid substances are distributed in particular locations of the membrane surface, or are distributed 'symmetrically'.

Table 4.2 LIPID DISTRIBUTION IN RED CELLS FROM VARIOUS MAMMALIAN SPECIES

	mg per ml packed cells	g per cell	Cholesterol	Total gangliosided	Other glycolipids	Phospholipid
Cat	6·04	$3·45 \times 10^{-13}$	26·8	8·8	3·1	61·3
Cow	4·44	$2·58 \times 10^{-13}$	27·5	5·5	2·2	64·8
Dog	5·76	$4·84 \times 10^{-13}$	24·7	11·8	10·9	52·6
Goat	6·14	$1·23 \times 10^{-13}$	26·2	5·7	17·9	50·2
Guinea pig	5·72	$4·41 \times 10^{-13}$	27·0	2·2	15·2	55·6
Horse	5·37	$2·58 \times 10^{-13}$	24·5	15·5	8·0	52·0
Pig	4·33	$2·52 \times 10^{-13}$	25·8	3·3	10·1	59·8
Rabbit	4·57	$4·15 \times 10^{-13}$	28·9	4·5	0·8	65·8
Rat	5·08	$3·15 \times 10^{-13}$	24·7	6·3	2·0	67·0
Sheep	4·91	$1·62 \times 10^{-13}$	26·5	7·8	2·5	63·2

After Rouser et al.[38]

Table 4.3 PHOSPHOLIPID DISTRIBUTION IN RED CELLS FROM VARIOUS MAMMALIAN SPECIES

	Rat	Rabbit	Pig	Dog	Horse	Sheep	Cow	Goat	Cat	Guinea Pig
PC	47·5	33·9	23·3	46·9	42·4	ND	ND	ND	30·5	41·1
PE	21·5	31·9	29·7	22·4	24·3	26·2	29·1	27·9	22·2	24·6
PS	10·8	12·2	17·8	15·4	18·0	14·1	19·3	20·8	13·2	16·8
PI	3·5	1·6	1·8	2·2	0·3	2·9	3·7	4·6	7·4	2·4
PA	0·3	1·6	0·3	0·5	0·3	0·3	0·3	0·3	0·8	4·2
Sph	12·8	19·0	26·5	10·8	13·5	51·0	46·2	45·9	26·1	11·1
LPC	3·8	0·3	0·9	1·8	1·7	ND	ND	ND	0·3	0·3
X	—	—	—	—	—	4·8	1·7	0·8	—	—

After Rouser et al.[38]

100

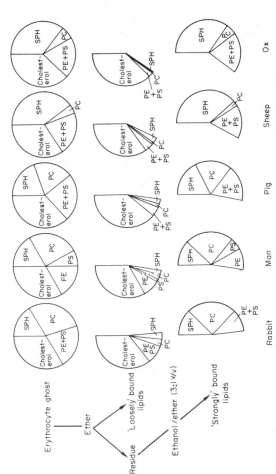

Figure 4.3. Scheme for the extraction of 'loosely' and 'strongly' bound lipids of red cell ghosts and their relative distribution over the corresponding fractions, presented for a series of five mammalian species. Abbreviations: SPH = sphingomyelin: PC = phosphatidyl choline; PS = phosphatidyl serine; PE = ethanolamine containing phosphoglycerides. Note that extraction of ox-red cell ghosts with ether removes cholesterol almost quantitatively, with minimal disturbance of the phospholipid fraction (After van Deenen[128], courtesy Prentice Hall Inc.)

101

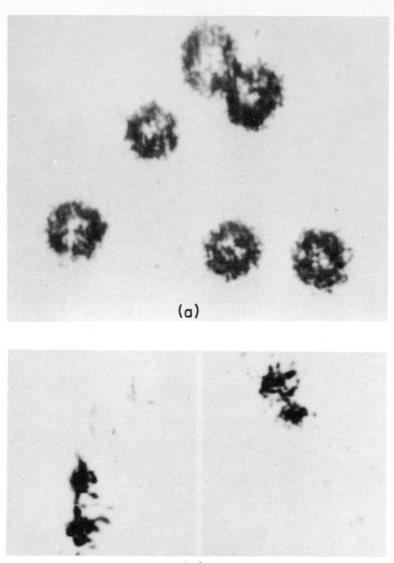

(a)

(b)

Figure 4.4. (a) *Photomicrograph of a radioautograph of human red cells containing cholesterol-7-α-³H.* (b) *Photomicrograph of a radioautograph of human red cells containing cholesterol-7α-³H, and cut in cross-section. The cross-section represents the middle of a cell. In as much as the cholesterol is entirely in the membrane, the relative absence of grains over the central area of the cell indicates that cholesterol is concentrated in the peripheral area of the membrane disc (From Murphy[47]; courtesy J. Lab. clin. Med.)*

By equilibrating cells with cholesterol-7α-^{3}H in plasma, and then examining the cells by radioautography, it was concluded that membrane cholesterol is concentrated around the periphery of the biconcave disc (see Figure 4.4)[47]. The relative absence of cholesterol from the concavities results in a more wettable surface and lower interfacial tension in these regions, and the asymmetric distribution of cholesterol may be a shape-determining factor of the red cell membrane. Partial removal[48] or replacement of membrane cholesterol with lecithin, or certain other steroidal substances[49], results in an increase of red cell osmotic fragility, presumably because removal of cholesterol causes the cell to sphere, and approach the threshold, or critical haemolytic volume. There is a selective loss of red cell lipids during the normal process of red cell ageing[3, 4], during which time individual cells become more spherical and more osmotically fragile.

Table 4.4 CHOLESTEROL: PHOSPHOLIPID MOLE RATIOS OF RED BLOOD CELLS OF SOME COMMON MAMMALS

Species	Sample No.	Cholesterol (mol per cell $\times 10^{16}$)	Phospholipid phosphorus (mol per cell $\times 10^{16}$)	Cholesterol: Phospholipid ratio
Cow	1A	1·79	1·95	0·92
	1B	1·88	1·99	0·94
Dog	2A	3·17	3·30	0·96
	2B	3·04	3·40	0·89
Goat	3A	0·80	0·82	0·98
	3B	0·85	0·80	1·07
Horse	4A	1·67	1·81	0·92
	4B	1·74	1·85	0·94
Pig	5	1·73	2·00	0·87
Rabbit	6	3·10	3·61	0·86
Rat	7	2·04	2·81	0·73
Sheep	8A	1·06	1·37	0·77
	8B	1·06	1·27	0·83

From Nelson[39].

Evidence suggestive of a symmetrical distribution of lipids in the red cell membrane comes from observations of a light membrane-like sediment obtainable from the plasma of patients with morphological evidence of red cell fragmentation. Lipid analyses of these fragments show that they contain all the red cell membrane lipids, in just the same proportion as they are found in the intact cell[50].

Future developments may be expected to show whether particular

Table 4.5 ALKALI AND CHLORIDE CONCENTRATIONS IN CELLS AND PLASMA

Species	K^+	Na^+	Cl'	K^+	Na^+	Cl'
	mequiv l^{-1}	cell water		mequiv l^{-1}	plasma	
Man	136	19	78	5·0	155	112
Baboon	145	24	78	4·7	157	115
Rabbit	142	22	80	5·5	150	110
Rat	135	28	82	5·9	152	118
Horse	140	16	85	5·2	152	108
Sheep	46	98	78	4·8	160	116
Ox	35	104	85	5·1	150	109
Cat	8	142	84	4·6	158	112
Dog	10	135	87	4·8	153	112

From Bernstein, R. E., *Science, N.Y.*, **120**, 459 (1954); courtesy *Science* lipid classes are equally distributed on the inner and outer surface. ^{35}S-sulphanilic acid diazonium salt, which does not penetrate red cells, can be used to identify components, both lipid and protein, that are situated on the outside of the permeability barrier,[5] and, no doubt, lipid-reactive reagents of more limited group specificity will be developed.

MEMBRANE PROTEINS

While the study of membrane lipids is now beginning to provide a clear picture of a structure–function relationship, the role of individual proteins in membranes is still far from understood. The main reason for this is that lipids, free or in association with membranes, take up their thermodynamically most favoured configurations on interacting with solvent water. This is why the study of model membranes has been so profitable. The proteins of cell membranes, however, find themselves in a special situation where hydrophobic and van der Waals bonding forces predominate over all aqueous interactions. In order to release the proteins, the techniques used (treatment with ionic detergents and organic solvents) are those that allow proteins to attain a stable, but in most cases, inactive structure. The aim is to achieve 'solubilisation', i.e. a close interaction of the proteins with the aqueous phase—and critical alterations in their tertiary structure are inevitable.

A commonly used criterion of a soluble preparation is one that is clear to the eye, and nonprecipitable after centrifugation for one hour at $100\,000 \times g$. Certain treatments that allow such a degree of solublisation are reversible. Thus, treatment of membranes with the nonionic detergent Triton X 100 in presence of glycerol (40%) allows the 'solubilisation' of phosphomonoester hydrolase together with much other protein. This is the enzyme that appears to mimic the terminal step of the Na^+- and K^+-dependent ATPase in membranes (see page 122, reference 52). The enzyme is no longer K^+-stimulated, as it is on the untreated membrane, but there is no evidence that the essential lipid–protein structure has been damaged by treatment with the detergent. The individual proteins are not separable from the other after this treatment, although they penetrate agarose gel having an exclusion limit of 1.5×10^6. If the detergent is removed, a particulate suspension having K^+-dependent phosphatase activity is recovered[53].

A number of enzymically active preparations have been isolated by 'gentle' procedures, such as treatment of the membranes with

E*

salt. It is likely that these enzymes are attached to the membrane surface, rather than forming part of the membrane structure, and may even be normally in equilibrium with unbound enzyme in the cell cytoplasm. In this way, acetylcholine esterase has been released and purified from membranes treated with the nonionic detergent Tween 20[54] or NaCl[55] and a number of glycolytic enzymes are found both as cytoplasmic components and bound to the membrane[56-60].

A single protein (about 20% of the total membrane protein) can be isolated on extraction by dialysis against EDTA and mercapto-ethanol. In the presence of divalent cations, it polymerises and forms coiled filaments and is thought to be involved in the maintenance of membrane structure. It has been called 'spectrin' on account of its association with 'ghosts'[61, 62]. This protein may be identical with the actin-like proteins that have previously been found in red cell ghosts[63], but a close identification with actin is not justified[64]. This protein may be responsible for the Ca^{2+}-dependent ATPase activity of red cell membranes[65a]. By reacting carefully prepared red cell ghosts with the nonpenetrating fluorescent amino-reactive substance 4-acetamido-4'-isothiocyanate-stilbene-2,2'-disulphonic acid ('SITS')[65b], before extraction with low-ionic-strength aqueous solutions in the absence and presence of EDTA, two groups of proteins emanating from the interior side of the membrane have been isolated and characterised[66]. The protein liberated accounts for 50% of the total ghost protein—none of it was labelled with the fluorescent marker. It is likely that a large part of the most easily liberated protein is normally washed away by the normal procedures of ghost preparation. Most of the antigenic material, sialic acid, and carbohydrate remains with the insoluble residue. Choline esterase was not liberated, but a considerable amount of an 'ATPase' requiring no added cations for activation, was present in both the soluble fractions and the residue, giving an apparent recovery of 150%.

The bulk of membrane proteins are released from well-washed ghosts into the aqueous phase by extraction with butanol[67] and some of these have been characterised[68]. The proteins so released are not properly separable by electrophoresis or gel chromatography, though two major peaks are observed on analytical ultra-centrifugation. A number of laboratories are now carrying out exhaustive studies in an attempt to isolate and characterise pure membrane proteins[69-71]. Analyses of total membrane protein amino acid composition have been published[69, 72, 71]. Apart from a large discrepancy in the determination of tryptophan, the agreement between laboratories using differing means of protein extraction from membranes is striking, but no special inferences concerning

106

membrane structure or function can be drawn from these analyses, which show the amino acid composition to be typical of proteins in general (see Table 4.6).

Table 4.6 AMINO ACID CONTENT OF PROTEINS FROM HUMAN RED CELL MEMBRANES. RESULTS EXPRESSED AS RESIDUES PER HUNDRED TOTAL RESIDUES

| Amino acid | Bakerman and Wasemiller[69] | Rosenberg and Guidotti[71] | Zwaal and van Deenen[72] | | |
			Ghosts	BuOH extracts	Pent OH
Lysine	4·77	5·21	5·0	4·9	5·0
Histidine	2·40	2·44	2·7	2·6	2·6
Arginine	4·82	4·53	5·1	5·2	5·2
Trytophan	0·37	2·49	†	†	†
Aspartic acid	8·17	8·49	8·3	8·0	7·9
Threonine	5·72	5·86	5·2*	5·0*	5·1*
Serine	7·04	6·26	6·3*	6·2*	6·1*
Glutamic acid	12·12	12·15	13·9	14·0	13·8
Proline	4·78	4·26	5·6	5·8	5·8
Glycine	6·57	6·73	6·8	7·0	7·0
Alanine	8·07	8·15	8·1	8·1	8·0
Cystine (half)	1·43	1·08	0·5	0·5	0·5
Valine	6·86	7·10	6·6	6·7	6·6
Methionine	2·43	2·02	2·1	2·2	2·2
Isoleucine	5·06	5·29	4·8	5·0	5·3
Leucine	12·35	11·34	11·6	11·6	11·7
Tyrosine	2·54	2·41	2·4	2·3	2·3
Phenylalanine	4·48	4·20	5·0	4·9	4·9

* Threonine and serine, uncorrected values.
† Tryptophan, not determined.

Many different techniques have been used to achieve a soluble preparation of red cell membranes. The following list was assembled by Hoogeveen et al.[66] who were themselves concerned with extraction of water-soluble proteins under conditions of low ionic strength, together with EDTA.

Use of non-aqueous solvents: references 69, 72–79.

Use of ionic detergents: references 51, 69, 73, 80–82.

Use of non-ionic detergents: references 69, 83–86.

Use of non-ionic detergents generally in combination with other reagents or treatments including urea: references 69–71, 82, 84, 87; mercaptoethanol: references 61, 62, 64, 69, 84, 87, 88; guanidine: references 69, 71; chelating agents: references 61, 62, 65, 71, 88; acid: references 82, 85, 87, 89, 90; alkali: references 62, 69, 88, 91; low ionic strength: references 62, 65, 83, 88, 92, 93, 94; high ionic strength: references 55, 63, 71; chemical modification: references 77, 82; and sonication: references 69, 71, 75, 81, 85, 95. Proteins which can be liberated without the use of detergents, but by manipulation of ionic strength, usually in the presence of chelating agents and sometimes mercaptoethanol, have been reported in references 55, 61, 62, 65, 68, 69, 71, 92, 93, 94.

MEMBRANE SULPHYDRYL GROUPS

Sulphydryl groups are important for the maintenance of membrane structure. Treatment of sheep red cells with organic mercurials (phenyl mercuric hydroxide and salyrgan) produces haemolysis[96]. Only 22% of the membrane —SH groups are attacked by p-chloromercuri benzene sulphonic acid (PCMBS)—the rest are masked to the reagent[97] and solubilisation of the membranes with n-butanol produces only a small increase (5%) in the number of unmasked —SH groups[68]. Exposed —SH groups on the exterior and interior membrane surfaces can be distinguished by the use of penetrating (chlormerodrin and p-chloromercuri benzoate) and non-penetrating reagents (PCMBS). Interaction of —SH groups on the exterior surface of the membrane with organic mercurials results in inhibition of glucose transport[97]. By using such 'side' specific reagents, and by relating the rate of interaction of various —SH reagents with membrane sulphydryl groups, it is possible to detect a correlation between certain classes of sulphydryl groups with particular membrane functions. Thus, 'readily reactive' groups, that combine with N-ethylmaleimide (NEM), organic mercurials, and $HgCl_2$ (7% of membrane —SH groups) are concerned with glucose transport, but are not involved in cation permeability or ATPase functions. 'Partially reactive' groups, that do not react with NEM (18% of membrane —SH groups) are involved in cation permeability and ATPase functions[78]. The remaining 75% of membrane —SH groups—the 'masked' groups—are only detectable by treatment with $HgCl_2$[98]. The idea of differently reacting groups of sulphydryl groups is supported by spectral studies with a spin-labelled sulphydryl-reactive compound[99]. Binding at one set of sites permits relatively free movement of the spin-labelled compound; binding at the other sites permits only restricted movement.

PROTEIN–LIPID INTERACTIONS

In order to learn about protein–lipid interactions in membranes, there are two general approaches. One can isolate the proteins and lipids to study the product of their recombination[25], and it should become possible to extend this procedure to the interaction of membrane proteins with bimolecular lipid model membranes. The functional interaction of spectrin with liposomes has been studied[100]. In the absence of purified membrane proteins, one can study the interaction of enzyme proteins, such as phospholipases, which are known to interact with membrane lipids[101]. Alternatively, it is possible to make use of spectroscopic techniques, and compare the results obtained with simple model systems, to obtain information on the generalised tertiary structure of membrane proteins, the degree of helicity, etc. It should be noted that such spectroscopic methods provide an average of information, and little can be concluded directly about structure–function relationships. For example, it is estimated that there is about one sodium pumping site per 1 μm^2 of membrane surface[102, 103]; and so perturbations at this level are hardly likely to produce useful alterations in the gross spectroscopic behaviour of the whole membrane. However, application of phospholipase C, phospholipase A, and sodium dodecyl sulphate, followed by spectroscopic investigation, leads to the conclusion that the overall membrane architecture is strongly dependent on hydrophobic, or lipid-sensitive, interactions[104–106] in agreement with all other experience of membrane structure. It can be stated that the spectroscopic approach is able to see real events in the membrane if the perturbations applied are large enough; the difficulty of the approach is in knowing what the real events are. Though we reserve judgement for the time being on the spectroscopic investigation of membrane structure, much can be expected in the future, especially when purified membrane enzymes are to hand.

ION-EXCHANGE PROPERTIES OF THE RED CELL MEMBRANE

When red cells or ghosts are suspended in saline or plasma, and subjected to an electric field, they migrate towards the anode. Therefore, they bear a net negative charge. The magnitude of this charge varies between species and can be drastically altered by treating the membranes with the enzyme neuraminidase—which cleaves 95–100% of the membrane-bound sialic acid[107]. By carrying out such experiments with red cells from a number of animal sources, a clear correlation was found between electrophoretic mobility, and hence the surface charge, and the surface density of sialic acid[108]. If the carboxyl group of sialic acid plays a major role in determining the net charge of the membrane, it is apparent that the phospholipids make a major contribution to the binding of cations. The array of ordered fixed charges, which the polar surface of phospholipids present, ensures a high degree of ion association on membrane surfaces[109]. The association of monovalent cations on living and non-living ion exchange surfaces has a much greater stability than the association between cations and simple monovalent anions in solution. Such associations exhibit selectivity between individual cations[110–114]. By using the negatively charged fluorochrome 1-anilinonaphthalene-8-sulphonate to measure ionic interaction with membranes, it has been found that the ion exchange selectivity of red cell membranes is in the order $Cs^+ > Rb^+ > K^+ > Na^+ > Li^+$ [115]. The association reactions are endothermic and are driven by a favourable entropy change that probably arises from the large aqueous entropy of monovalent cation solvation. Interactions of membranes and phospholipid miscelles with divalent cations are complicated by the appearance of a number of apparent binding sites[115, 116]. It appears that the binding of divalent cations induces alterations in the organisation of the surface groups, and this results in abrupt transitions during the course of titration. Binding of divalent cations at tight sites is exothermic but is endothermic at the weakest site. None of these reactions, which were tested with human red cell membranes and rat brain microsomal membranes, were sensitive to treatment with neuraminidase.

MECHANICAL PROPERTIES OF THE RED CELL

The typical biconcave shape of the red cell is not maintained if the hydrostatic pressure difference between inside and outside exceeds 1 mm water[117]. There is a uniform interfacial tension, or surface energy, of about $3 \times 10^{-5} \, J \, m^{-2}$.

Red cells swell in hypotonic media. The rate of swelling has been related to the difference between the internal (p) and the external (P) osmotic pressures (or water activities), so that rate of swelling, $dv/dt = KA(p - P)$, where A is the surface area and K a measure of water permeability[118]. However, K has now been shown to depend on the osmolarity of the solution, so that water moves through the membrane of a shrunken cell more slowly than through the membrane of a swollen cell[119]. The swelling causes the shape to become spherical, and at still lower tonicities stretching of the membrane makes it temporarily permeable to haemoglobin, the escape of which provides a means of measuring the extent of haemolysis.

Exposure of red cells to extreme conditions, such as media of low tonicity, causes loss of internal salt[120], and so they do not expand exactly to the volume that would be predicted if they were behaving as true osmometers. The presence of internal colloid is also important because its effect on water activity can only be compensated by adding a large, non-penetrating molecule to the medium (unless metabolically driven active salt extrusion is taking place). The internal colloid osmotic pressure may be balanced by external sucrose[121].

The critical volume (V_H) at which haemoglobin is released can be calculated from the equation[122, 123].

$$P_c = \frac{0.58 V_0}{V_H - V_0}$$

where V_0 is the initial volume at 310 milliosmolar and P_c is the critical osmotic pressure. The osmotically inert volume is taken as 42% of the total volume. V_H can also be determined from the relation

$$V_H = 6 \times A^{\frac{3}{2}} \pi^{\frac{1}{2}}$$

(A is the surface area). The salt concentration at which haemolysis commences is given by $C_H = (0.0032 P_c - 0.015)g\ 1^{-1}$.

In the osmotic haemolysis experiment, cells that initially approximate to spherical in shape, and are therefore closer to the critical haemolytic volume, will haemolyse at higher salt concentrations than cells that are initially biconcave[6, 124]. One reason for the preferential haemolysis of old red cells is that they are more spherical than young cells[125-127]. This is due to the loss of membrane fragments without loss of intracellular colloid material[50]. Calculation of membrane tension shows that the membrane can withstand tensions up to a maximum of 2×10^{-2} N m^{-1} for short periods. The membrane substance has a Young's modulus of 10^5–10^7 N m^{-2} and a viscosity of 10^9–10^{12} cP, and may be thought of as a tough viscoelastic solid[13]. Sphered cells, and cells with solid inclusions, such as occur in certain pathological states (e.g. Heinz body anaemias) are particularly prone to mechanical haemolysis.

Many other factors can cause haemolysis. Haemolysis may be induced by many lipid-soluble and surface-active substances, though a large number of these compounds, at very low concentrations (such as alcohol and phenothiazine anaesthetics) will protect the cell against osmotic haemolysis. Higher concentrations of these substances are immediately lytic. At drug concentrations that afford protection against lysis, the critical haemolytic volume and corresponding surface area are increased[129, 130a].

REVERSAL OF HAEMOLYSIS, GHOST CELLS

When exposed to solutions less concentrated than the critical value, the cell membrane becomes permeable to large molecules and much of the internal haemoglobin emerges, to equilibrate with the medium. The amount of haemoglobin retained is in proportion to the ratio cell water : total water, which is normally very small. The haemoglobin-free cells have a very pale, translucent appearance, whence the term 'ghosts'. The ghosts can be made to shrink and regain their original permeability characteristics, by resuspending them in salt solutions of physiological tonicity at 37 °C for about 30 min. One may add selected enzyme proteins, metabolic intermediates, inhibitors, etc., to the suspending medium, and by washing the ghost preparation afterwards, these substances can be incorporated within the reconstituted ghost interior[130b, 162], and an extracellular medium of alternative selected composition can be established. Thus one can study the interactions of the red cell membrane with cytoplasmic and extracellular environments of one's own choice.

There is an alternative method of re-establishing the intracellular composition that does not involve cell lysis. Red cells become permeable to alkali cations when they are suspended in non-electrolyte solution (lactose, about 0·17 M is preferred), and to this solution is added the electrolyte substances that it is desired to present to the intracellular compartment. The original permeability characteristics of the cell can be regained on addition of Ca^{2+} (2·5 mM), but there is one difference: the permeability of the cells is now absolutely dependent on the continued presence of Ca^{2+} in the extracellular fluid[130c].

ION REGULATION

In order to maintain the steady state, the activity of water inside the cell must be equal to that outside the cell. If the red cells were inert, semi-permeable sacs containing haemoglobin, they would take up water to the point of haemolysis. Haemoglobin is present in the red cell at a concentration of 7 mmol per kilogramme of water. The balance of water activity can be maintained only if sufficient external non-penetrant (such as sucrose) is added to the suspending medium[131] or by the operation of the sodium pump that dynamically maintains the total concentration difference of penetrating ions across the membrane. The ionisation of the internal haemoglobin also plays a part in osmotic balance, because its contribution as an anion is subtracted from the number of small, singly charged anions (Cl^-, HCO_3^-) that would be required to neutralise the internal cations (Table 4.5). The ionised haemoglobin normally provides 35–50 mequiv. of anion to be associated with cations per litre of cell water[132]. The degree of ionisation depends on the pH and the extent of oxygen association of haemoglobin. Oxyhaemoglobin is a stronger acid than deoxyhaemoglobin.

Careful studies of the relation between osmolarity and cell volume suggested that only 80% of the cell water acts as a solvent[133]. Thus, the cell does not shrink as much as would be expected when put in a solution of higher than normal osmolarity. This is due to the anomalous behaviour of haemoglobin, which makes a contribution to osmotic pressure about three times greater than would be predicted on the basis of its concentration alone[134]. Cell volume (v) and osmotic pressure (π) may be related by the empirical relationship ·

$$\pi(v - b)^3 = 0.683\,(v - b)^2 + 0.009\,58\,(v - b) + 0.009\,03$$

where b is the sum of the volumes occupied by internal solutes, and which is conveniently put as $0.855 \times (Hb\,\mathrm{g\,ml^{-1}})$[135]. In isotonic solution, $b = 0.286$ ml per millilitre of cells. This anomalous behaviour of haemoglobin is due to its reduced isonisation as the concentration increases. Thus, the shrunken cell with its very high haemoglobin concentration requires additional small anions to balance its cations, which are no longer neutralised by haemoglobin. Since shrinkage leads to a net gain of total solute, as well as to a loss of water, the water movement is less than would otherwise be the case[136].

It is primarily the haemoglobin that sets the Donnan distribution of those penetrating ions that are not subject to an active pump mechanism. Direct measurement with microelectrodes shows the transmembrane potential to be about 8 mV (inside negative)[137]. (See Figures 4.5 and 4.6.) The membrane potential is reduced in old red cells[138]. This difference, which is small, agrees approximately with potential differences calculated using the Nernst equation for internal and external chloride concentration. The haemoglobin itself makes only a small contribution to the potential difference. A Donnan equilibrium distribution holds for the highly mobile ions Cl^-, HCO_3^-, and H^+, so that

$$\frac{[Cl^-]_i}{[Cl^-]_e} = \frac{[HCO_3^-]_i}{[HCO_3^-]_e} = \frac{[OH^-]_i}{[OH^-]_e} = \frac{[H^+]_e}{[H^+]_i}$$

(see references 139, 140, 132, 141, 142). The resulting transmembrane potential could contribute to the driving force, causing downhill

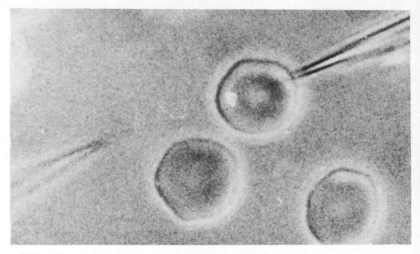

Figure 4.5.   A red cell being impaled at the rim by a KCl-filled glass microelectrode with a tip diameter of less than 0·25 *μm (From Jay and Burton*[138]; *courtesy* Biophys. J.)

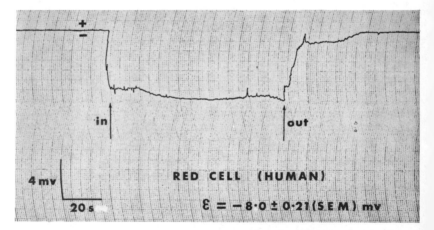

Figure 4.6.   A typical potential profile obtained in a measurement of the electrical potential difference across the human red cell membrane. Measurements were only considered acceptable if the potential remained constant during 5 s, and microscopic observation made it certain that the cell had been penetrated without contact of the needle with the glass cavity in which the red cells were placed (From Jay and Burton[137]; *courtesy Biophys. J.)*

114

movements of the cations[143], and it appears that it does so in a complex manner. By partial substitution of external salt by sucrose to control the Donnan ratio, it has been found that the relationship of cation efflux to potential shows very sharp inflections, at 45 and 170 mV, suggesting an all-or-none shift from one membrane state to another, but this is more readily understood on the basis of a gradual increase in permeability above 45 mV, with permeability independent of potential below this figure[144].

OBLIGATORY CATION EXCHANGE

When cations are transferred across the red cell membrane there is an obligation for electroneutrality to be maintained, and it has been shown that this occurs by the movement of an alternative cation in the contrary direction[145]. Such obligatory cation exchange is illustrated by experiments with ionophorous antibiotic substances, and the results are generally similar to those with liposomes (see Chapter 1) and mitochondria (see Chapter 8). Treatment of cells with valinomycin or the actins renders them permeable to K^+ and Rb^+, as can be shown with isotopic tracers (see Figure 4.7)[146].

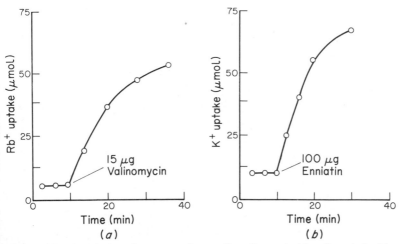

Figure 4.7. Stimulation of cation exchanges by valinomycin (a) and enniatin (b). In (a), valinomycin is seen to induce exchange between internal K^+ and external $^{86}Rb^+$: the medium contained isotopically labelled RbCl (75 mM), plus choline chloride (70 mM) and tris chloride (6 mM). In (b), enniatin is seen to induce exchange between internal K^+ and external $^{42}K^+$; the medium contained isotopically labelled KCl (75 mM), plus choline chloride (70 mM) and tris chloride (6 mM). 0·75 ml cells in 7·6 ml medium at 30°C (Henderson et al.[146]; courtesy Biochem. J.)

115

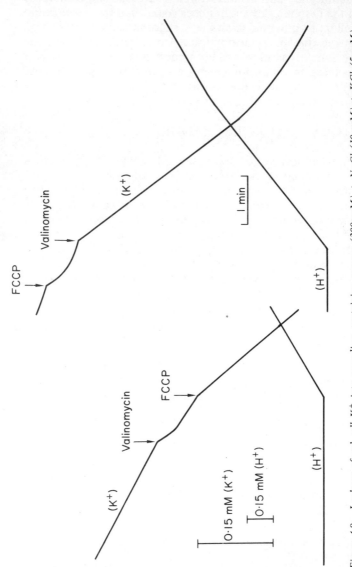

Figure 4.8. *Leakage of red cell K^+ into a medium containing sucrose (300 mM) + $NaCl$ (10 mM) + KCl (5 mM). At 'Valinomycin', 4 μg valinomycin was added, and at 'FCCP' trifluoromethoxycarbonyl cyanide was added to 5 mM. The traces show the responses of K^+ and H^+ sensitive glass electrodes monitoring concentrations in the medium. Note that addition of val or FCCP alone has only a transient effect whereas addition of the two agents together leads to an extensive K^+ for H^+ exchange*

However, the rate of leak of K^+ into a solution of low ionic strength, or into a solution of isotonic NaCl, is only transiently increased by application of these substances. If the cells are now made permeable to H^+ by the addition of mitochondrial uncoupling agents (e.g. FCCP), or to Na^+, by the addition of gramicidin or monensin, then K^+ will emerge rapidly (see Figure 4.8). With gramicidin addition, internal K^+ can be discharged into a high Na^+ medium, or internal Na^+ can be discharged in exchange for K^+ (see Figure 4.9). Addition of monensin to suspensions of human red cells in a

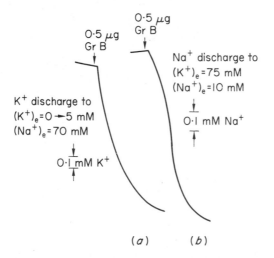

Figure 4.9. The effects of gramicidin. In (a) red cells are suspended in a medium containing NaCl, and the loss of internal K^+ to the medium following addition of gramicidin (0·5 µg) is measured with a K^+ sensitive glass electrode. In (b), loss of internal Na^+ into a medium containing a relatively high concentration of KCl following addition of gramicidin, is measured with a Na^+ sensitive glass electrode. Each experiment with 0·5 ml human red cells in 10 ml medium (Harris, unpubl.)

high Na^+ medium appears first to bring about a K^+ efflux in exchange for H^+, and then Na^+ entry with H^+ efflux[146]. A direct exchange between internal K^+ and H^+ is obtained with nigericin (Figure 4.10); the entry of protons causes internal acidification and the cells collapse because the protons combine with buffering anions and so do not make a compensating osmotic contribution for the K^+ that has been lost. Dog red cells that naturally have a high Na^+ content will exchange internal Na^+ for external K^+ when a suitable ion transporting substance such as gramicidin is present

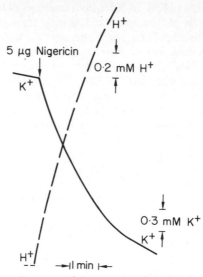

Figure 4.10. Exchange of internal K^+ for protons induced by addition of nigericin to a suspension of human red cells. The traces from K^+ and H^+ sensitive glass electrodes show the increase in K^+ and the decrease in H^+ in the medium. The medium contained sucrose (300 mM), tris chloride (20 mM) and NaCl (10 mM). 0·5 mM human red cells in 10 mM medium (After Harris and Pressman[145]; courtesy Nature, Lond.)

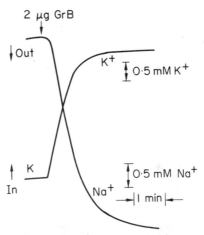

Figure 4.11. Exchange of internal Na^+ for external K^+ promoted by addition of gramicidin B to a suspension of dog red cells in a medium containing sucrose (260mM), NaCl (10mM), KCl (30mM). The traces are from glass electrodes monitoring the concentrations of K^+ and Na^+ in the medium. 0·4 ml dog red cells in 10 ml medium (After Harris and Pressman[145]; courtesy Nature, Lond.)

118

(Figure 4.11). It will be clear from these experiments that in no case do anions move to maintain electroneutrality when the cation balance of red cells is disturbed.

STOICHIOMETRY OF CATION TRANSPORT

The red cells of most species contain K^+ at about 100 mM, while the plasma is about 150 mM in Na^+ salt. The difference in composition is maintained by an active outward movement of Na^+[147, 148, 149] associated with an ATPase enzyme located in the membrane[150]. Cells that are deprived of glucose will only extrude Na^+ in proportion to their endogenous ATP content (plus ATP precursors such as 2,3-diphosphoglycerate)[151]. In a gross sense, the ATP-dependent extrusion of Na^+ requires the presence of external K^+[152] and this led to the proposal of a tight coupling between the Na^+ and K^+ transport processes[153, 31]. Na^+ for Na^+ exchange processes that are dependent on the presence of ATP but that occur without detectable ATP hydrolysis have been described[154]. The molar ratios of Na^+ and K^+ transport to ATP consumed (in human red cells) have been estimated to be $3Na^+$: $2K^+$: 1 ATP[155, 149, 31, 156], though some reports favour a 1 : 1 exchange of Na^+ and K^+[157, 86]. The products of ATP hydrolysis, ADP and inorganic phosphate, are liberated within the cell[156, 158].

Cardiac glycosides, when present in the external medium, interfere with the ATP-dependent Na^+ for K^+ exchange; this inhibition can be partly overcome by raising the external K^+ concentration[157, 159]. Ghosts that have been loaded with Na^+, Mg^{2+}, and ATP show a reduction in the rate of Na^+ output if a cardiac glycoside is present, and this shows that the inhibiting effect is directly on the transport system, rather than on the metabolic processes of the cell[160].

The kinetic course of adjustment from a high cell Na^+ content to a low level, coupled with hydrolysis of ATP, is like that of a first-order rate process (for equation, see Chapter 2). With the cell Na^+ between 7 and 15 mequiv l^{-1}, a steady state is reached with the output balancing the input, so the content remains steady. If the cell Na^+ content is made to fall (by exposure to a medium of low Na^+ concentration), the efflux rate decreases more than would be predicted from the cell Na^+ concentration. Over a certain range of intracellular Na^+ concentration, the rate of efflux can be linearly related to the cube of intracellular Na^+ concentration, which is similar to the flux–$[Na^+]$ relation seen in frog muscle cells. It has been suggested that packets of $3Na^+$ ions are moved together[161].

119

SODIUM- AND POTASSIUM-DEPENDENT ATPASE

Red cell membranes suspended in a solution containing Mg^{2+} will catalyse the hydrolysis of added ATP. Studies on cation activation and inhibition, and selective inhibition studies with cardiac glycosides, have indicated that two main types of ATPase exist in red cell membranes[162]. The ratio between the activities of the two ATPase components has been shown to depend on the concentration of Ca^{2+} ions[163], and the method of preparation of the enzyme extract[76, 163, 164]. It is possible that the presence of two ATPase systems is due to a variable degree of coupling between ATP hydrolysis and ion transport[165], rather than the existence of two separate enzymes or functional sites. Because of the apparent existence of two membrane ATPase enzymes, they are distinguished descriptively as the Na^+- and K^+-sensitive ATPase, and the Na^+- and K^+-insensitive ATPase. For both enzymes, the inorganic phosphate and ADP derived from the substrate ATP are released into the interior of the cell[156, 166].

Working with membrane fragments, it has been shown that the ATPase reaction has a dependence upon the three metal ions, Na^+, K^+, and Mg^{2+}. Magnesium ion is an absolute requirement, and activates the enzyme in the absence of Na^+ and K^+. In the presence of magnesium, addition of either Na^+ or K^+ alone has no effect; when added together, there is a large activation of the enzyme[163]. The Michaelis constants of Na^+ and K^+ activation of ATP hydrolysis by membrane fragments, and of active transport, are very close[76]. Similarly, the K_i for ouabain inhibition of ATP hydrolysis by membrane fragments ($= 0.1$ mM) is the same as the K_i for inhibition of active transport in the intact cells[163]. These observations fulfil the criteria for identification of an enzyme-catalysing ATP-dependent cation transport[167, 168].

The process of Na^+ and K^+ transport coupled to ATP hydrolysis is reversible. By exposing red cell ghosts containing potassium to a sodium medium (containing no potassium) with ^{32}P-phosphate, under conditions in which glucose metabolism is inhibited with iodoacetate, it is possible to demonstrate labelling of ATP. This reverse reaction is inhibited by cardiac glycosides, as is the active transport of sodium, and the sodium- and potassium-dependent ATPase. This has been interpreted as indicating the reversal of the sodium pump to form ATP[169]. and conditions under which it is possible to demonstrate net synthesis of ATP coupled to reverse ion movements have now been described[170]. ATP synthesis can be coupled to transmembrane electron flow when ferricyanide is added externally[171].

120

Most experience with cation transport and cation-dependent ATPase indicates that the cations which activate are the same as those that are actively transported[158, 172, 173], i.e. binding of the activating cation at the receptor is inextricably linked with movement and ejection from the opposite face. However, this association of activities may not be absolute. While K^+ can be replaced by Rb^+ or Cs^+ in both transport and activator functions on the exterior of the ghost membrane, Li^+ can replace both K^+ on the exterior and Na^+ on the interior as an activator[158] but there is no evidence of active Li^+ extrusion from Li^+-loaded cells[174].

A study of the steady-state kinetics of the red cell membrane ATPase suggests that the enzyme has two sites with high affinity for monovalent cations. At one site, the affinity for Na^+ is six to eight times greater than the affinity for K^+, but these sites can interact, as gross excess of either cation is inhibitory[168, 175] (see Chapter 2).

METABOLIC CONSEQUENCES OF CATION TRANSPORT

In whole cells, the conversion of ATP to ADP by the Na^+ extrusion mechanism activates glycolysis, since it provides ADP to interact with 1,3-diphosphoglycerate (phosphoglycerate kinase catalysed reaction)[158, 176, 177]. Relevant to long-term incubation experiments with whole cells is the observation that glycolytically produced lactic acid activates the destruction of 2,3-diphosphoglycerate[178], so depleting the cell of an alternative metabolic precursor of ATP.

Further, conversion of nondiffusible 2,3-diphosphoglycerate into diffusible pyruvate and inorganic phosphate will necessitate some reorganisation of the anion content of the cells to maintain electrochemical neutrality. Also, 2,3-diphosphoglycerate (but not its monophosphorylated analogues) binds divalent cations at physiological pH ($K'_{assoc} = 2.6 \, mM^{-1}$ for Ca^{2+} at 37 °C, pH 7.2 decreasing to $0.5 \, mM^{-1}$ in presence of KCl, $160 \, mM^{179}$). Clearly, the loss of 5 mM 2,3-diphosphoglycerate that occurs during prolonged experiments on intact cells can have a multiplicity of effects, direct and indirect (i.e. through metabolic control processes) on the transport and accumulation properties of the red cell.

Another alternative substrate for red cell glycolysis is the ribose of nucleosides such as adenosine. Deoxyribose is also effective. The glycolytic process is controlled by the availability of ADP, whether glucose[180] or nucleoside[181] is the substrate.

MOLECULAR BASIS OF THE TRANSPORT ENZYME SYSTEM

By labelling red cell ghosts with tritiated digoxin to form a stable complex, it has been shown that there are about 200 Na^+- and K^+-dependent ATPase active sites per cell[103, 102], giving a catalytic site turnover of 10.5×10^3 min^{-1} for the enzyme. This figure agrees well with estimates of turnover based on labelling with ATP-γ-^{32}-P[182]. Radiation inactivation techniques have been used to estimate the molecular weight of the enzyme complex—current data suggest a value around 0.25×10^6 (reference 186).

The first considerable step in gaining an understanding of the molecular mechanism of the Na^+- and K^+-dependent ATPase and the coupled transport system was the discovery of the phospho-protein cycle[152]. Using terminally labelled ATP, it was found that membrane phosphoprotein takes up the label in a Na^+- and Mg^{2+}-dependent reaction, and loses the label in a separate K^+ and Mg^{2+}-dependent reaction. The second reaction is inhibited by ouabain[183].

$$\text{Protein} + \text{ATP} \underset{\text{Phosphoproteinkinase}}{\xrightarrow{Na^+ \; Mg^{2+}}} \text{Phosphoprotein} + \text{ADP}$$

$$\text{Phosphoprotein} \underset{\text{Phosphoprotein phosphatase}}{\xrightarrow{K^+ \; Mg^{2+}}} \text{Protein} + \text{phosphate}$$

Net reaction

$$\text{ATP} \xrightarrow{Na^+ \; K^+ \; Mg^{2+}} \text{ADP} + \text{phosphate}$$

All membranes able to catalyse this series of reactions also carry a phosphatase activity that can be measured using model substrates such as p-nitrophenyl phosphate. The phosphatase is Mg^{2+}- and K^+-dependent and ouabain-inhibited, and is thought to mimic the final reaction of the ATPase reaction complex.

By using N-ethylmaleimide to inhibit the reaction, a second intermediate form of the phosphoprotein, $E-P_2$, has been identi-fied[184]. The phosphate attachment to the protein (presumably $E-P_2$) is thought to be to the γ-carboxyl group of a glutamate residue[185].

An intimate understanding of the mechanism of action of this enzyme must await its purification. It is likely that the 'enzyme' is in truth a complex of closely coupled enzymes and ion carriers. Unlike the case with redox-coupled reactions of mitochondria, spectroscopic investigation of the intact enzyme to investigate the various steps is not possible. Like mitochondria, however[187], it is to be hoped that eventually the separate entities may be dismantled from the whole, and reconstituted in simpler parts so that the separate steps of the reaction can be studied in isolation. At this time, even the more optimistic hope that relatively simple 'iono-phorous' carrier molecules might be discovered in mammalian membranes[188] is still unfulfilled, and it can be added that the whole concept of phosphoprotein involvement in ion transport is not without its critics[189].

In the meantime, kinetic and molecular model mechanisms of active ion transport proliferate. One difficulty is to devise a model which suggests a critical experiment (see references 174, 185, 190–195 —the authors cited cannot all be correct).

SOME BIOLOGICAL AND CLINICAL ASPECTS OF CATION TRANSPORT

Biological and clinical observations are pertinent to the under-standing of the molecular basis of cation transport. Na^+- and K^+-dependent ATPase may be induced and repressed in the red cells of rats by suitable control of the K^+ level in the diet[196a]. Among sheep and cattle, the intracellular K^+ concentrations, and the Na^+- and K^+-dependent ATPase activity show genetic and ontogenetic variation. The young animals all have a high K^+ concentration and high enzyme activity. All adult cattle, and some sheep, have low intracellular K^+ concentration and a low level of membrane enzyme activity. Low intracellular K^+ concentrations are always associated with a low activity of Na^+- and K^+-dependent ATPase[196b]. A single-locus, two-allele system is responsible for the polymorphism in sheep, with the low-potassium (LK) low-ATPase gene dominant[197, 7]. The high-potassium (HK) gene is constantly associated with the M blood-group gene[198] but the gene products do not appear to be immunologically related. By immunising HK sheep with LK red cells, it is possible to produce an anti-L serum [199, 200]. Addition of anti-L serum to LK cells produces a large stimulation of K^+ influx and of the Na^+- and K^+-dependent ATPase[200], but has no effect on membrane acetyl choline esterase nor on the electrophoretic properties of the cells. The antibody therefore appears to interact rather specifically with the membrane cation transport system. The L and M blood-group antigens are very weak in or absent from foetal cells, and appear together with the adult red cells at the time when the adult (i.e. LK or HK) K^+ levels are being established and the adult haemoglobin makes its appearance[201]. Preliminary experiments suggest that the L sub-stance controls the effective number, but not the quality of the cation-specific active sites because (1) treatment of LK cells with anti-L serum increases the extent of digoxin binding and (2) the K^+ affinity of LK ATPase is not affected by anti-L serum[202].

It is reported the ATPase activity of red cell membranes (and sarcoplasmic vesicles) of patients suffering from Duchenne muscular dystrophy (and of white Pekin ducks with a spontaneously developed myopathy) is activated by ouabain at $10^{-4} M$[203]. Activations up to 68% are recorded, compared with between 30% and 80% inhibition by ouabain of the ATPase of normal red cells. A similar effect of ouabain is observed when normal red cells are incubated with the serum of patients with Duchenne dystrophy[204], suggesting that it is some component in the dystrophic serum that affects the membrane enzyme rather than a structural disorder of the membrane itself.

In the inherited condition known as hereditary spherocytosis, the red cells have an increased leakiness to Na^+; this is accompanied by an increased specific activity of Na^+- and K^+-dependent ATPase, and the rate of glycolysis is increased because of the consequent release of metabolic control. There is an increased rate of turnover of membrane phospholipids and a progressive diminution in membrane surface area, which causes the cells to sphere and become fragile[205]. It has been suggested that the disease arises from a mutation at a control gene locus[206]. In two other very rare inherited conditions, the functions of an increased leak pathway without a compensated increase in ATPase activity[207a,b] and of a deficiency of ATPase activity[208] are separately affected.

DIVALENT CATIONS

The red cell membrane is asymmetric with respect to divalent cations[162,209]. The normal human cell population contains Mg^{2+} at a concentration of about $3.5 mM$[210]: Ca^{2+} concentration is less than $0.04 mM$[174]. There is a Mg^{2+}- and Ca^{2+}-dependent ATPase system in the red cell membrane[211,212] and inhibitors of the Mg^{2+}- and Ca^{2+}-dependent ATPase (such as mersalyl and ethacrynic acid) also inhibit Ca^{2+} extrusion[213]. The divalent cation-sensitive enzyme is probably distinct from the Na^+- and K^+-dependent ATPase. Certainly, the divalent alkaline earth cations other than the universally required Mg^{2+}, which inhibit the Na^+- and K^+-dependent ATPase, are all activators of the Mg^{2+}- and Ca^{2+}-dependent enzyme[214]. Similarly, it can be expected that K^+ and other activators of the Na^+- and K^+-dependent enzyme would be apparent inhibitors of the divalent cation-stimulated enzyme owing to their calculable effect on the activity coefficients of solvated divalent cations. Demonstration of inhibition by monovalent cations at the enzyme catalytic site would have to take this consideration into account. Some inhibitors (atebrin, salyrgan, and possibly ouabain) affect both enzymes, but guanidine, amytal, and 2,4-dinitrophenol are specific for the divalent cation-dependent enzyme[214], although treatment with 2,4-dinitrophenol results in a marked increase in the passive permeability to Na^+. Ca^{2+}-dependent ATPase may be the enzyme protein 'spectrin' that, when isolated, forms fibrils in presence of Ca^{2+} [61, 62, 65]. After treatment with non-electrolyte solution, external Ca^{2+} is required to keep the passive permeability of Na^+ and K^+ at their normal low value[130c, 215].

It has been reported that the Mg^{2+}- and Ca^{2+}-dependent ATPase is depressed in the red cells of patients with cystic fibrosis[216].

INORGANIC ANIONS

Monovalent inorganic anions move very rapidly across the red cell membrane when compared with ions such as sulphate and phosphate. As a result, sulphate and phosphate movements are much more accessible to experimental study, and more is known about their mechanism of transfer.

Chloride efflux from ox and human red cells is a first-order rate process with a rate constant of $3.1\ s^{-1}$ at $25\ °C$[217] rising to $6.3\ s^{-1}$ at $38\ °C$[217–219]. This is about half a million times faster than K^+ movement and between 70 and 200 times slower than water movement (depending on whether water movement is measured by an osmotic relaxation procedure[220] or with isotopic tracers[221]).

Sulphate ions are present in blood plasma (normal range in humans is about 1–2 mM); they slowly attain a Donnan equilibrium with other anions (Cl^-, HCO_3^-) in the cells[222] that is pH-dependent. Since sulphate ions require two monovalent cations for neutralisation, the uptake of sulphate is marked by a shrinkage of the cell. The movement of slowly transferred ions, such as sulphate and phosphate, can also be measured by following the alteration in pH that arises because of the compensatory shifts of OH^- and Cl^-, which move rapidly to maintain electroneutrality[223, 224]. The external acidification can be measured with a pH electrode; the interior alkalinisation can be measured spectroscopically using methaemoglobin as an intrinsic indicator.

The kinetics of this process are species-dependent[225]. Half-time of sulphate exchange at $20\ °C$ is 10 min for pig cells and is a first-order process. Half-time for ox cells is 40 min and is a second-order process[226]. The rate of exchange is inhibited above pH 8 and below pH 5. Addition of Ca^{2+} ions prevents sulphate movement due to precipitation of $CaSO_4$[227].

Phosphate also exchanges with internal Cl in an exponential process[223]. The half-times of phosphate exchange vary widely: $t_{\frac{1}{2}}$ is 11 min for pig cells and 145 min for ox cells, with an apparent energy of activation of 113 kJ mol^{-1}[226]. Half-time for phosphate exchange in human red cells is 14 min[229,230], giving a first-order rate constant of 6.7×10^{-3} s^{-1}, with an apparent activation energy of 83 kJ mol^{-1}. The high apparent activation energies observed have been used to argue that the transport of phosphate is a carrier-facilitated process. The initial rate of incorporation of tracer phosphate into the terminal position of ATP in metabolising cells is strikingly similar to the rate of phosphate penetration[231]. Phosphate entry into metabolising cells is inhibited by iodoacetate[232] and by phlorizin, a steroid inhibitor of glucose movement[233]. However, the idea that phosphate movement is other than a diffusion process arises from confusion between steady-state observations on metabolising cells and true equilibrium measurements. Phosphate enters the red cell by a diffusion process, as the following reasons make clear. (1) There is no evidence of saturation kinetics over a wide concentration range[234]. (2) When true equilibrium conditions are established, and taking account of the distribution of ionic species at physiological pH, it is found that interior and exterior phosphate concentrations follow the Donnan distribution[235]. (3) The high energy of activation for phosphate transfer, which has been used to argue for an active or carrier-facilitated process, probably arises from constriction of ion movement in positively charged pores of approximately molecular dimensions. There is the possibility of single-file movement[236] and increase of temperature would result in greatly accelerated ion movement on this account. The activation energy for phosphate movement is of the same order as for sulphate movement—which is clearly acknowledged to move by a diffusion process[237]. Recent measurements of phosphate movements in ghosts confirm that HPO_4^{2-} (the main phosphate species at pH 7·4) enters by a diffusion process, but raise the possibility that $H_2PO_4^-$ movements display saturation kinetics possibly involving a mobile equilibrating carrier[238].

TRANSPORT OF GLUCOSE AND MONOSACCHARIDES

The monosaccharide transport problem has been excellently and exhaustively reviewed by workers active in this field of research[239,240]. Here, only the main findings and some of the current problems are presented.

TECHNIQUES

There are two main techniques used to measure the rate of monosaccharide transport through red cell membranes. Either one follows the outflow of an isotopically labelled molecule, or one uses an osmotic relaxation technique[241–243]. Red cells, equilibrated in a medium of high sugar concentration, are suspended in a medium of lower sugar concentration, which causes them to swell very rapidly as water moves inwards. The rate of recovery to normal dimensions that accompanies the re-equilibration of sugar across the membrane can be measured by following changes in the light-scattering properties of the cell suspension. The isotopic method of measurement provides rate data under equilibrium conditions; the osmotic method provides rate data under the influence of a concentration gradient. Both techniques can be used to relate rate of transport to the conditions (concentration of sugar, inhibitor, pH, temperature, etc.) of the experiment.

The rate of glucose movement into human red cells (at physiological glucose concentrations) is about 250 times faster than the metabolic requirement[244]. There are wide species and ontogenetic variations in transport rates, with red cells of non-primate adults being almost impermeable[245]. The foetal cells of many non-primates display rapid glucose movements[246]. Measurement of penetration rate as a function of concentration shows clear evidence of a saturation phenomenon. The movement of sugar is not dependent on the presence of insulin[247], nor upon a transphosphorylation reaction involving ATP[248]. Glucose movement is not inhibited by iodoacetate—a potent glycolytic poison[249]. For these reasons, evidence of a mobile carrier for hexoses has long been sought. The non-metabolic, equilibrating movement of hexoses, displaying saturation kinetics, is referred to as a facilitated diffusion process.

SUBSTRATE SPECIFICITY

By studying the substrate specificity of the transport system, one can infer certain things about the geometry of the transport receptor site. The hexose ring can assume two main strainless configurations, which have been called the 'chair' and 'boat' forms. Hexose sugars all have one hydrogen (small) and one hydroxyl (large) substituent at each of carbon atoms 1, 2, 3, and 4, with a hydroxymethyl substituent at carbon atom 5. With boat form structures, the possibility arises of the close approach in space of two large substituents from different ring carbons. The staggering of the non-cyclic bonds in the chair forms decreases the possibility of this occurring and so, except in some special cases, the chair forms are inherently more

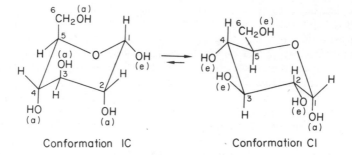

Conformation IC   Conformation CI

Figure 4.12. The two chair conformations of α-D-glucose. In the 1C conformation, the hydroxyl groups at C-2, C-3, C-5, and the hydroxymethyl group at C-5 are all axial and give rise to compression strains which result in a low stability. If the front of the 1C chair is raised and the back lowered, conformation C1 results. By this transformation all axial groups become equatorial and vice versa

stable. While the various boat forms are readily interconvertible, the energy of activation for the interconversion of the two chair forms, called C1 and 1C, is very great, since large deformations of the tetrahedral valence angles of the ring carbons would be required (see Figure 4.12). For any particular hexose molecule, the favoured chair configuration is that which places the maximum number of large substituents in equatorial positions (i.e. projecting approximately in the plane of the ring). Thus, there is an equilibrium contribution of both the C1 and 1C forms to the total. These are separated by a very high energy barrier, and the rate of attainment of equilibrium, should one of the forms—but not the other—be consumed or removed by a metabolic process, is very slow indeed.

Measurement of the concentration dependence of hexose transport by red cells shows a clear correlation between carrier-complex stability and the stability of the C1 conformation (see Figure 4.13). Thus, the α and β anomers of D-glucose (which favours the C1 conformation) have $K_M \sim 5$ mM. L-glucose, which takes up the 1C conformation, has $K_M > 3$ M.

The affinity of the carrier has a clear demand for the C1 conformation, but may be modified by those factors which determine

128

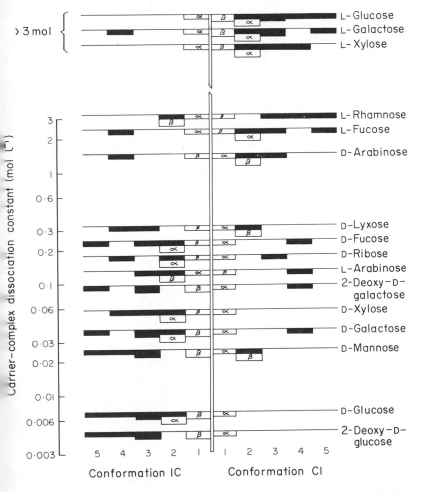

Figure 4.13. Conformational stabilities of aldopyranoses in relation to affinity for the human red cell sugar transport system. The blocks represent 'instability factors' as weighted by Reeves[266] at each ring carbon as numbered at the bottom, in each of the two alternative chair conformations. Open blocks mark instability factors which alter between the α- and β-anomers. Thus, the stability of the β-anomer in the C1 conformation is greater than α-anomer, as the C1 carbon of the β-anomer is equatorial. Since the α ⇌ β equilibrium is rapidly established, this has no apparent effect on carrier complex stability (After LeFevre[239]; courtesy Pharmac. Rev.)

129

F

conformational stability. Thus, axially substituted methyl groups in C1 hexoses drastically reduce the carrier-complex stability compared with the parent substance, while equatorial substitution results in compounds having reasonably high affinities for the carrier system.

It is concluded that a fairly rigid three-point receptor assembly in the membrane could account for the affinity series of the hexose receptor in red cell membranes. Binding of glucose to the receptor site is an exothermic reaction, probably involving the formation of several hydrogen bonds[243]. Attempts have been made to isolate a glucose receptor molecule from solublised preparations of red cell ghosts[250, 251] but, as with so many 'soluble' preparations involving treatment with non-ionic detergents, it is not clear if any measure of purification has been achieved.

THE INTERNAL TRANSFER MODEL

In the search for possible glucose carriers, little work has so far been done with model systems. There is evidence that red cell phospholipids enhance considerably the partition of glucose in the organic phase of chloroform–water mixtures[252] and increase the mobility of glucose through water–chloroform–water three-layer systems[253]. The rate of glucose diffusion from liposomes is enhanced when they are treated with the red cell membrane protein, spectrin[100]. Most ideas about the nature of the carrier systems are derived from kinetic studies, and information at the molecular level is scanty. The reason for this, it is now proposed, is that no mobile carrier system exists[254, 255]. Detailed analysis of the kinetic data from non-equilibrium experiments presents a number of discrepancies that cannot be explained on the basis of a simple mobile carrier mechanism on which the assumptions leading to the acceptance of the Michaelis equation can be made. The carrier model has been progressively refined. Corrections have to be made because the Michaelis constant K_M obtained from rate measurements cannot be equated with the dissociation constant of the glucose receptor complex except under the limiting conditions where the mobility of the carrier in the membrane is infinitesimally small[256]. It is now thought that the mobility of the carrier may depend on whether or not it is loaded, and that the mobility of the loaded carrier compared with the unloaded carrier may be greatly enhanced at low temperatures[257]. Such considerations lead to a satisfactory resolution of the problems of the carrier model in terms of data from non-equilibrium measurements. But a fundamental paradox remains.

In one laboratory[258a,b], non-equilibrium (osmotic relaxation) and equilibrium exchange (isotopic) techniques have been used to measure the steady-state kinetic parameter of glucose transport. All experiments were carried out at 20 °C, pH 7·35. Non-equilibrium measurements gave $K_M = 1·8$ mM, $V_{max} = 104$ mmol per minute per cell unit. Equilibrium measurements gave $K_M = 38$ mM, $V_{max} = 260$ mmol per minute per cell unit (1 cell unit is the quantity of cells whose water volume is 1 litre under isotonic conditions). Further, while both techniques provide values of K_M that decrease with increase of temperature, non-equilibrium measurements require that carrier-complex formation be exothermic (i.e. ΔH negative)[243, 257], but equilibrium measurements require that complex formation be endothermic[259]. Owing to the impossibility of developing a mobile carrier model to explain all these observations without introducing arbitrary assumptions and parameters, Lieb and Stein[255] have now resolved this problem in terms of a tetrameric transmembrane protein, having noninteracting strong and weak binding sites at both surfaces. On this basis, 2^4 model occupancy states are defined, but consideration of the experimental data, and exclusion of impossible occupancy states, remove all but four from further consideration.

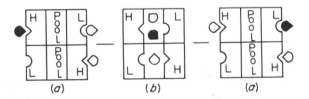

Figure 4.14. Illustration of the details of a transport event in the internal transfer model having an internal pool. The case considered is where $K_H \ll K_L$. The substrate bound initially at the left-hand side is shaded. In this example the transformation from state (b) back to state (a) can with equal probability result in the substrate being redistributed as at the extreme left (no transport) or at the extreme right (two molecules exchanged) (After Lieb and Stein[255]; courtesy Biophys. J.)

The protein is assumed to be fixed in the membrane, the glucose moving between the subunits on either side (see Figure 4.14). The tetrameric transfer complex has two conformations—one in which the binding sites are exposed to the solvent on either side of the membrane, the other in which the binding sites are in communication with each other across a central pool. In the second conformational state, the sugar molecules are redistributed according to their affinities for the various binding sites that are now in communication with each other. The rate of transition between conformational states, and hence the rate of transfer between the subunits, depends on the state of occupancy of the receptor sites on both side of the membrane. The binding affinity at the various sites is unaffected by binding at other sites.

The fit of the predictions of this internal transfer model with most of the experimental data is very good, and certainly it overcomes the central paradox of the mobile carrier hypothesis of monosaccharide transport. The authors indicate how models of this kind—using the concepts of multiple subunit interacting enzyme proteins (in this case, the Koshland model, see reference 260)—can be extended to a wide variety of membrane transport processes, though active cation transport is excluded from their discussion (see also reference 254).

CONSIDERATION OF A NON-CARRIER MODEL FOR CATION TRANSPORT

It should be asked whether the success of the mobile carrier concept in describing the kinetics of active transport[261] and the behaviour of the ionophorous antibiotics in membranes has not excluded from consideration alternative model systems that may be equally relevant. In particular, the failure so far to isolate and identify a small molecular translocator from mammalian membranes as predicted[189], in spite of the availability of powerful analytical methods[267], suggests that the time is ripe for consideration of other mechanisms. A further problem of the mobile carrier model is to account for the necessary asymmetric quality of active transport whereby the carrier must have a high affinity for the carried ion at one surface and a low affinity at the other. A development of an asymmetric internal transfer model would be attractive. Transport, active or passive, would be controlled by the affinity of cations for receptor sites both on the exterior of the membrane and in a central pool. Affinity of the various sites would be the key to such a system, and it is suggested that this would arise from the charge density properties of the ion exchange surface of the receptors according to the Eisenmann rules[111]. The charge density could be controlled by varying the degree of compression of the molecules contributing to the receptor site ion exchange surface—this depending on the simultaneous presence of ATP, Na^+, and K^+. In this case, hydrolysis of ATP could be a purely secondary reaction, allowing for relaxation between the compression states of the ion exchange surface, or the compression process itself could be actuated by transphosphorylation from ATP. Control of charge density at the ion exchange surface could also arise from ATP-dependent transphorylation.

This model has certain advantages over mobile carrier systems: it provides a distinct operational role for anionic fixed groups that are such a dominant feature of membrane structure; control of cation specificity can be developed along the lines of physical chemistry[114]; there is no possibility of loss of control due to disturbance of the equilibrium between the free carriers and carriers in the membrane; the role of ATP can be expressed in terms of chemical notions: no energy linked process is called for[263]; a large number of the well-known carrier substances under investigation at the present time are antibiotics—toxic not only to bacterial cells but to mammalian cells, too. On the other hand, it is easier to comprehend the obligatory exchange of cations on the basis of an association between cation and negatively charged carrier—the neutral cation carrier complex so formed being the only mobile species.

CONCLUSION

At present, no final judgments can be made concerning the nature of the transport systems of the red-cell membrane. The very low activity of the red cell as a mover of ions raises the question of whether the red cell is the best place in which to search for the molecular apparatus of ion transport, though an eightfold purification of Na^+- and K^+-dependent ATPase from red cell membranes treated with sodium dodecyl sulphate has been reported[264].

The success of studies with red cells has been in the field of kinetics. Certainly, the approach to the molecular level of comprehension is closer with certain bacterial transport systems[265, 266]. In eucaryotic cells, the best approach so far to the description of a transport system at the molecular level is of the vitamin-D-induced calcium-binding protein of the intestinal mucosal cells[267]. The isolation of the molecular apparatus of ion transport still has a long way to advance before it can help towards an understanding of the mechanisms of the transport processes, but it will be an essential step. 'The process of transport will be understood when, as in enzyme catalysis, all of the parts of some transport system are isolated, and their structures and interactions have been clarified. Kinetics is always a limited approach—it at best eliminates hypotheses, but by itself can never provide final proof—owing to the complexity of biological systems with their numerous unknown parts.'[265]

REFERENCES

1. GOMPERTS, B. D., 'The biochemistry of red blood cells in health and disease', in *The Biological Basis of Medicine* (ed. E. E. BITTAR), Academic, London, vol. 3 (1969)

2. MARIKOVSKY, Y., DANON, D., and KATCHALSKY, A., 'Agglutination by polylysine of young and old red blood cells', *Biochim biophys, Acta,* **124**, 154 (1966)

3. WESTERMAN, M. P., PIERCE, L. E., and JENSEN, W. N., 'Lipid quantification in normal young and normal aged circulating erythrocytes', *J. clin. Invest.,* **38**, 1054 (1959)

4. WESTERMAN, M. P., PIERCE, L. E., and JENSEN, W. N., 'Erythrocyte lipids. A comparison of normal young and normal old populations'. *J. Lab. clin. Med.,* **62**, 394 (1963)

5. BERNSTEIN, R. E., 'Alterations in metabolic energetics and cation transport during ageing of red cells. *J. clin. Invest.,* **38**, 1572 (1959)

6. PRANKERD, T. A. J., 'The ageing of red cells', *J. Physiol.,* **143**, 325 (1958)

7. BREWER, G. J., EATON, J. W., BECK, C. C., FEITLER, L., and SHREFFLER, D. C., 'Sodium–potassium-stimulated ATPase activity of mammalian hemolysates: clinical observations and dominance of ATPase deficiency in potassium polymorphism of sheep', *J. Lab. clin. Med.,* **71**, 744 (1968)

8. BEUTLER, E., 'Cellular mosaicism and heterogeneity in red cell disorders'. *Am. J. Med.,* **41**, 724 (1966)

9. ONI, S. B., OSUNKOYA, B. O., and LUZZATTO, L., 'Paroxysmal nocturnal haemoglobinuria: evidence for monoclonal origin of abnormal cells', *Blood,* **36**, 145 (1970)

10. AUDITORE, J. V., and HARTMANN, R. C., 'Paroxysmal nocturnal haemoglobinuria. II Erythrocyte acetyl choline esterase defect', *Am. J. Med.,* **27**, 401 (1959)

11. SURGENOR, D. M., 'The transport of oxygen and carbon dioxide', in *The Red Blood Cell* (eds C. BISHOP and D. M. SURGENOR), Academic, New York (1964)

12. HENDERSON, J. F., and LE PAGE, G. A., 'Transport of adenine-8-C^{14} among mouse tissues by blood cells', *J. biol. Chem.,* **234**, 3219 (1959)

13. RAND, R. P., 'Mechanical properties of the red cell membrane. II, Visco-elastic breakdown of the membrane', *Biophys. J.,* **4**, 303 (1964)

14. PERUTZ, M. F., 'Submicroscopic structure of the red cell', *Nature, Lond.,* **161**, 204 (1948)

15. GORTER, E., and GRENDEL, F., 'On bimolecular layers of lipids on the chromocytes of the blood', *J. exp. Med.,* **41**, 439 (1925)

16. DANIELLI, J. F., and HARVEY, E. N., 'The tension at the surface of mackerel egg oil with remarks on the nature of the cell surface', *J. cell. comp. Physiol.,* **5**, 483 (1934)

17. DANIELLI, J. F., in *Surface Phenomena in Chemistry and Biology,* (eds J. F. DANIELLI, K. G. A. PANKHURST, and A. C. RIDDIFORD), London, Pergamon, 246 (1958)

18. MUELLER, P., RUDIN, D. O., TIEN, H. T., and WESCOTT, W. C., 'Reconstitution of excitable cell membrane structure *in vitro*', *Circulation* **26**, 1167 (1962)

19. MUELLER, P., and RUDIN, D. O., 'Resting and action potentials in experimental bimolecular lipid membranes', *J. theoret. Biol.,* **18**, 222 (1968)

20. HUANG, C., WHEELDON, L., and THOMPSON, T. E., 'The properties of lipid bilayer membranes separating two aqueous phases: formation of a membrane of of simple composition', *J. molec. Biol.,* **8**, 148 (1964)

21a. KORN, E. D., 'Structure of biological membranes', *Science,* **153**, 1491 (1966)

21b. BAR, R. S., DEAMER, D. W., and CORNWELL, D. G., *Science*, **153**, 1010 (1966)
22a. ROBERTSON, J. D., 'The ultrastructure of cell membranes and their derivatives', *Biochem. Soc. Symp.*, **16**, 3 (1959)
22b. ROBERTSON, J. D., 'Origin of the unit membrane concept', in *Biophysics and Physiology of Biological Transport* (eds L. BOLIS, V. CAPRARO, K. R. PORTER, and J. D. ROBERTSON), Springer Verlag, New York, 218 (1967)
23. WINTROBE, M. M., *Clinical Haematology*, 5th edn, Kimpton, London (1961)
24. ZWAAL, R. F. A., and VAN DEENEN, L. L. M., 'Interactions between proteins and lipids from human red cell membranes', *Chem. & Phys. Lipids*, **4**, 311 (1970)
25. ZAHLER, P., and WEIBEL, E. R., Reconstitution of membranes by recombining proteins and lipids derived from erythrocyte stroma', *Biochim. biophys. Acta*, **219**, 320 (1970)
26a. DEAMER, D. W., 'An alternative model for molecular organization in biological membranes', *J. Bioenergetics*, **1**, 237 (1970)
26b. HENDLER, R. W., 'Biological membrane ultrastructure', *Physiol. Rev.*, **51**, 66 (1971)
27. CHRISTENSEN, J. J., and IZATT, R. M., 'Thermodynamics of proton dissociation in dilute aqueous solution. II. Heats of proton dissociation from ribonucleotides and related compounds determined by a thermometric titration procedure', *J. Phys. Chem.*, **66**, 1030 (1962)
28. PONDER, E., 'Present concepts of the structure of the mammalian red cell', *Blood*, **9**, 227 (1954)
29. BAYER, M. E., 'Elektronenoptische Untersuchungen an der Membran des Erythrocyten', *Pflügers Arch. ges. Physiol.*, **270**, 323 (1960)
30. MCLELLAN, W. L., and LIONETTI, F., 'Hypoxanthine production from nucleosides in erythrocyte ghosts', *J. biol. Chem.*, **234**, 1622 (1959)
31. POST, R. L., and JOLLY, P. C., 'The linkage of sodium, potassium and ammonium active transport across the human erythrocyte membrane', *Biochim. biophys. Acta*, **25**, 118 (1957)
32. STEIN, W. D., 'Permeability of erythrocyte ghosts', *Expl. Cell. Res.*, **11**, 232 (1956)
33. LE FEVRE, P. G., 'Persistance in erythrocyte ghosts of mediated sugar transport', *Nature, Lond.*, **191**, 970 (1961)
34. O'DONNELL, J. M., and ELLORY, J. C., 'Shape, volume and integrity of reconstituted human red cell ghosts', *Experientia*, **25**, 529 (1969)
35. PONDER, E., and PONDER, R. V., 'The electrophoretic velocity of human red cells, of their ghosts and mechanically produced fragments, and of certain lipid complexes', *J. gen. Physiol.*, **43**, 503 (1960)
36. DODGE, J. T., MITCHELL, C., and HANAHAN, D. J., 'The preparation and chemical characterization of hemoglobin-free ghosts of human erythrocytes', *Arch. biochem. biophys.*, **100**, 119 (1963)
37. VAN DEENEN, L. L. M., and DE GIER, J., 'Chemical composition and metabolism of lipids in red cells of various species', in *The Red Blood Cell* (eds C. BISHOP and D. M. SURGENOR), Academic, New York (1964)
38. ROUSER, G., NELSON, G. J., FLEISCHER, S., and SIMON, G., 'Lipid composition of animal cell membranes, organelles and organs', in *Biological Membranes* (ed D. CHAPMAN), Academic, London (1968)
39. NELSON, G. J., 'Composition of neutral lipids from erythrocytes of common mammals', *J. Lipid Res.*, **8**, 374 (1967)
40. BALLANTINE, R. S., and PARPART, A. K., 'The action of lipase on the red cell surface', *J. cell. comp. Physiol.*, **16,** 49 (1940)
41. NEERHOUT, R. C., 'Abnormalities of erythrocyte stromal lipids in hepatic disease', *J. Lab. clin. Med.*, **71**, 438 (1968)
42. FARNSWORTH, P., DANON, D., and GELHORN, A., 'The effect of fat free diet on the rat erythrocyte membrane', *Brit. J. Haematol.* **11**, 200 (1965)

RED BLOOD CELLS

43. HAGERMAN, J. S., and GOULD, R. G., 'The *in vitro* interchange of cholesterol between plasma and red cells', *Proc. Soc. exp. Biol. Med.*, **78**, 329 (1951)

44. LONDON, I. M., and SCHWARTZ, H., 'Erythrocyte metabolism. The metabolic behaviour of the cholesterol of human erythrocytes', *J. clin. Invest.*, **32**, 1248 (1953)

45. SAKAGAMI, T., MINARI, O., and ORII, T., 'Interaction of individual phospholipids between rat plasma and erythrocytes *in vitro*', *Biochim. biophys. Acta*, **98**, 356 (1965)

46. SILBER, R., WINTER, R., and KAYDEN, H. J., 'Tocopherol transport in the rat erythrocyte', *J. clin. Invest.*, **48**, 2089 (1969)

47. MURPHY, J. R., 'Erythrocyte metabolism. V. Cell shape and the location of cholesterol in the erythrocyte membrane', *J. Lab. clin. Med.*, **65**, 756 (1965)

48. MURPHY, J. R., 'Erythrocyte metabolism. III. Relationship of energy metabolism and serum factors to the osmotic fragility following incubation', *J. Lab. clin. Med.*, **60**, 86 (1962)

49. BRUCKDORFER, K. R., DEMEL, R. A., DE GIER, J., and VAN DEENEN, L. L. M., 'The effect of partial replacements of membrane cholesterol by other steroids on the osmotic fragility and glycerol permeability of erythrocytes', *Biochim. biophys. Acta,* **183**, 334 (1969)

50. WEED, R. I., and REED, C. F., 'Membrane alterations leading to red cell destruction', *Am. J. Med.*, **41**, 681 (1966)

51. BERG, H. C., 'Sulfanilic acid diazonium salt: a label for the outside of the human erythrocyte membrane', *Biochim. biophys. Acta,* **183**, 65 (1969)

52. JUDAH, J. D., AHMED, K., and MCLEAN, A. E. M., 'Phosphoproteins and sodium transport', *Nature, Lond.,* **196**, 484 (1962)

53. GOMPERTS, B. D., Unpublished data

54. ZITTLE, C. A., DELLAMONICA, E. S., and CUSTER, J. H., 'Purification of human red cell acetylcholinesterase', *Arch. Biochem. Biophys.*, **48**, 43 (1954)

55. MITCHELL, C. D., and HANAHAN, D. J., 'Solubilization of certain proteins from the human erythrocyte stroma', *Biochemistry*, **5**, 51 (1966)

56. GREEN, D. E., MURER, H., HULTIN, H. O., RICHARDSON, S. H., SALMON, B., BRIERLEY, G. P., and BAUM, H., *Arch. Biochem. Biophys.,* **112**, 635 (1965)

57. SCHRIER, S. L., 'Organisation of enzymes in human erythrocyte membranes', *Am. J. Physiol.*, **210**, 139 (1966)

58. RONQUIST, G., and AGREN, G., 'Formation of adenosine triphosphate by human erythrocyte ghosts', *Nature, Lond.,* **209**, 1090 (1966)

59. NILSSON, O., and RONQUIST, G., 'Enzyme activities and ultra structure of a membrane fraction from human erythrocytes', *Biochim. biophys. Acta,* **183**, 1 (1969)

60. SCHRÖTER, W., and NEURIANS, M., 'Membrane bound 2,3-diphosphoglycerate phosphatase of human erythrocytes', *J. Membrane Biol.*, **2**, 31 (1970)

61. MARCHESI, V. T., and STEERS, E., 'Selective solubilisation of a protein component of the red cell membrane', *Science,* **159**, 203 (1968)

62. MARCHESI, S. L., STEERS, E., MARCHESI, V. T., and TILLACK, 'Physical and chemical properties of a protein isolated from red cell membranes', *Biochemistry*, **9**, 51 (1970)

63. OHNISHI, T., 'Extraction of actin and myosin-like proteins from erythrocyte membrane', *J. Biochem. Tokyo*, **52**, 307 (1962)

64. STEERS, E., and MARCHESI, V. T., 'Studies on a protein component of guinea pig erythrocyte membranes', *J. gen. Physiol.*, **54**, 65 S (1969)

65a. ROSENTHAL, A. S., KREGENOW, F. M., and MOSES, M. C., 'Some characteristics of a Ca^{++} dependent ATPase activity associated with a group of erythrocyte membrane proteins which form fibrils', *Biochim. biophys. Acta,* **196**, 254 (1970)

65b. MADDY, A. H., 'A fluorescent label for the outer components of the plasma membrane', *Biochim. biophys. Acta*, **88**, 390 (1965)

66. HOOGEVEEN, J. TH., JULIANO, R., COLEMAN, J., and ROTHSTEIN, A., 'Water soluble proteins of the human red cell membrane', *J. Membrane Biol.*, **3**, 156 (1970)

67. MADDY, A. H., 'The solubilization of the protein of the oxerythrocyte ghost', *Biochim. biophys. Acta*, **88**, 448 (1964)

68. REGA, A. F., WEED, R. I., REED, C. F., BERG, G. G., and ROTHSTEIN, A., 'Changes in the properties of human erythrocyte membrane protein after solubilization by butanol extraction', *Biochim. Biophys. Acta,* **147**, 297 (1967)

69. BAKERMAN, S., WASEMILLER, G., 'Studies on structural units of human erythrocyte membrane [1] Separation, isolation and partial characterization', *Biochemistry*, **6**, 1100 (1967)

70. LAUF, P. K., and POULIK, M. D., 'Solubilization and structural integrity of the human red cell membrane', *Br. J. Haemat.,* **15**, 191 (1968)

71. ROSENBERG, S. A., and GUIDOTTI, G., 'The protein of human erythrocyte membranes. I. Preparation, solubilization and partial characterization', *J. biol. Chem.*, **243**, 1985 (1968)

72a. BLUMENFELD, O. G., 'The proteins of the erythrocyte membrane obtained by solubilization with aqueous pyridine solution', *Biochim. biophys. Res. Commun.*, **30**, 200 (1968)

72b. ZWAAL, R. F. A., and VAN DEENEN, L. L. M., 'The solubilization of human erythrocytes by n-pentanol', *Biochim. biophys. Acta*, **150**, 323 (1968)

73. BLUMENFELD, O. G., GOLDSTEIN, M., and SEIFTER, S., 'The protein of the human erythrocyte membrane', *Fedn Proc. Fedn Am. Socs exp. Biol.*, **26**, 381 (1967)

74. MADDY, A. H., 'The properties of the protein of the plasma membrane of ox erythrocytes', *Biochim. biophys. Acta,* **117**, 193 (1966)

75. MORGAN, T. E., and HANAHAN, D. J., 'Solubilization and characterization of a lipoprotein from erythrocyte stroma', *Biochemistry*, **5**, 1050 (1966)

76. POST, R. L., MERRITT, C. R., KINSOLVING, C. R., and ALBRIGHT, C. D., 'Membrane adenosine triphosphatase as a participant in the active transport of sodium and potassium in the human erythrocyte', *J. biol. Chem.*, **235**, 1976 (1960)

77. POULIK, M. D., and LAUF, P. K., 'Heterogeneity of water soluble structural components of human red cell membrane', *Nature, Lond.*, **208**, 874 (1965)

78. REGA, A. F., ROTHSTEIN, A., and WEED, R. I., Erythrocyte membrane sulphydryl groups and the active transport of cations', *J. cell. comp. Physiol.*, **70**, 45 (1967)

79. WINZLER, R., 'Determination of serum glycoproteins', in *Methods in Biochemical Analysis*, Interscience, New York, vol. 2, 279 (1955)

80. GREEN, F. A., 'Disaggregated human erythrocyte membranes and blood group activity', *J. Immunol.*, **99**, 56 (1967)

81. ROSENBERG, S. A., and GUIDOTTI, G., 'Fractionation of the protein components of human erythrocyte membranes', *J. biol. Chem.*, **244**, 5118 (1969)

82. ROSENBERG, S. A., and MCINTOSH, J. R., 'Erythrocyte membranes: effects of sonication', *Biochim. biophys. Acta,* **103**, 285 (1968)

83. MAZIA, D., and RUBY, A., 'Dissolution of erythrocyte membranes in water and comparison of the membrane protein with other structural proteins', *Proc. natn. Acad. Sci. U.S.A.*, **61**, 1005 (1968)

84. SCHNEIDERMAN, L. J., 'Solubilization and electrophoresis of human red cell stroma', *Biochem. biophys. Res. Commun.*, **20**, 763 (1965)

85. SCHNEIDERMAN, L. J., and JUNGA, I. G., 'Isolation and partial characterization of structural protein derived from human red cell membrane', *Biochemistry*, **7**, 2281 (1968)

86. TOSTESON, D. C., HOFFMAN, J. F., 'Regulation of cell volume by active cation

F*

transport in high and low potassium sheep red cells', *J. gen. Physiol.*, **44**, 169 (1960)

87. AZEN, E. A., ORR, S., and SMITHIES, D., 'Starch gel electrophoresis of erythrocyte stroma', *J. Lab. clin. Med.*, **65**, 440 (1965)

88. TILLACK, T. W., MARCHESI, S. L., MARCHESI, V. T., and STEERS, E., 'A comparative study of spectrin: a protein isolated from red blood cell membranes', *Biochim. biophys. Acta*, **200**, 125 (1970)

89. POULIK, M. D., 'Preparation, isolation and charactization of soluble red cell membrane proteins', in *Metabolism and Membrane Permeability of Erythrocytes and Thrombocytes*, (eds E. DEUTCHE, E. GERLACH, and K. MOSER), Thieme, Stuttgart, 360 (1968)

90. POULIK, M. D., and LAUF, P. K., 'Some physico-chemical and serological properties of isolated protein components of red cell membranes', *Clin. & exp. Immunol.*, **4**, 165 (1969)

91. PRANKERD, T. A. J., ALTMAN, K. I., and ANDERSON, J. R., 'Electrophoresis of human red cell stroma', *Nature, Londn,* **174**, 1146 (1954)

92. HARRIS, J. R., 'Release of a macromolecular protein component from human erythrocyte ghosts', *Biochim. biophys. Acta,* **150**, 534 (1968)

93. HARRIS, J. R., 'The isolation and purification of a macromolecular protein component from the human erythrocyte', *Biochim. biophys. Acta*, **188**, 31 (1969)

84. MOSKOWITZ, M., and CALVIN, M., 'On the components and structure of the human red cell membrane', *Exptl Cell. Res.*, **3**, 33 (1952)

95. ANDERSON, H. M., and TURNER, J. C., 'Relation of hemoglobin to the red cell membrane', *J. clin. Invest.*, **39**, 1 (1960)

96. BENESH, R. E., and BENESH, R., 'Relation between erythrocyte integrity and sulphydryl groups', *Arch. Biochem. Biophys.*, **48**, 38 (1954)

97. VAN STEEVENINK, J., WEED, R. I., and ROTHSTEIN, A., 'Localization of erythrocyte membrane sulphydryl groups essential for glucose transport', *J. gen. Physiol.*, **48**, 617 (1965)

98. SUTHERLAND, R. M., ROTHSTEIN, A., and WEED, R. I., 'Erythrocyte membrane sulphydryl groups and cation permeability', *J. cell. comp. Physiol.*, **69**, 185 (1967)

99. SANDBERG, H. E., BRYANT, R. G., and PIETTE, L. A., 'Studies on the location of sulphydryl groups in erythrocyte membranes with magnetic resonance spin probes', *Arch. Biochem. Biophys.,* **133**, 144 (1969)

100. SWEET, C., and ZULL, J. E., 'Interaction of the erythrocyte-membrane protein, spectrin, with model membrane systems', *Biochem. biophys. Res. Commun.*, **41**, 135

101. DAWSON, R. M. C., 'The nature of the interaction between protein and lipid during the formation of lipoprotein membranes', in *Biological Membranes,* (ed. D. CHAPMAN), Academic, London (1968)

102. HOFFMAN, J. F., 'The interaction between tritiated ouabain and the Na^+-K^+ pump in red blood cells', *J. gen. Physiol.*, **54**, 343 S (1969)

103. ELLORY, J. C., and KEYNES, R. D., 'Binding of tritiated digoxin to human red cell ghosts', *Nature, Lond.*, **221**, 776 (1969)

104. WALLACH, D. F. H., 'Membrane lipids and the conformation of membrane proteins', *J. gen. Physiol.*, **54**, 3S (1969)

105. GORDON, A. S., WALLACH, D. F. H., and STRAUS, J. H., 'The optical activity of plasma membranes and its modification by lysolecithin, phospholipase A and phospholipase C', *Biochim. biophys. Acta,* **183**, 405 (1970)

106. LENARD, J., and SINGER, S. J., 'Structure of membranes: Reaction of red blood cell membranes with phospholipase C', *Science,* **159**, 738 (1968)

107. COOK, G. M. W., HEARD, D. H., and SEAMAN, G. V. H., 'Sialic acids and the electro-kinetic charge of the human erythrocyte', *Nature, Lond.*, **191**, 44 (1961)
108. EYLAR, E. H., MADOFF, M. A., BRODY, O. V., and ONCLEY, J. D., 'The contribution of sialic acid to the surface charge of the erythrocyte', *J. biol. Chem.*, **237**, 1992 (1962)
109. LING, G. N., *A Physical Theory of the Living State*. Blaisdell, New York, ch. 2 (1962)
110. LING, G. N., *A Physical Theory of the Living State*, Blaisdell, New York, ch. 9 (1962)
111. EISENMANN, G., 'On the elementary atomic origin of equilibrium ionic specificity', in *Symposium on Membrane Transport and Metabolism* (eds A. KLEINZELLER and A. KOTYK), Academic, London (1960)
112. EISENMANN, G., 'Cation selective glass electrodes and their mode of operation. *Biophys. J.*, **2**, 259 (1962)
113. REICHENBERG, D., 'Ion exchange selectivity', in *Ion Exchange* (ed. J. A. MARINSKY). Dekker, New York, vol. 1 (1966)
114. DIAMOND, J. M., and WRIGHT, E. M., 'Biological membranes. The physical basis of ion and non-electrolyte selectivity', *A. Rev. Physiol.*, **31**, 581 (1969)
115. GOMPERTS, B. D., LANTELME, F., and STOCK, R., 'Ion association reactions of biological membranes studied with the fluorescent dye, 1-anilino-8-naph-thalene sulfonate', *J. Membrane Biol.*, **3**, 24 (1970)
116. ROTHSTEIN, A., and HAYES, A. D., 'The relationship of the cell surface to meta-bolism. XIII. The cation binding properties of the yeast cell surface', *Arch. Biochem. Biophys.*, **63**, 87 (1956)
117. RAND, R. P., 'Some biophysical considerations of the red cell membrane', *Fedn Proc. Fedn Am. Socs exp. Biol.*, **26**, 1780 (1967)
118. PONDER, E., *Hemolysis and Related Phenomena*, Churchill, London (1948)
119. RICH, G. T., SHAAFI, R. I., ROMUALDEZ, A., and SOLOMON, A. K., 'Effect of osmolality on the hydraulic permeability coefficient of red cells', *J. gen. Physiol.*, **52**, 941 (1968)
120. RUMMEL, W., and WILBRANDT, W., 'Über scheinbare Hemmung der osmotischen Hämolyse durch Zucker', *Pflügers Arch. ges. Physiol.*, **253**, 194 (1951)
121. WILBRANDT, W., 'Der Kompensationtest der kolloidosmotischen Hämolyse', *Helv. physiol. pharmac. Acta*, **6**, 234 (1948)
122. CANHAM, P., 'Curves of osmotic fragility calculated from the isotonic areas and volumes of individual erythrocytes', *J. cell. comp. Physiol.*, **74**, 203 (1969)
123. CANHAM, P. B., and BURTON, A. C., 'Distribution of size and shape in populations of normal human red cells', *Circulation Res.*, **22**, 405 (1968)
124. SILVER, M. M., MCMILLAN, C. C., and SILVER, M. D., 'Haemolytic anaemia in cholesterol-fed rabbits', *Br. J. Haemat.*, **10**, 271 (1964) '
125. CHALFIN, D., 'Difference between young and mature rabbit erythrocytes', *J. cell. comp. Physiol*, **47**, 215 (1956)
126. VAN GASTEL, C., VAN DEN BERG, J., DE GIER, J., and VAN DEENEN, L. L. M., 'Some lipid characteristics of normal red blood cells of different age', *Brit. J. Haemat.*, **11**, 193 (1965)
127. CANHAM, P. B., 'The difference in geometry of young and old human erythrocytes explained by a filtering mechanism', *Circulation Res.*, **23**, 39 (1969)
128. VAN DEENEN, L. L. M., 'Membrane lipids and lipophylic proteins', in *The Mole-cular Basis of Membrane Function* (ed. D. C. TOSTESON), Prentice Hall, New York (1969)
129. SEEMAN, P., KWANT, W. O., SAUKS, T., and ARGENT, W., 'Membrane expansion of intact erythrocytes by anaesthetics', *Biochim. biophys. Acta*, **183**, 490 (1969)
130a. SEEMAN, P., 'Membrane expansion and stabilization by anaesthetics and other

drugs', in *Permeability and Function of Biological Membranes* (eds L. BOLIS, A. KATCHALSKY, R. D. KEYNES, W. R. LOEWENSTEIN, and B. A. PETHICA). North Holland Publishing Co., Amsterdam (1970)

130b. STRAUB, F. B., 'Über die Akkumulation der Kaliumionen durch menschliche Blutkörperchen', *Acta. physiol. hung.*, **4**, 235 (1953)

130c. BOLINGBROKE, V., and MAIZELS, M., 'Calcium ions and the permeability of human erythrocytes', *J. Physiol.*, **149**, 563 (1959)

131. WILBRANDT. W.. 'Osmotische Natur sogenanntner nichtosmotische Hämolysen, [Kolloid-osmotische Hämolyse]', *Pflügers Arch. ges. Physiol.*, **245**, 22 (1941)

132. HARRIS, E. J., 'Distribution of ions in suspensions of human erythrocytes', *J. Physiol.*, **118**, 40 (1952)

133. PONDER, E., 'The paracrystalline state of the red cell', *J. gen. Physiol.*, **29**. 89 (1945)

134. MCCONAGHEY, P. D., and MAIZELS, M., 'The osmotic coefficients of haemoglobin in red cells under varying conditions', *J. Physiol.*, **155**. 28 (1961)

135. DICK, D. A. T., and LOWENSTEIN, L., 'Osmotic phenomena in red cells studied with the aid of immersion refractometry', *Proc. Roy. Soc. B*, **148**, 241 (1958)

136. GARY-BOBO, C. M., and SOLOMON, A. K., 'Properties of hemoglobin solutions', *J. gen. Physiol.*, **52**, 825 (1968)

137. JAY, A. W. L., and BURTON, A. C., 'Direct measurement of potential difference across the human red blood cell membrane', *Biophys. J.*, **9**, 115 (1969)

138. JAY, A. W. L., and BURTON, A. C., 'Membrane potentials in young and old human red cells', *Biophys. J.*, **14**, Abstracts 77a WPM-K7 (1970)

139. DONNAN, F. G., 'The theory of membrane equilibria', *Chem. Rev.*, **1**, 73 (1924)

140. BOLAM, T. R., *The Donnan Equilibria and Their Application to Chemical, Physiological and Technical Processes*, Bell. London (1932)

141. BROMBERG, P. A., THEODORE, J., ROBIN, E., and JENSEN, W., 'Anion and hydrogen ion distribution in human blood', *J. Lab. clin. Med.*, **66**, 464 (1965)

142. FUNDER, J., and WIETH, J. O., 'Chloride and hydrogen ion distribution between human red cells and plasma', *Acta physiol. scand.*, **68**, 234 (1966)

143. WILBRANDT, W., and SCHATZMANN, H. J., 'Changes in the passive cation permeability of erythrocytes in low electrolyte media', in *Ciba Fdn Study Grp Symp.*, No. 5, *Regulation of the Inorganic Ion Content of Cells*, Churchill, London, 340 (1960)

144. DONLON, J. A., and ROTHSTEIN, A., 'The cation permeability of erythrocytes in low ionic strength media of various tonicities', *J. Membrane Biol.*, **1**, 37 (1969)

145. HARRIS, E. J., and PRESSMAN, B. C., 'Obligate cation exchanges in red cells', *Nature, Lond.*, **216**, 918 (1967)

146. HENDERSON, P. J., MCGIVAN, J. D., and CHAPPELL, J. B., 'The action of certain antibiotics on mitochondrial, erythrocyte and artificial phospholipid membranes', *Biochem. J.*, **111**, 521 (1969)

147. HARRIS, J. E., 'The influence of the metabolism of human erythrocytes on their potassium content', *J. biol. Chem.*, **141**, 579 (1941)

148. DANOWSKI, T. S., 'The transfer of potassium across the human blood cell membrane', *J. biol. Chem.*, **139**, 693 (1941)

149. FLYNN, F., and MAIZELS, M., 'Cation control in human erythrocytes', *J. Physiol.*, **110**, 301 (1949)

150. CLARKSON, E. M., and MAIZELS, M., 'Distribution of phosphatases in human erythrocytes', *J. Physiol.* **116**, 112 (1952)

151. WHITTAM, R., 'Potassium movements and ATP in human red cells', *J. Physiol.*, **140**, 479 (1958)

152. HARRIS, E. J., and MAIZELS, M., 'The permeability of human erythrocytes to Na', *J. Physiol.*, **113**, 506 (1951)

153. HARRIS, E. J., 'Linkage of sodium and potassium active transport in human erythrocytes', *Symp. Soc. exp. Biol.*, **8**, 228 (1954)
154. GARRAHAN, P. J., and GLYNN, I. M., 'Uncoupling the sodium pump', *Nature, Lond.*, **207**, 1098 (1965)
155. MAIZELS, M., 'Cation control in human erythrocytes', *J. Physiol.*, **108**, 247 (1959)
156. SEN, A. K., and POST, R. L., 'Stoichiometry and localization of adenosine triphosphate dependent sodium and potassium transport in the erythrocyte', *J. biol. Chem.*, **239**, 345 (1964)
157. GLYNN, I. M., 'The ionic permeability of the red cell membrane', *Progr. Biophys. biophys. Chem.*, **8**, 241 (1957)
158. WHITTAM, R., and AGER, M. E., 'Vectorial aspects of adenosine triphophatase activity in erythrocyte membranes', *Biochem. J.*, **93**, 337 (1964)
159. BOWER, B. F., 'Site of cardiac glycoside inhibition of cation transport', *Nature, Lond.*, **204**, 786 (1964)
160. SCHATZMANN, H. J., 'Effect of cardiac glycosides on active Na–K transport', in *Biophysics and Physiology of Biological Transport* (eds L. BOLIS, V. CAPRARO, K. R. PORTER, and J. D. ROBERTSON), Springer Verlag, New York, 136 (1967)
161. KEYNES, R. D., 'Some further observations on the sodium efflux in frog muscle', *J. Physiol.*, **178**, 305 (1965)
162. HOFFMAN, J. F., 'Cation transport and the structure of the red-cell plasma membrane', *Circulation*, **26**, 1201 (1962)
163. DUNHAM, E. T., and GLYNN, I. M., 'Adenosine triphosphatase activity and the active movements of alkali metal cations', *J. Physiol.*, **156**, 274 (1961)
164. YOSHIDA, H., and FUGISAWA, H., 'Influence of subcellular structures on the activity of Na^+ and K^+-activated adenosine triphosphatase in brain', *Biochim. biophys. Acta*, **60**, 443 (1962)
165. ASKARI, A., and FRATANTONI, J. C., 'Apparent uncoupling of the Na^+ and K^+ activation of the human erythrocyte membrane adenosine triphosphatase', *Biochim. biophys. Acta*, **71**, 232 (1963)
166. MARCHESI, V. T., and PALADE, G. E., 'The localisation of Mg–Na–K-activated adenosine triphosphatase on red cell ghost membranes', *J. Cell. Biol.*, **35**, 385 (1967)
167. SKOU, J. C., 'Enzymatic aspects of active linked transport of Na and K through the cell membrane', *Prog. Biophys. biophys. Chem.*, **14**, 131 (1964)
168. SKOU, J. C., 'The role of membrane ATPase in the active transport of ions', in *The Molecular Basis of Membrane Functions* (ed. D. C. TOSTESON) Prentice Hall, New York, 455 (1969)
169. GARRAHAN, P. J., and GLYNN, I. M., 'The incorporation of inorganic phosphate into adenosine triphosphate by reversal of the sodium pump', *J. Physiol.*, **192**, 237 (1967)
170. LEW, V. L., GLYNN, I. M., and ELLORY, J. C., 'Net synthesis of ATP by reversal of the sodium pump', *Nature, Lond.*, **225**, 865 (1970)
171. MISHRA, R. K., and PASSOW, H., 'Induction of intracellular ATP synthesis by extracellular ferricyanide in human red blood cells', *J. Membrane Biol.*, **1**, 214 (1969)
172. GARRAHAN, P. J., and GLYNN, I. M., 'The stoichiometry of the sodium pump'. *J. Physiol.*, **192**, 217 (1967)
173. HOKIN, L. E., 'On the molecular characterization of the sodium potassium transport adenosine triphosphatase', *J. gen. Physiol.*, **54**, 327s (1969)
174. MAIZELS, M., 'Effect of sodium content on sodium efflux from human red cells suspended in sodium-free media containing potassium, rubidium, caesium or lithium chloride', *J. Physiol.*, **195**, 657 (1968)
175. AHMED, K., JUDAH, J. D., and SCHOLEFIELD, P. G., 'Interaction of sodium and

141

potassium with a cation dependent adenosine triphosphatase from rat brain', *Biochim. biophys. Acta,* **120**, 351 (1966)

176. ECKEL, R. E., RIZZO, S. C., LODISH, H., and BEGGREN, A. B., 'Potassium transport and control of glycolysis in human erythrocytes', *Am. J. Physiol.,* **210**, 737 (1966)

177. PARKER, J. C., and HOFFMAN, J. F., 'The role of membrane phosphoglycerate kinase in the control of glycolytic rate by active cation transport in human red blood cells', *J. gen. Physiol.,* **50**, 893 (1967)

178. GOMPERTS, B. D., 'Metabolic changes in human red cells during incubation of whole blood *in vitro*', *Biochem. J.,* **102**, 782 (1967)

179. GOMPERTS, B. D., Unpublished data

180. MINAKAMI, S., KAKINUMA, K., and YOSHIKAWA, H., 'The control of erythrocyte glycolysis by active cation transport', *Biochim. biophys. Acta,* **90**, 434 (1964)

181. WILEY, J. S., 'Effect of ouabain on deoxynucleoside metabolism in hereditary spherocytic human erythrocytes', *Biochim. biophys. Acta,* **135**, 1071 (1967)

182. BADER, H., POST, R. L., and BOND, G. H., 'Comparison of sources of a phosphory-lated intermediate in transport ATPase', *Biochim. biophys. Acta,* **150**, 41 (1968)

183. JUDAH, J. D., and AHMED, K., 'The biochemistry of sodium transport', *Biol. Rev.,* **39**, 160 (1963)

184. POST, R. L., KUME, S., TOBIN, T., ORCUTT, B., and SEN, A. K., 'Flexibility of an active center in sodium–potassium adenosine triphosphatase', *J. gen. Physiol.,* **54**, 306S (1969)

185. KAHLENBERG, A., GALSWORTHY, P. R., and HOKIN, L. E., 'Studies on the characteri-zation of the sodium–potassium transport adenosine triphosphatase. II. characterization of the acyl phosphate intermediate as an L-glutamyl-γ-phosphate residue', *Arch. Biochem. Biophys.,* **126**, 331 (1968)

186. KEPNER, G. R., and MACEY, R. I., 'Molecular weight estimation of membrane bound ATPase by *in vacuo* radiation inactivation', *Biochem. biophys. Res. Commun.,* **30**, 582 (1968)

187. RACKER, E., *Mechanisms in Bioenergetics,* Academic Press (1965)

188. PRESSMAN, B. C., HARRIS, E. J., JAGGER, W. S., and JOHNSON, J. H., 'Antibiotic mediated transport of alkali ions across lipid barriers', *Proc. Natn. Acad. Sci. U.S.A.,* **58**, 1949 (1967)

189. HEINZ, E., 'Transport through biological membranes', *A. Rev. Physiol.,* **29**, 21 (1967)

190. OPIT, L. J., and CHARNOCK, J. S., 'A molecular model for a sodium pump', *Nature, Lond.,* **208**, 471 (1965)

191. JARDETSKY, O., 'Simple allosteric model for membrane pumps', *Nature, Lond.,* **211**, 969 (1966)

192. KATCHALSKY, A., 'Membrane thermodynamics', in *The Neurosciences,* (eds G. QUARTON, TH. MELNECHUK, and F. O. SCHMITT), Rockefeller University Press, New York, 326 (1967)

193. CALDWELL, P. C., 'Factors governing movement and distribution of inorganic ions in nerve and muscle', *Physiol. Rev.,* **48**, 1 (1968)

194. LOWE, A. G., 'Enzyme mechanism for the active transport of sodium and potassium ions in animal cells', *Nature, Lond.,* **219**, 934 (1968)

195. MIDDLETON, H. W., 'Kinetics of ion activation of the $[Na^+ + K^+]$ dependent adenosine triphosphatase and a model for ion translocation and its inhibition by cardiac glycosides', *Arch. Biochem.,* **136**, 280 (1970)

196a. CHAN, P. C., and SANSLONE, W. R., 'The influence of a low potassium diet on rat erythrocyte membrane adenosine triphosphatase', *Arch. biochem. biophys.,* **134**, 48 (1969)

196b. TOSTESON, D. C., 'Active transport, genetics and cellular evolution', *Fedn Proc. Fedn Am. Socs exp. Biol.*, **22**, 19 (1963)

197. EVANS, J. V., KING, J. W. B., COHEN, B. L., HARRIS, H., and WARREN, F. L., 'Genetics of haemoglobin and blood potassium differences in sheep', *Nature, Lond.*, **178**, 849 (1956)

198. RASMUSEN, B. A., and HALL, J. G., 'Association between potassium concentration and serological blood type of sheep red blood cells', *Science*, **151**, 1551 (1966)

199. RASMUSEN, B. A., *Genetics*, **65**, 549 (1970)

200. ELLORY, J. C., and TUCKER, E. M., 'Stimulation of the potassium transport system in low potassium type sheep red cells by a specific antigen-antibody reaction', *Nature, Lond.*, **222**, 477 (1969)

201. ELLORY, J. C., and TUCKER, E. M., 'Active potassium transport and the development of m antigen on the red cells of LK type lambs', *J. Physiol.*, **204**, 161 (1969)

202. ELLORY, J. C., and TUCKER, E. M., 'A specific antigen-antibody reaction affecting ion transport in sheep LK erythrocytes', in *Permeability and Function of Biological Membranes* (eds L. BOLIS, A. KATCHALSKY, R. D. KEYNES, W. R. LOEWENSTEIN, and B. A. PETHICA), North Holland Publishing Co., Amsterdam (1970)

203. BROWN, H. E., CHATTOPADHYAY, S. K., and PATEL, A. B., 'Erythrocyte abnormality in human myopathy', *Science*, **157**, 1577 (1967)

204. PETER, J. B., WORSFOLD, M., and PEARSON, C. M., 'Erythrocyte ghost adenosine triphosphatase in Duchenne dystrophy', *J. Lab. clin. Med.*, **74**, 103 (1969)

205. JACOB, H. S., 'Abnormalities in the physiology of the erythrocyte membrane in hereditary spherocytosis', *Am. J. Med.*, **41**, 734 (1966)

206. WILEY, J. S., 'Inheritance of an increased sodium pump in red cells', *Nature, Lond.*, **221**, 1222 (1969)

207a. ZARKOWSKY, H. S., OSKI, F. A., SHA'AFI, R., SHOHET, S. B., and NATHAN, D. G., 'Congenital hemolytic anemia with high sodium, low potassium red cells 1. Studies of membrane permeability', *New Engl. J. Med.*, **278**, 593 (1968)

207b. OSKI, F. A., NAIMAN, J. L., BLUM, S. F., ZARKOWSKY, H. S., WHAUN, J., SHOHET, S. B., GREEN, A., and NATHAN, D. G., 'Congenital hemolytic anemia with high sodium, low potassium red cells', *New Engl. J. Med.*, **280**, 909 (1969)

208. HARVALD, B., HANEL, K. H., SQUIRES, R., and TRAP-JENSEN, J., 'Adenosine-triphosphatase deficiency in patients with non-spherocytic haemolytic anaemia', *Lancet II*, **18** (1964)

209. WASSERMAN, R. H., and KAHLFELZ, F. A., 'Transport of calcium across biological membranes', in *Biological Calcification* (ed. H. SCHRAER), North Holland Publishing Co., Amsterdam, 313 (1970)

210. HUNT, B. J., and MANERY, J. F., 'Use of ion-exchange resin in preparing erythrocytes for magnesium determinations', *Clin. Chem.*, **16**, 269 (1970)

211. PASSOW, H., 'Metabolic control of passive cation permeability in human red cells', in *Cell Interface Reactions* (ed. H. D. BROWN), New York Scholar's Library, New York, 58 (1963)

212. SCHATZMANN, H. J., 'ATP dependent Ca^{++} extrusion from human red cells', *Experientia*, **22**, 364 (1966)

213. VICENZI, F. F., and SCHATZMANN, H. J., 'Some properties of Ca-activated ATPase in human red cell membranes', *Helv. physiol. pharmac. Acta*, **25**, CR 233 (1967)

214. WINS, P., and SCHOFFENIELS, E., 'Studies on red cell ghost ATPase systems: properties of a $[Mg^{++} + Ca^{++}]$-dependent ATPase', *Biochim. biophys. Acta*, **120**, 341 (1966)

215. MAIZELS, M., 'Calcium ions and permeability of human red cells', *Nature, Lond.*, **184**, 366 (1959)

RED BLOOD CELLS

216. HORTON, C. R., COLE, W. Q., and BADER, H., 'Depressed Ca^{++} transport ATPase in cystic fibrosis erythrocytes', *Biochem. biophys. Res. Commun.*, **40**, 505 (1970)
217. TOSTESON, D. C., 'Halide transport in red blood cells', *Acta physiol. scand.*, **46**, 19 (1959)
218. DIRKEN, M. N. J., and MOOK, H. W., 'The rate of gas exchange between blood cells and serum', *J. Physiol.*, **73**, 349 (1931)
219. LUCKNER, H., 'Die Temperaturabhängigkeit des Anionenaustausches roter Blutkörperchen', *Pflügers Arch. ges. Physiol.*, **250**, 303 (1948)
220. SIDEL, V. A., and SOLOMON, A. K., 'Entrance of water into human red cells under an osmotic pressure gradient', *J. gen. Physiol.*, **41**, 243 (1957)
221. PAGANELLI, C. V., and SOLOMON, A. K., 'The rate of exchange of tritiated water across the red cell membrane', *J. gen. physiol.*, **41**, 259 (1957)
222. RICHMOND, J. E., and HASTINGS, A. B., 'Distribution equilibrium of sulphate *in vitro* between red blood cells and plasma', *Am. J. Physiol.*, **199**, 821 (1960)
223. WILBRANDT, W., 'Die Kinetik des Ionenaustausches durch selektiv ionen-permeable Membranen'. *Pflügers Arch. ges. Physiol.*, **246**, 274 (1942)
224. SCHWEITZER, C. H., and PASSOW, H., 'Kinetik und Gleichgewichte bei der langsamen Anionenpermeabilität roter Blutkörperchen, *Pflügers Arch. ges. Physiol.*, **256**, 419 (1953)
225. MOND, R., and GERTZ, H., 'Vergleichende Untersuchungen über Membranstruktur und Permeabilität der roten Blutkörperchen verschiedender Säugetiere', *Pflügers Arch. ges. Physiol.*, **221**, 623 (1929)
226. DUNKER, E. and PASSOW, H., 'Zwei Arten des anionenaustausches bei den roten Blutkörperchen verschiedener saügetiere', *Pflügers Arch. ges. Physiol.*, **256**, 446 (1953)
227. PARPART, A. K., 'The permeability of the erythrocyte for anions', *Cold Spring Harb. Symp. quant. Biol.*, **8**, 25 (1940)
228. HOLTON, F. A., 'The alleged effect of acid phosphate on the anion permeability of erythrocytes', *Biochem. J.*, **52**, 506 (1952)
229. PRANKERD, T. A. J., and ALTMAN, K. I., 'A study of the metabolism of phosphorus in mammalian red cells', *Biochem. J.*, **58**, 622 (1954)
230. GERLACH, E., FLECKENSTEIN, A., and GROSS, E., 'Der intermediäre Phosphatstoffwechsel des Menschen-Erythrocyten', *Arch. ges. Physiol.*, **266**, 528 (1958)
231. GOMPERTS, B. D., 'The time course of phosphorus incorporation in human red cells', *Biochim. biophys. Acta*, **177**, 144 (1969)
232. JONAS, H., and GOURLEY, D. R. H., 'Effect of ATP, magnesium and calcium on phosphate uptake by rabbit erythrocytes', *Biochim. biophys. Acta*, **14**, 535 (1954)
233. HARRIS, E. J., and PRANKERD, T. A. J., 'Phloridzin and red cell phosphate turnover', *Experientia*, **14**, 249 (1958)
234. ZIPURSKY, A., and ISRAELS, L. G., 'Transport of phosphate into erythrocytes', *Nature, Lond.*, **189**, 1013 (1961)
235. VESTERGAARD-BOGIND, B., and HESSELBO, T., 'The transport of phosphate ions across the human red cell membrane. I. The distribution of phosphate ions in equilibrium at comparatively high phosphate concentrations', *Biochim. biophys. Acta*, **44**, 117 (1960)
236. HODGKIN, A. L., and KEYNES, R. D., 'The potassium permeability of a giant nerve fibre', *J. Physiol.*, **128**, 61 (1955)
237. PASSOW, H., 'Ion and water permeability of the red blood cell', in *The Red Blood Cell* (eds C. BISHOP and D. M. SURGENOR), Academic, New York (1964)
238. SCHRIER, S. L., 'Transfer of inorganic phosphate across human erythrocyte ghost membranes', *J. Lab. clin. Med.*, **75**, 422 (1970)
239. LE FEVRE, P. G., 'Sugar transport in red blood cells: structure activity relationship in substrates and antagonists', *Pharmac. Rev.*, **13**, 39 (1961b)

144

240. STEIN, W. D., *The Movement of Molecules Across Cell Membranes*, Academic, London (1967)
241. MELDAHL, K. F., and ØRSKOV, S. L., 'Photoelektrische Methode zur Bestimmung der Permeierungsgeschwindigkeit von Anelektrolyten durch die Membran von roten Blutkörperchen. Untersuchung über die Gültigkeit des Fick'schen Gesetzes für die Permeierungsgeschwindigkeit', *Skand. Arch. Physiol.*, **83**, 266 (1940)
242. ØRSKOV, S. L., 'Eine Methode zur fortlaufenden photographischen Aufzeichung von Volumänderungen der roten Blutkörperchen', *Biochem. Z.*, **279**, 241 (1935)
243. SEN, A. K., and WIDDAS, W. F., 'Determination of the temperature and pH dependence of glucose transfer across the human erythrocyte membrane measured by glucose exit.', *J. Physiol.*, **160**, 392 (1962)
244. WIDDAS, W. F., 'Facilitated transfer of hexoses across the human erythrocyte membrane', *J. Physiol.*, **125**, 163 (1954)
245. KOZAWA, S., 'Beiträge zum arteigenen Verhalten der roten Blutkörperchen', *Biochem. Z.*, **60**, 231 (1914)
246. WIDDAS, W. F., 'Hexose permeability of foetal erythrocytes', *J. Physiol.*, **127**, 318 (1955)
247. PARK. C. R., POST. R. L., KALMAN, C. F., WRIGHT, J. H., JOHNSON, L. H., and MORGAN, H. E., 'The transport of glucose and other sugars across cell membranes and the effect of insulin', *Ciba Fdn Colloq. Endocrin.*, **9**, 240 (1956)
248. LE FEVRE, P. G., 'Evidence of active transfer of certain non-electrolytes across the human red cell membrane', *J. gen. Physiol.*, **31**, 505 (1948)
249. WILBRANDT, W., GRENSBERG, E., and LAUENER, H., 'Der Glucoseeintritt durch die Erythrocytenmembran', *Helv. physiol. pharmac. Acta*, **5**, C20 (1947)
250. BONSALL, R. B., and HUNT, S., 'Solubilization of a glucose binding component of the red cell membrane', *Nature, Lond.*, **211**, 1368 (1966)
251. BOBINSKY, H., and STEIN, W. D., 'Isolation of a glucose binding component from human erythrocyte membrane', *Nature, Lond.*, **211**, 1366 (1966)
252. JUNG, C. Y., CHANEY, J. E., and LE FEVRE, P. G., 'Enhanced migration of glucose from water into chloroform in presence of phospholipids', *Arch. Biochem. Biophys.*, **126**, 664 (1968)
253. LE FEVRE, P. G., JUNG, C. Y., and CHANEY, J. E., 'Glucose transfer by red cell membrane phospholipids in $H_2O/CHCl_3/H_2O$ three layer systems', *Arch. Biochem. Biophys.*, **126**, 677 (1968)
254. STEIN, W. D., 'Intraprotein interactions across a fluid membrane as a model for biological transport', *J. gen. Physiol.*, **54**, 81 S (1969)
255. LIEB, W. R., and STEIN, W. D., 'Quantitative predictions of a non-carrier model for glucose transport across the human red cell membrane', *Biophys. J.*, **10**, 585 (1970)
256. DAWSON, A. C., and WIDDAS, W. F., 'Variations with temperature and pH of the parameters of glucose transfer across the erythrocyte membrane in the foetal guinea pig', *J. Physiol.*, **172**, 107 (1964)
257. BOLIS, L., LULY, P., PETHICA, B. A., and WILBRANDT, W., 'The temperature dependence of the facilitated transport of D[+]-glucose across the human red cell membrane', *J. Membrane Biol.*, **3**, 83 (1970)
258a. MILLER, D. M., 'The kinetics of selective biological transport. II. Equations for induced uphill transport of sugars in human erythrocytes', *Biophys. J.*, **5**, 417 (1965)
258b. MILLER, D. M., 'The kinetics of selective biological transport. III. Erythrocyte-monosaccharide transport data', *Biophys. J.*, **8**, 1329 (1968)
259. LEVINE, M., and STEIN, W. D., 'The kinetic parameters of the monosaccharide

transfer system of the human erythrocyte'. *Biochim. biophys. Acta,* **127**, 179 (1966)

260. KOSHLAND, D. E., and NEET, K. E., 'The catalytic and regulatory properties of enzymes', *A. Rev. Biochem.,* **37**, 359 (1968)

261. CALDWELL, P. C., 'Energy relationships and the active transport of ions', in *Current Topics in Bioenergetics* (ed. D. R. SANADI), Academic, London (1969)

262. MUELLER, P., and RUDIN, D. O., 'Bimolecular lipid membranes: techniques of formation, study of electrical properties and induction of ionic gating phenomena', in *Laboratory Techniques in Membrane Biophysics,* (eds H. PASSOW and R. STRÄMPFLI), Springer Verlag, New York (1970)

262. BALFE, J. W., COLE, C., SMITH, E. K. M., GRAHAM, J. B., and WELT, L. G., 'Hereditary sodium transport defect in human red blood cells', *J. clin. Invest.,* **47**, 4a abstract 11 (1968)

263. BANKS, B. E. C., and VERNON, C. A., 'Reassessment of the role of ATP *in vivo*', *J. theor. Biol.,* **29**, 301 (1970)

264. DUNHAM, P. B., and HOFFMAN, J. F., 'Partial purification of the ouabain-binding component and of Na, K-ATPase from human red cell membranes', *Proc. Natn. Acad. Sci., U.S.A.,* **66**, 936 (1970)

265. PARDEE, A. B., 'Biochemical studies on active transport', *J. gen. Physiol.,* **52**, 279S (1968)

266. REEVES, R. E., 'Cuprammonium-glycoside complexes', *Adv. Carbohyd. Chem.,* **6**, 107 (1951)

266. ROSEMAN, S., *J. gen. Physiol.,* **54**, 138 (1969)

267. WASSERMAN, R. H., CORRADINO, R. A., and TAYLOR, A. N., 'Binding proteins from animals with possible transport function', *J. gen. Physiol.,* **54**, 114 S (1969)

5

TRANSPORT ACROSS THE EPITHELIAL MEMBRANE WITH PARTICULAR REFERENCE TO THE TOAD BLADDER

GEOFFREY W. G. SHARP
The Massachusetts General Hospital, Boston, U.S.A.

GENERAL INTRODUCTION

The electrical activity of isolated amphibian skin has been a source of interest and much physiological study since the early reports of DuBois-Reymond[1] and Galeotti[2]. Ussing and Zerahn[3], by their development of the short-circuiting technique, determined the nature of the electrical activity. They demonstrated that the electrical current, which was present under short-circuited conditions, was equal to the net transport of sodium ions from the outside to the inside of the skin. As the technique was simple and allowed instantaneous and continuous measurement of sodium transport, intensive investigations into the properties of transepithelial sodium transport became possible.

Skins and urinary bladders from frogs and toads transport sodium ions. As the toad bladder is histologically simpler than the amphibian skins, much of this chapter will be devoted to its properties. In the toad the urinary bladder acts as a functional extension of the kidney and is in many respects similar to the distal convoluted portion and collecting duct of the mammalian nephron. Its epithelial cells reabsorb sodium salts from the urine and return them to the body. It acts as a storage organ for water so that water may be reabsorbed during periods of water deprivation, and controls the reabsorptive process by secretion of anti-diuretic hormone (ADH). This hormone, in the toad, is arginine oxytocin (vasotocin). In addition to its effect upon water reabsorption, the hormone also stimulates the reabsorption of sodium ions and alters the permeability of the cells to urea, acetamide and certain other small molecules. The bladder also responds with an increased rate of

147

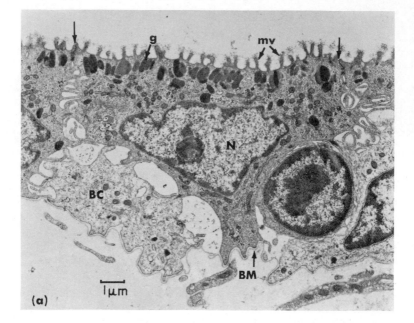

Figure 5.1. Cross-section of cells in the wall of the urinary bladder of the toad. The epithelial cell layer lining the mucosal or urinary surface of the bladder is responsible for the transport functions of the bladder. It is supported on a connective tissue stroma containing blood vessels and smooth muscle. A thin serosal cell layer lines the body surface. The mucosal cell layer contains four types of cells, namely, goblet or mucous secreting cells, basal cells (probably germinal), granular cells, and mitochondria-rich cells. Only the last two are thought to be directly involved in sodium transport and only the granular cells in water movement. The electronmicrographs show: (a) a granular cell and a basal cell; (b) a mitochondria-rich cell bordered by a granular cell and a goblet cell. Symbols: g, granule; mv, microvilli; mvb, multivesicular body; N, nucleus; BM, basement membrane; BC, basal cell; GC, goblet cell; GrC, granular cell; the arrows indicate the tight junctions at adjacent cell boundaries. (Electronmicrographs made by Dr Donald R. DiBona, Department of Medicine, the Massachusetts General Hospital, Boston, Mass.)

148

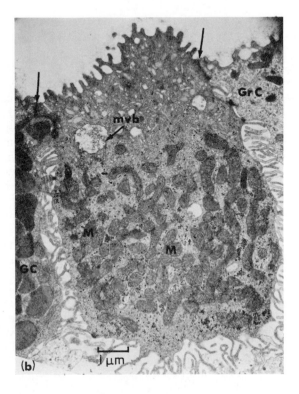

sodium reabsorption to aldosterone, which is secreted by the inter-renal glands of the toad. Aldosterone does not have any effect upon water flow. Reviews on different aspects of transepithelial sodium and water transport exist in the literature [4-9].

MORPHOLOGY OF THE BLADDER

The bladder consists of an epithelial cell layer on the mucosal surface, which is supported by a submucosa consisting of loosely structured connective tissue, smooth muscle bands, and blood vessels. The muscle bands are discontinuous and probably serve to control bladder size proportional to its urine volume. Urination is accomplished by contraction of the muscles of the abdominal wall. The serosal cell layer lies on the basal surface of the submucosa and apparently has no transport functions. It has a high permeability to salts and water and a low electrical resistance. The bladder is distensible and occupies much of the abdominal cavity, so that, when the toad is well hydrated, as much as half of the abdominal space may be taken up by the storage of 10–30 ml of dilute urine.

The cell layer of interest to transport function is the mucosal epithelial cell layer (Figure 5.1). This has been described histo-logically by several people [10-13]. It contains four cell types, the most common being the granular cell, which occupies most of the lumenal surface of the bladder. Other cells which have access to the lumenal surface are the mitochondria-rich cells, comprising 5–10% of the total, and the mucous secreting goblet cells. The basal cells which underly the surface layer make a discontinuous second cell layer. Functionally, however, the bladder may be described as a single cell-layer. By careful serial sectioning of the toad bladder, DiBona, Civan, and Leaf [14] found that the granular cells are shaped so that the apical membranes cover a relatively large area of the surface and the basal membranes always come close at some point to the basement membrane. Similarly, the mitochondria-rich cells, despite their oval shape and small area of exposure at the apical membrane, make contact with the apical surface (Figure 5.2). As the basal cells do not have access to the lumen, and as the goblet cell function is the secretion of mucous at the apical surface, it appears that the transport and permeability characteristics of the bladder may be ascribed to the granular cells and the mitochondria-rich cells. Evidence is available, and will be presented later, that the water flow through the bladder in response to anti-diuretic hormone is a property of the granular cells and not the mitochondria-rich cells.

The characteristic morphology of the individual cells can be seen

150

in the electronmicrograph in Figure 5.1. The granular cells are named because of the numerous dense oval granules situated at the apical surface just beneath the small villi which project from the surface. They contain few mitochondria. The mitochondria-rich

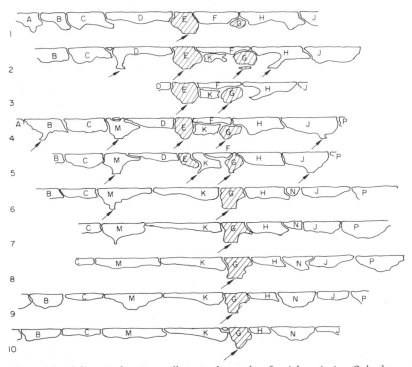

Figure 5.2. Schematic drawing to illustrate the results of serial sectioning. Only the outlines of granular (clear) and mitochondria-rich cells (cross-hatched) are shown; basal cells, omitted for clarity, occupied space between granular cells and basement membrane. The basement membrane is indicated by arrows at those places at which a cell makes contact. (Contact regions are drawn as flattened margins.) The natural convolutions of the bladder are not included in the figure which is, therefore, only approximately accurate to scale. The mucosal margin of cell E (in the first view) measures about 5 μm and the average separation between adjacent views is of the order of ½ μm. Note that cells E and G (mitochondria-rich) make contact with the basement membrane over a fairly broad region. This is as expected from the frequency with which basement membrane contacts are found in randomly selected sections. Of the granular cells shown, cells B, D, H, J, K, and M are seen to make contact. The other granular cells are incompletely followed. Cells A and P are seen only rarely and are included simply to delineate the field; cell C was missed, unfortunately, in sections 3, 6, 7; cells F and N are seen only at their peripheries; cell F is terminating as the sequence begins; cell N is only beginning to appear as the sequence ends (ca. ×6000) (After DiBona, Civan, and Leaf[14]; courtesy J. Cell. Biol.)

cells are barrel-shaped and contain numerous mitochondria. Villus projections are also more numerous, for a given surface area, than those of the granular cells.

The surface cells are fused close to the apical border by tight junctions, the *zona occludens*, which provide a morphological base for the concept of transcellular, rather than an intercellular, transport of water and ions across the epithelial layer. Intercellular spaces below the tight junctions exist as narrow, convoluted channels under basal conditions, but may open to form pools or intercellular lakes during the reabsorption of water. The nature of the 'tight junctions' which apparently bind the cells under normal conditions is not known. It is of interest that removal of calcium ions from the bathing medium causes the cells to come apart from each other, and as a result the cell layer becomes highly permeable[15]. Another interesting property of the tight junctions is that they provide a low-resistance pathway between adjacent cells[16]. Interconnection between cells by means of these low-resistance pathways was found to be a property of some, but not all, neighbouring cells, so that a tortuous star-shaped pattern of communication exists. These low-resistance pathways should allow the easy passage of ions between cells, and thus permit several cells to operate in unison on the same intracellular transport pool of ions.

A general model for the transport of sodium and water is shown in Figure 5.3 and will be fully discussed in subsequent sections. It is assumed that the granular and mitochondria-rich cells are capable of active transport of sodium, and the process comprises a passive, but specific, interaction with the apical surface, whereby the sodium ion is recognised and enters the transport pool. Active sodium transport then occurs across the basal or lateral surfaces of the cells by the activity of Na^+, K^+ activated ATPase. Water flow is thought to occur through the granular cells which, depending upon the rate of water movement, may swell during the transport process.

WATER FLOW

Ewer[17] showed that the bladder acts as a water storage organ and reabsorbs water under the combined influence of an osmotic gradient and neurohypophysial hormones. In the absence of hormone, water loss from the bladder is extremely low even in the presence of high osmotic gradients. Bentley[18] described a simple and effective method for the determination of water flow *in vitro*. He mounted the bladders as small bags on glass tubes. By filling the bags with dilute Ringer's solution and immersing them in full-

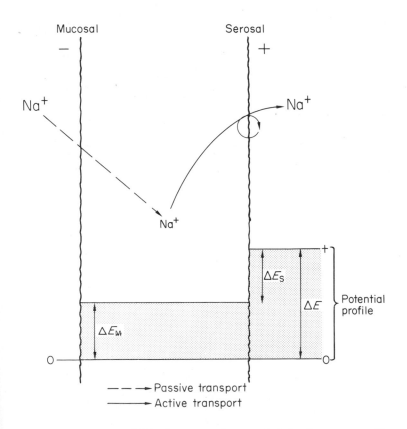

Figure 5.3. *A current model for the process of transepithelial sodium transport. The model assumes that sodium moves from urine to the cell interior passively, down a chemical gradient, and that a specific interaction of the sodium ion takes place at the apical membrane. Active transport takes place as the sodium ions are extruded into the serosal medium. The extrusion occurs without either an obligatorily coupled exchange for cations in the serosal medium or a paired movement of anions into the serosal medium. Thus, the process is considered electrogenic. The potential step at the apical membrane is thought to be a sodium diffusion potential. The potential step at the basal membrane is thought to arise from the separation of charge resulting from sodium ion extrusion (After Sharp and Leaf[8]; courtesy Recent Prog. Horm. Res.)*

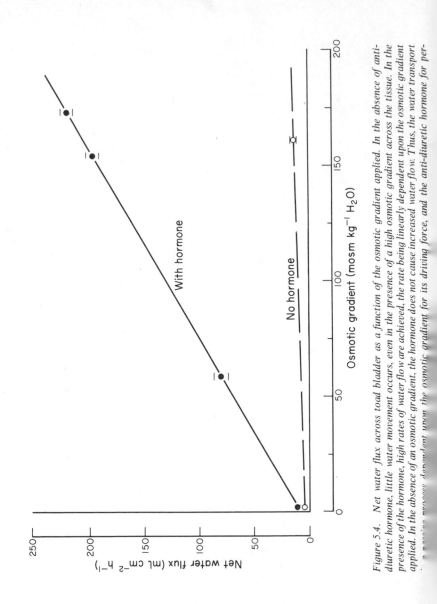

Figure 5.4. Net water flux across toad bladder as a function of the osmotic gradient applied. In the absence of anti-diuretic hormone, little water movement occurs, even in the presence of a high osmotic gradient across the tissue. In the presence of the hormone, high rates of water flow are achieved, the rate being linearly dependent upon the osmotic gradient applied. In the absence of an osmotic gradient, the hormone does not cause increased water flow. Thus, the water transport is a passive process dependent upon the osmotic gradient for its driving force, and the anti-diuretic hormone for per-

154

strength solution, an osmotic gradient may be set up across the bladder. Simple weighing of the bags at specific time intervals enables one to determine the weight loss (water flow) from the bag. When anti-diuretic hormone is added to the serosal bathing medium, but not to the mucosal medium, a high rate of water flow is seen. It has been shown that the water flow under such conditions is a linear function of the osmotic gradient (Figure 5.4) except at gradients above 200 mosm kg^{-1}. Therefore, the movement of water across the tissue is a passive process, which is determined by the gradient and the permeability characteristics of the membrane. As water flow occurs even in the absence of sodium in the mucosal medium, active sodium transport is not essential for the process[19].

The anatomic site of the permeability barrier to water is located at the apical surface of the mucosal epithelial cells. Evidence for this has been derived from the application of hypotonic solutions to one side or the other of the bladder. When such a solution or even water is applied to the urinary surface of the bladder, swelling of the cells does not occur. On the other hand, when similar solutions are applied to the serosal side of the bladder, swelling occurs in all the four types of cell in the mucosal epithelial layer. Furthermore, with a hypotonic solution on the urinary surface, the addition of anti-diuretic hormone results in high rates of water flow and swelling of the granular cells[20]. The specificity of the cell swelling which occurs only in the granular cells, under these conditions, is remarkable when one considers that all cells swell under conditions of serosal hypotonicity. It supports the theory that under the influence of anti-diuretic hormone, water passes through the cells. If it were to pass between the cells, the mitochondria-rich cells would also swell. Thus anti-diuretic hormone specifically alters the permeability characteristics of the apical membranes of the granular cells to allow the bulk flow of water through them. The mechanism whereby anti-diuretic hormone activates adenyl cyclase to produce cyclic AMP, the mediator of this permeability effect, will be discussed later in connection with the mechanism of action of the hormone.

SODIUM TRANSPORT

When a frog skin or toad bladder is mounted in glass or lucite chambers, so that the Ringer's solution bathing the two surfaces is separated, a potential difference of up to 100 mV is found, serosa positive to mucosa. Using the technique of Ussing and Zerahn[3] to short-circuit the membrane, the transbladder potential can be reduced to zero by the application of current from a battery placed

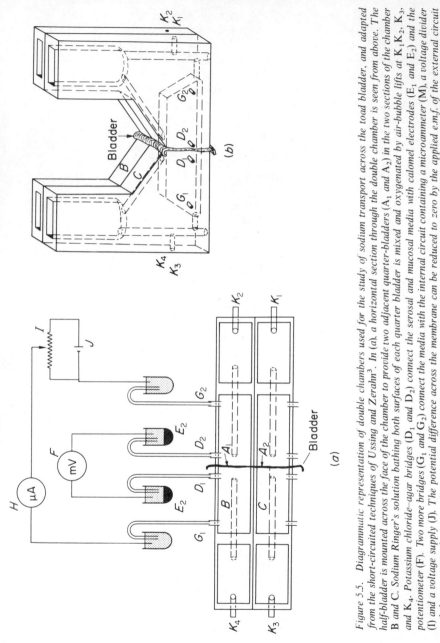

Figure 5.5. Diagrammatic representation of double chambers used for the study of sodium transport across the toad bladder, and adapted from the short-circuited techniques of Ussing and Zerahm³. In (a), a horizontal section through the double chamber is seen from above. The half-bladder is mounted across the face of the chamber to provide two adjacent quarter-bladders (A₁ and A₂) in the two sections of the chamber B and C. Sodium Ringer's solution bathing both surfaces of each quarter bladder is mixed and oxygenated by air-bubble lifts at K₁, K₂, K₃, and K₄. Potassium chloride–agar bridges (D₁ and D₂) connect the serosal and mucosal media with calomel electrodes (E₁ and E₂) and the potentiometer (F). Two more bridges (G₁ and G₂) connect the media with the internal circuit containing a microammeter (M), a voltage divider (I) and a voltage supply (J). The potential difference across the membrane can be reduced to zero by the applied e.m.f. of the external circuit and the current in the external circuit recorded on the microammeter. Only one circuit is shown in the diagram, that connected to chamber B. An identical circuit is, of course, connected to chamber C. In (b), the double chamber is shown in side view, exposing the fluid compartments

156

in an external circuit. With identical solutions on the two sides, which will eliminate the possibility of chemical gradients, and with no electrical driving force under conditions of zero potential, net ion movement should be zero in the absence of active ion transport. Under these short-circuited conditions, however, ion movement does take place and is measured as the short-circuit current (see Figure 5.5). By measuring the fluxes of ^{22}Na and ^{24}Na from serosa to mucosa and mucosa to serosa, respectively, it is clear that for frog skin and toad bladder the short-circuit current is equal to the net movement of sodium from mucosa to serosa (Table 5.1). Thus,

Table 5.1 COMPARISON OF THE NET SODIUM FLUX AND THE SHORT-CIRCUIT CURRENT THROUGH THE ISOLATED URINARY BLADDER OF THE TOAD, *Bufo marinus*

Periods		*Mean* Na *flux* (μA cm^{-2})		A	B
No.	Duration (min)	M → S	S → M	Mean net Na transport	Mean short-circuit current
16	60	57·5	15·3	42·2	43·1

Sodium flux was measured from mucosal to serosal surfaces (M → S) by means of Na^{22} and simultaneously in the opposite direction (S → M) with Na^{24}. The short-circuit current was monitored continuously throughout the period of flux measurements.
Mean difference A − B = −0·9 μA cm^{-2}
S.E. of mean difference = ±2·9 μA cm^{-2}
(After Leaf[22])

the sodium ion is actively transported from urine to body fluids, no other ion is being actively transported and the energy for the process must be derived from cell metabolism. This active sodium transport system of the skin and bladder, like the distal portion of the mammalian nephron, can move sodium against high concentration gradients. Active sodium transport will persist even when the concentration of sodium in the serosal medium is 100 mequiv l^{-1} and that in the mucosal medium is 1 mequiv l^{-1}. When the bladder is open-circuited, as, for instance, *in vivo*, then the potential difference set up by the active transport process will exert a force for the reabsorption of anions such as chloride or for the secretion of cations. The chloride permeability of the bladder is low and exerts a limiting drag effect upon the rate of sodium transport in the open-circuited condition.

It is clear from the morphology of the mucosal epithelial cell layer that the movement of sodium across the cells allows for a division of function into at least two parts: an entry process across the apical membrane and an exit process across the basal or lateral cell membranes. The available evidence suggests that the entry process is essentially passive and occurs down a chemical gradient for sodium, while the active extrusion process takes place at the serosal boundary of the cell.

THE PASSIVE SODIUM ENTRY PROCESS

Three possible mechanisms exist which would move sodium ions across the apical surfaces of the cells and into the transport pool: (1) an electrical potential gradient, (2) a chemical gradient, and (3) active transport.

The possibility that an electrical gradient was responsible was ruled out by direct measurement of the electrical potential profile of the bladder. Frazier[23], by inserting micropipettes into the cells from the apical side, found that, in the open-circuited state, the interior of the cell is electrically positive relative to the urine. As the probe enters the cell, an initial positive step in potential is seen at the apical membrane. As the probe is pushed through and out of the cell, a second positive step in potential is found (Figure 5.6). Thus, the electrical profile is such as to act against, rather than for, sodium entry into the cell from the mucosal side. The two-step profile is a consistent finding of these experiments and suggests that the transepithelial transport of sodium is a two-step process.

Evidence for the existence of a chemical gradient across the apical surface, sufficient for the sodium entry process, was derived from ^{24}Na pool studies. In these experiments various concentrations of sodium chloride were placed in the mucosal Ringer's solution and labelled with ^{24}NaCl. After sufficient time for equilibration, the amount of isotopic sodium in the tissue was determined and compared with that in the mucosal medium. At all concentrations from $114 \text{ mequiv l}^{-1}$ to $0.4 \text{ mequiv l}^{-1}$ the concentration of sodium in the tissue was less than that in the medium (Figure 5.7, Table 5.2). Thus, a passive entry of sodium could account for the initial step in the process, and an active energy-requiring mechanism is not necessary. Unfortunately, it is not possible to interpret these results rigorously, because of criticism which may be directed at the experimental technique. For a rigorous interpretation, it is necessary to know that the sodium measured in the tissue is contained in the cells and is located between the two permeability barriers, i.e. between the apical and basal membranes of the mucosal cells. As the experiments were performed with the whole bladder it is not

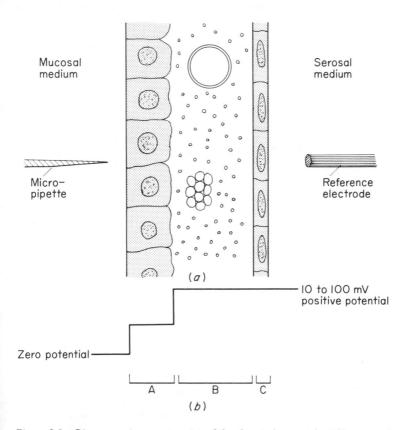

Figure 5.6. Diagrammatic representation of the electrical potential profile across the mucosal epithelial cell layer of the toad bladder. (a) A cross-section of the toad bladder is shown with the mucosal cells (on the left) separated from the serosal cell layer (on the right) by connective tissue. Active sodium transport is a property of the mucosal cells. (b) In the non-short-circuited state, measurement of the electrical potential profile by introduction of micropipettes into and through the cells demonstrates the existence of two potential steps, both positive with respect to the mucosal side of the bladder. The potential steps are thought to occur at the apical and basal membranes of the epithelial cells. Thus, electrical driving forces cannot be responsible for sodium ion movement from urine to cell interior nor from cell interior to serosa. No potential changes were seen in zones B and C, the connective tissue and serosal cell layers, respectively (From Frazier and Leaf [24]; courtesy Medicine)

159

possible to say where the sodium was in the tissue, It is possible that some of the radioactive sodium was trapped after it had left the cells, perhaps in the connective tissue layer, perhaps in the intracellular spaces. Then, again, any compartmentalisation of the sodium transport pool with reduction of volume would increase the calculated concentration of sodium in the pool. Future studies on transport pool size in isolated cells should be useful in elucidating these points.

Table 5.2 THE TISSUE SODIUM CONTENT AS A FUNCTION OF THE MUCOSAL SODIUM CONCENTRATION

Concentration in mucosal medium	Tissue content	No. of measurements
114	9·3 ± 1·3 (S.E.M.)	10
60	9·3 ± 1·3 (S.E.M.)	10
20	5·6 ± 1·1 (S.E.M.)	8
5	2·6 ± 0·32 (S.E.M.)	8
1·18	0·64	1
0·98	0·42	1
0·61	0·59	1
0·45	0·43	1
0·40	0·20	1

The mucosal medium was Ringer's solution diluted with sodium-free choline Ringer's solution to obtain the indicated concentrations of sodium. The serosal medium was a sodium-free choline Ringer's solution. The sodium pool was estimated with radioactive sodium added to the mucosal medium. Chamber area, 7·07 cm². Volume of medium bathing each surface, 20 ml. Duration of incubation, 60 min. Vasopressin was present in all experiments (From Frazier, Dempsey and Leaf[25])

Assuming, for the moment, that sodium entry is a passive process down a chemical gradient, the interaction of the ion with the apical membrane is of great interest. Specific interaction of the sodium ion and apical membrane (Figure 5.8) can be seen from the competition experiments of Frazier[27]. When the rate of sodium transport is measured as a function of sodium concentration in the mucosal medium, apparent saturation of the process occurs (Figure 5.9). Several lines of investigation point to the entry step as the site of saturation. It is thought that the main controlling factor in the determination of specificity is the field strength of negative charges on the membrane[28]. The action of anti-diuretic hormone to stimulate sodium transport occurs at the apical membrane[29]. Amiloride and triamterene, two potassium-sparing diuretic agents, also act at this membrane to inhibit the sodium entry process[30, 31].

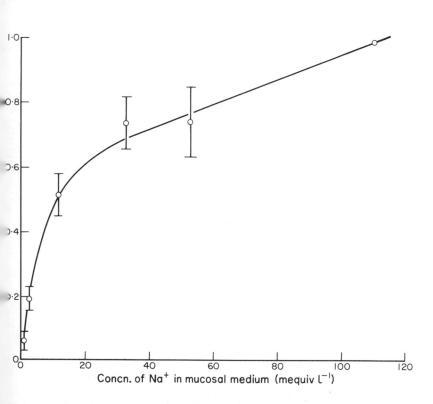

Figure 5.7. The mucosal-to-serosal flux of Na²⁴ as a function of concentration of stable Na in the mucosal medium. The bladder was maintained in the short-circuited state in a lucite chamber. Flux periods were 15 min in length and the concentration of stable Na in the mucosal medium was increased from an initial value of $1\cdot2$ mequiv l^{-1} by the addition of small volumes of $4\cdot8$ M NaCl. For each bladder, the flux at $1\cdot2, 3, 13, 33,$ and 53 mequiv l^{-1} of Na in the mucosal medium was compared to that at 114 mequiv l^{-1}. The normalised fluxes from 11 experiments were then averaged at each concentration of Na in the mucosal medium. The vertical bars represent ±1 s.e.m. In each case, the serosal bathing medium was ordinary Ringer's solution. Vasopressin was not present (From Frazier and Leaf [24]; courtesy Medicine)

161

G

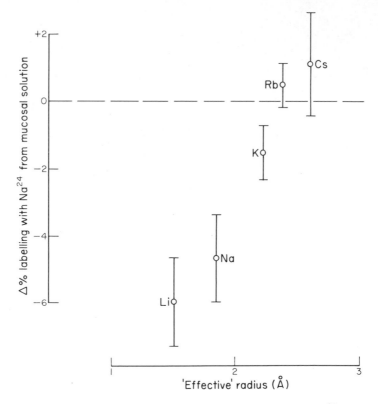

Figure 5.8. The effect of alkali metal ions on tissue labelling with Na^{24} from the mucosal medium. The experiments were performed in a dual chamber which permitted chemical and electrical isolation of the two halves of the same bladder preparation. The mucosal media contained 12·4 mM Na, labelled with Na^{24}. The experimental mucosal medium also contained the chloride of the alkali metal in a concentration of 0·10 M; the control mucosal medium was adjusted to the same osmotic activity by the addition of sucrose. The serosal bathing medium was sodium-free choline Ringer's solution. A negative value for $\Delta\%$ labelling means that the labelling of the experimental half was less than that of the control. The vertical bars represent $\pm$ 1 s.e.m. The 'effective' radius for each cation comes from a compilation by Latimer[26] (From Frazier and Leaf[24]; courtesy Medicine)

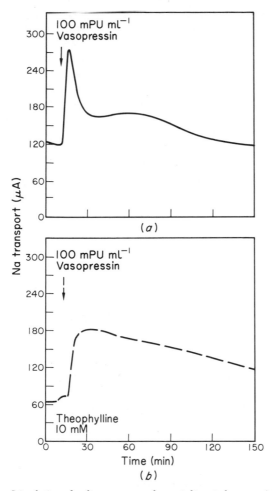

Figure 5.9.   Stimulation of sodium transport by anti-diuretic hormone (a) and by anti-diuretic hormone in the presence of theophylline (b). Sodium transport is expressed as the short-circuit current on the ordinate. The response to the hormone is prompt; transport reaches peak values within a few minutes, and then falls away. Increased rates of sodium transport may persist at a reduced level for two or three hours. Theophylline usually stimulates sodium transport weakly (b) but potentiates and prolongs the response to anti-diuretic hormone

THE ACTIVE SODIUM TRANSPORT PROCESS

Using histochemical techniques, Keller[32] showed that ATPase activity was present along the lateral and basal border of the mucosal epithelial cells, whereas no activity was detected at the apical border. This evidence is in accord with the concept of passive entry of sodium ions and active extrusion at the basal and lateral surfaces of the cells. Farquhar and Palade[33], in studies on the frog skin, have localised ATPase activity on the lateral borders of the cells, but not on the basal or apical borders. Several reviews of the relationship between Na^+, K^+-activated ATPase and sodium transport exist in the literature and the topic need not be discussed here[34-36]. The nature of the intermediate stages of sodium transport between entry at the apical membrane and exit at the lateral or basal membranes remains obscure. Whether this is by simple diffusion is not known. Similarly, the size and concentration of the sodium pool cannot be determined with certainty at the present time.

THE NATURE OF THE PERMEABILITY BARRIER

Permeability changes in response to anti-diuretic hormone are not restricted to sodium and water. The permeability to urea and some other uncharged amides and some alcohols is increased by the hormone[37,38]. The reaction is specific for certain molecules; for instance, it has been found that while urea permeability increases markedly, thiourea is unaffected (Figure 5.10), despite the fact that movement of these molecules through the bladder occurs by passive diffusion, and that solvent drag effects can be seen on urea movement during water transfer.

The effects of anti-diuretic hormone on water and sodium transport occur at, or near, the apical surface of the mucosal layer of cells. This raises the problem that the tissue can remain selective to sodium and solute molecules of small dimensions (e.g. urea), at a time when large channels would seem necessary to accommodate the net transport of water across the tissue. This apparent difficulty may be overcome by assuming that the permeability barrier is not a simple homogeneous structure, but rather a complex system of at least two barriers with different properties in series[39]. Thus, a dense diffusion barrier in series with an underlying porous barrier could theoretically account for the permeability properties of the bladder. The dense barrier might screen out most solutes, but allow water, urea, and sodium to pass. Modification of the porosity of the deeper barrier by ADH could then provide the specific changes in permeability induced by the hormone.

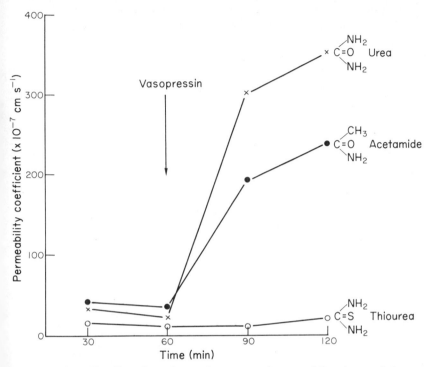

Figure 5.10. The effect of anti-diuretic hormone on the permeability characteristics of toad bladder with respect to urea, thiourea, and acetamide. Permeability coefficients were determined by the addition of radioactively labelled compounds to one side of the bladder and determination of the rate of appearance on the other side. Thirty-minute flux periods were used. Anti-diuretic hormone caused a marked increase in permeability to urea and acetamide. A striking specificity of effect is seen in the lack of permeability change to thiourea (From Maffly et al.[37]; courtesy J. clin. Invest.)

Evidence for the existence of such a dual barrier has been provided by Lichtenstein and Leaf[40, 41], who made the observation that amphotericin B added to the mucosal medium bathing the toad bladder will separate, functionally, the two barriers. Amphotericin B is a polyene antibiotic known to react with membranes containing sterols. For this reason it is toxic to fungi, but not to bacteria, which do not contain sterols in their cell walls. By adjusting the amounts of amphotericin B used in experiments with the toad bladder, it was possible to remove functionally the selective permeability barrier leaving the second barrier relatively intact. Figure 5.11 shows that amphotericin B, when added to the mucosal bathing medium, results in a stimulation of sodium transport (short-circuit current), which was not further augmented by vasopressin. By contrast, net

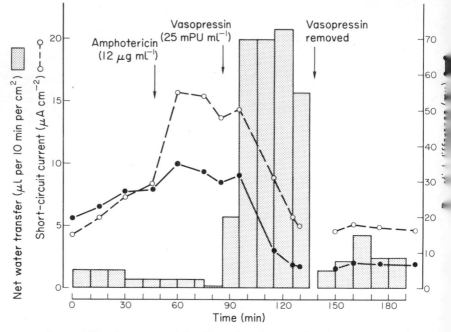

Figure 5.11. Comparison of the effects of amphotericin B and vasopressin on short-circuit current (broken line), potential difference (continuous line), and net water movement (shaded area) across the toad bladder. Addition of amphotericin B to the mucosal medium produced a large increase in short-circuit current without affecting net movement of water. Subsequent addition of vasopressin failed to augment the short-circuit current, but produced its usual large effect on transport of water. When vasopressin was removed, the net transport of water was reduced. In this experiment, sodium Ringer's solution was used as the serosal medium and sodium Ringer's solution (diluted 1:4) as the mucosal bathing medium (From Lichtenstein and Leaf[40]; courtesy J. clin. Invest.)

transport of water across the tissue was not significantly affected by the concentration of amphotericin B used in this experiment, but was increased in typical fashion in response to vasopressin. Thus amphotericin functionally removes the selective permeability barrier of the bladder. A diagrammatic respresentation of such a barrier is shown in Figure 5.12.

166

THE ACTION OF HORMONES ON TRANSPORT

THE ACTIONS OF ANTI-DIURETIC HORMONE

Several attempts have been made in the past to explain the action of this hormone. Ginetzinsky[42a] examined the mucopolysaccharide complexes in the intercellular cement of the basement membrane in

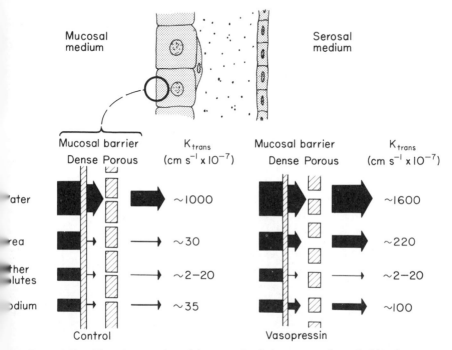

Figure 5.12.   The urinary surface of the mucosal cells is represented as a dual barrier, a dense diffusion, and a porous barrier, in series. All substances including water are retarded at the diffusion barrier. Vasopressin enhances the permeability of this tissue to urea and sodium by an effect on the dense diffusion barrier and to water by an effect on the porous barrier (From Lichtenstein and Leaf[40]; courtesy J. clin. Invest.)

distal portions of the nephron by using histochemical techniques and he estimated hyaluronidase activity in urine; he found that the mucopolysaccharide complexes in the distal nephron disappeared, and that hyaluronidase activity increased in the urine during anti-diuresis. He proposed that hyaluronidase was secreted by the renal epithelium in response to anti-diuretic hormone. This enzyme then digested the intercellular cement and basement membrane, making the renal tubule permeable to water by providing a pathway for water to move between the epithelial cells.

Another hypothesis was that anti-diuretic hormone interacted with cell receptor sites, resulting in the formation of covalent bonds by thioldisulfide exchange reactions, which in turn affected the porosity of the responsive membrane. The cysteine in the ring structure of the hormone was thought to provide the essential disulphide group for these exchange reactions[42b]. When a synthetic polypeptide lacking a disulphide group was subsequently tested and found to be active, this hypothesis of the hormonal action was retracted[43].

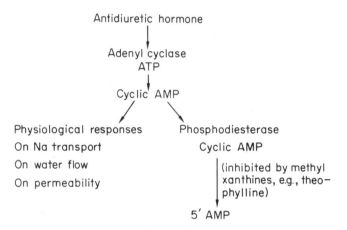

Figure 5.13

More recently, it has become clear that the hormone stimulates production of adenosine 3',5'-phosphate (cyclic 3',5'-AMP) by an action on adenyl cyclase (see Figure 5.13), and that this compound is the intracellular mediator of the hormonal effect[44]. 3',5'-AMP is known to serve as an intracellular mediator of the action of several other hormones[45]. The major evidence in support of this hypothesis is the finding of Orloff and Handler that cyclic 3',5'-AMP added in high concentrations to the medium bathing the toad bladder will mimic the known actions of ADH on permeability to water and sodium. They also found that theophylline which inhibits the hydrolysis of cyclic 3',5'-AMP by the cyclic 3',5'-nucleotide phosphodiesterase, has the same effect as ADH on the toad bladder *in vitro* and enhances the effects of cyclic 3',5'-AMP[46, 47]. Furthermore,

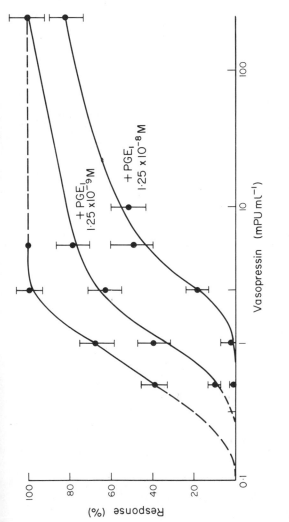

Figure 5.14. Dose-response characteristics of the action of vasopressin on osmotic water flow in toad bladder and the inhibitory effect of prostaglandin E_1 at concentrations of $1 \cdot 25 \times 10^{-9}$ M and $1 \cdot 25 \times 10^{-8}$ M. PGE_1 does not have any effect upon water flow when present alone, but markedly inhibits the response to anti-diuretic hormone. The inhibition can be overcome by high concentrations of ADH, a finding which suggests a competitive type of inhibition. A non-competitive type of inhibition is suggested by its ability to inhibit the theophylline-induced water flow (After Lipson and Sharp[51]; courtesy Am. J. Physiol.)

G*

ADH increases the concentration of cyclic 3',5'-AMP in the toad bladder[48], a finding in agreement with that of Brown *et al.*[49], that ADH increases the accumulation of cyclic 3',5'-AMP by the particulate fraction of a homogenate of dog kidney.

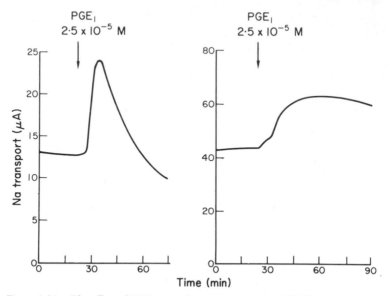

Figure 5.15.   The effect of PGE_1 on sodium transport in toad bladder. Stimulation is produced by concentrations of PGE_1 from 10^{-8} to 10^{-5} M. The onset of effect is rapid and the subsequent response varies from bladder to bladder. Two examples, of short and prolonged responses to PGE_1, are shown in the figure (After Lipson and Sharp[51]; courtesy Am. J. Physiol.)

The evidence, that cyclic 3',5'-AMP mimics the actions of ADH, and that its concentration is increased in tissue exposed to ADH, constitutes persuasive evidence that it is in fact an intermediary in the action of this hormone. How cyclic AMP induces the permeability change is not known. Furthermore, evidence is available that two adenyl cyclases exist in the bladder, one responsible for the effects of anti-diuretic hormone on water flow and one responsible for sodium transport[50]. It is known that when calcium or magnesium salts are present in the Ringer's solution at concentrations of 8–10 mM, the effect of anti-diuretic hormone on water flow is inhibited, while the effect on sodium transport is unaffected. Similarly, prostaglandin E_1 at concentrations as low as 10^{-9} M has a marked inhibitory effect upon water flow responses to ADH

(Figure 5.14) without effect on sodium responses[51, 42]. In fact, prostaglandin E_1 at high concentrations stimulates sodium transport (Figure 5.15). That these results are due to actions on the enzyme adenyl cyclase is shown by the lack of inhibitory effect of Ca^{2+}, Mg^{2+}, or prostaglandin E_1 on cyclic AMP induced water flow.

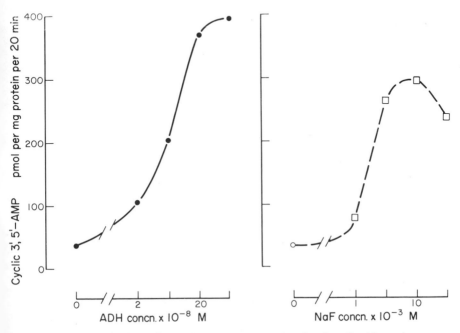

Figure 5.16. *The effects of ADH (arginine vasopressin) and sodium fluoride on the activity of adenyl cyclase in a homogenate of mucosal epithelial cells from toad bladder. Samples of homogenate were incubated at room temperature for 20 min with ATP-α-³²P, 0·1 mM; tris-HCl buffer (30 mM) at pH 7·5; MgCl₂, 5 mM; cyclic AMP, 0·1 mM; phosphoenol pyruvate kinase 40 μg ml⁻¹; myokinase 20 μg ml⁻¹. The activity of the adenyl cyclase was determined by the amount of cycle AM³²P produced during the incubation. The method has been described in detail elsewhere*[53, 54]. *Activation by ADH is usually greater than the activation by fluoride (After Hynie and Sharp*[54]; *courtesy* Biochim. biophys. Acta)

Adenyl cyclase activity in toad bladder has been measured by the method of Krishna, Weiss, and Brodie[53]. The method is essentially an incubation of homogenised mucosal epithelial cells with ³²P-labelled ATP. The ³²P-labelled cyclic AMP which is formed by the enzyme is separated from the reaction mixture and assayed by counting the radioactivity.

It has been shown that anti-diuretic hormones such as arginine vasopressin, arginine oxytocin, and lysine vasopressin all stimulate adenyl cyclase *in vitro*[54]. Sodium fluoride and Mg^{2+} also stimulate the enzyme and the maximal hormonal stimulation is usually greater than the fluoride stimulation (Figure 5.16). The Km for ATP was approximately 10^{-4} M under all conditions and the optimum pH $= 7.5$. All the activity was associated with particulate fractions of the cells and the enzyme was specific for ADH (see Table 5.3). Divalent cations had marked effects on the enzyme activity. Mg^{2+} at concentrations from 1 to 25 mM had a stimulatory effect upon the enzyme and upon stimulation by hormone and fluoride. Greater concentrations had an inhibitory effect.

Table 5.3. STIMULATION OF ADENYL CYCLASE FROM MUCOSAL EPITHELIAL CELLS OF TOAD BLADDER BY DIFFERENT HORMONES

Hormone	Concentrations tested	Response
Arginine oxytocin	$2 \times 10^{-12} - 2 \times 10^{-6}$ M	+
Arginine vasopressin	$2 \times 10^{-8} - 2 \times 10^{-6}$ M	+
Oxytocin	$2 \times 10^{-8} - 2 \times 10^{-6}$ M	+
Lysine vasopressin	$1 \times 10^{-7} - 3 \times 10^{-5}$ M	+
*Adrenocorticotropic hormone	$1 \times 10^{-10} - 1 \times 10^{-6}$ M	−
Glucagon	$3 \times 10^{-9} - 3 \times 10^{-5}$ M	−
Insulin	$3 \times 10^{-11} - 3 \times 10^{-7}$ M	−
Parathormone	$1 \times 10^{-10} - 1 \times 10^{-6}$ M	−
†Norepinephrine	$1 \times 10^{-8} - 1 \times 10^{-3}$ M	−
†Epinephrine	$1 \times 10^{-8} - 1 \times 10^{-3}$ M	−
†Serotonin	$1 \times 10^{-7} - 1 \times 10^{-3}$ M	−
Histamine	$1 \times 10^{-7} - 1 \times 10^{-3}$ M	−

+ = stimulation
− = no stimulation.
* $CaCl_2$ (10^{-5} M) added to the incubation medium.
† Dissolved in a solution containing 100 μg/ml ascorbic acid.
(After Hynie and Sharp[54])

Ca^{2+} was inhibitory at all concentrations above 10^{-5} M. Mn^{2+} was stimulatory only in the presence of fluoride and was inhibitory to the action of anti-diuretic hormone. The stimulation of adenyl cyclase is specific, as is shown by the lack of effect of parathyroid hormone, insulin, ACTH, and glucagon, a group of hormones known to affect adenyl cyclase in other tissues. A similar lack of stimulatory effect occurs for serotonin, catecholamines, and histamine. Norepinephrine and isoproterenol have slight inhibitory effects upon the enzyme.

The affinities of different hormones with anti-diuretic activity, namely arginine oxytocin, arginine vasopressin, vasopressin (Pitressin), lysine vasopressin, and oxytocin, as measured by activation of adenyl cyclase, are found to be in the same order as their relative potencies in stimulating water flow and sodium transport in the intact tissue.

It is of considerable interest that the intrinsic activity of these hormones is not the same. Arginine oxytocin has a greater maximal stimulating effect upon adenyl cyclase than any of the other hormones. Arginine vasopressin has lower intrinsic activity than arginine oxytocin but higher than vasopressin (Pitressin). Similarly, oxytocin has less intrinsic activity than arginine oxytocin, despite the fact that in the intact tissue, both hormones can stimulate sodium transport and water flow to the same extent. The combined observations give rise to the possibility that arginine oxytocin (and perhaps all fully effective anti-diuretic hormones) can stimulate the production of more cyclic AMP in the tissue than is necessary for a maximal physiological response. The possibility that spare receptors exist in the tissue needs to be tested experimentally by the measurement of cyclic AMP levels in the cells after administration of maximal and supramaximal concentrations of the hormones.

While activation of the enzyme in this tissue is very specific, inhibition of enzyme activity can be produced by several agents. For instance, the water permeability effect of ADH in toad bladder is inhibited by epinephrine and norepinephrine, the inhibition being demonstrable in the presence of submaximal concentrations, and not supramaximal concentrations, of ADH. The inhibitory effects are blocked by phentolamine, an α-adrenergic blocking agent, but not by propanolol, a β-adrenergic blocking agent. Because the catecholamines do not block the action of cyclic AMP on water flow it has been concluded that catecholamines affected adenyl cyclase[55]. This conclusion has been partially substantiated by demonstration of catecholamine inhibition of the adenyl cyclase in the broken cell preparation when stimulated by anti-diuretic hormone and by blockade of this inhibitory catecholamine effect by tolazoline. The character of the inhibitory effect is unknown and the situation complicated by the fact that isoproterenol, a β-sympathomimetic agent, also inhibits the effect of ADH on adenyl cyclase activity. Further studies will be necessary for complete elucidation of this effect.

It is clear, however, that there is broad agreement between physiological responses to ADH in the tissue and the activity of the enzyme adenyl cyclase under a variety of different conditions.

THE ACTIONS OF ALDOSTERONE

There have been several demonstrations of the stimulation of sodium transport by adrenocortical steroids on amphibian epithelial membranes[56-63]. However, it was from the demonstration by Crabbe[63] of an effect of aldosterone *in vitro* on sodium transport in toad bladder, and subsequent studies in which the methods were modified to yield more reproducible stimulation of transport[64, 21], that progress in understanding the molecular action of the hormone began.

Figure 5.17 shows a typical response of the toad bladder to aldosterone administered *in vitro*. After a lag period of 30–120 minutes a gradual increase in short-circuit current occurs in the tissue exposed to aldosterone. This increase in short-circuit current after exposure to aldosterone is accounted for by an equivalent rise in the rate of sodium transport, as determined by double isotope flux experiments to measure net sodium transport[62]. In separate measurements, the unidirectional flux of sodium from serosal to mucosal surface has been found unaffected by aldosterone[64].

Since the action of aldosterone is to stimulate the active transport, it might be expected that its effect would depend upon the availability of metabolisable substrates to supply the energy. When aldosterone is added to a bladder shortly after its removal from the animal, the expected stimulation of sodium transport occurs after the usual latent period. Addition of substrate, such as glucose, to such fresh tissue has no effect on the hormonal response. On the other hand, if the tissue is incubated in the absence of substrate for 12 to 24 hours the subsequent addition of aldosterone will cause little or no stimulation of sodium transport until substrate is added to the incubating medium[65, 8]. Sodium transport will then increase almost immediately, and the hormonal effect is evident; tissue similarly incubated in the absence of aldosterone shows little or no response to addition of substrate alone (see Figure 5.18). Thus, either the absence of aldosterone or the restricted availability of substrate as an energy source may be rate-limiting on sodium transport.

Only those substrates which yield acetyl-coenzyme A in the course of their metabolism serve as readily effective energy sources to support the aldosterone stimulation of sodium transport, although aldosterone-stimulated transport has been demonstrated under anaerobic conditions[55]. Thus, glucose, lactate, pyruvate, acetoacetate, and β-hydroxybutyrate are effective, while most other substrates, including acetate and members of the tricarboxylic acid cycle, prove ineffective[66, 67]. In the case of the latter, their low rate

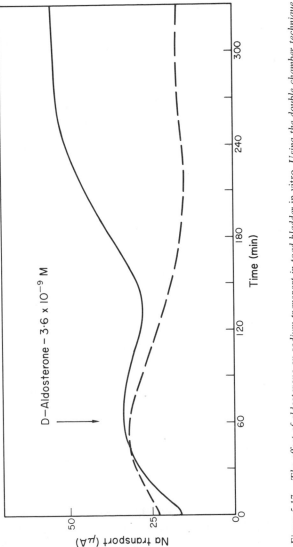

Figure 5.17. The effect of aldosterone on sodium transport in toad bladder in vitro. Using the double chamber technique described in the text, the addition of a low concentration of D-aldosterone to the serosal bathing solute results in a stimulation of sodium transport (measured as short circuit current). The response is characterised by a latent period of approximately 1 hour, during which no effect is seen, followed by an increase of transport, which reaches peak values at 3–4 hours after the aldosterone was given (After Sharp and Leaf[21]; courtesy Nature)

175

of penetration is the probable reason for their ineffectiveness, since the mitochondria, isolated from the mucosal epithelial cells, can be shown to oxidise succinate, despite the failure of this compound to support sodium transport when added to the medium bathing the intact tissue.

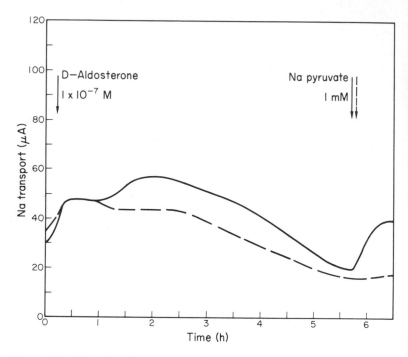

Figure 5.18.   The effect of pyruvate on sodium transport in the presence and absence of aldosterone. The addition of aldosterone to the serosal medium bathing one quarter-bladder results in stimulation of sodium transport after the usual one hour latent period. With the tissues mounted in simple Ringer's solution, the bladder becomes depleted of endogenous substrate and rates of transport in both control and hormone-treated tissue decline. At approximately 6 hours, the rates of transport are almost the same. However, if pyruvate is added to both quarter-bladders, a marked stimulation of transport occurs only in the tissue treated with aldosterone. Little or no effect occurs in control tissues not given aldosterone

The lowest concentration at which aldosterone has been found effective is 3.3×10^{-10} M, and maximal responses are obtained with concentrations of D-aldosterone close to 10^{-7} M. Other steroids such as cortisol, corticosterone, cortexolone, 18-hydroxy-corticosterone, 9α-fluorocortisol, deoxycorticosterone, and pred-nisolone, also stimulate sodium transport.

BINDING OF ALDOSTERONE

In recent years, there has been a rapid development of the theory that the physiological effects of steroid hormones are produced by regulation of the induction of protein synthesis, and evidence is available that aldosterone, also, stimulates sodium transport by an effect upon RNA and protein synthesis. With this background, several studies on the cell receptors for aldosterone have been made. The toad bladder has been shown to concentrate aldosterone, and this tissue was used to investigate the relationship between aldosterone binding and the latent period before onset of action. In the toad bladder, a latent period, following aldosterone, of 40–120 minutes occurs before sodium transport is increased. The length of the latent period seems to be independent of the concentration of aldosterone in the medium and, therefore, of the rate at which aldosterone reaches the site of the receptors[63]. Furthermore, aldosterone uptake into the toad bladder is complete at 30–45 minutes in bladders, which do not begin to increase their transport rate for 60 minutes or more.

Using a technique that should distinguish bound hormone in the tissue from hormone present that is not bound (see Figure 5.19), and by applying a 'Scatchard plot' to the data obtained at different concentrations of aldosterone, the affinity of the sites for aldosterone, and the maximal number of sites having the observed affinity, have been calculated. Two sets of sites were found[69]. The sites with the higher affinity for aldosterone were detected at concentrations of ^{3}H-D-aldosterone less than 10^{-10} M. This small set of sites with higher affinity binds only 9×10^{-4} mol aldosterone per g tissue, whereas the sites of lower affinity bind 3×10^{-12} mol aldosterone per g tissue.

The physiological significance of this binding was examined by the displacement, by other compounds, of aldosterone from sites in the tissue. It was found that mineralocorticoids, in addition to aldosterone, and structural competitive antagonists of aldosterone displace ^{3}H-D-aldosterone from the sites, whereas inactive steroids would not. Aldosterone, deoxycorticosterone, cortisol and spironolactone all displaced aldosterone, and their displacing ability is of the order to be predicted from their activity in the tissue. Furthermore, the correspondence of the physiological concentration range of the hormone with the degree of saturation of the sites comprising the larger set; and the displacement studies with other compounds lend support to the hypothesis that attachment of aldosterone at these sites is the initiating reaction of the hormonal response.

177

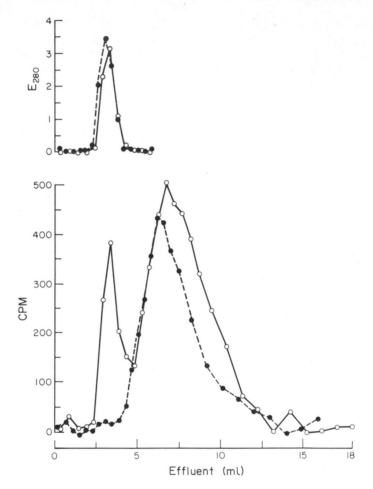

Figure 5.19. Elution patterns of the supernatant obtained after sonication of crude nuclear preparations of toad bladder. The protein was eluted between 2 and 5 ml (upper figure). Paired tissues were incubated with 10^{-8} M ^{3}H-D-aldosterone (continuous line) or 10^{-8} M ^{3}H-D-aldosterone + 10^{-5} M deoxycorticosterone (broken line) to displace ^{3}H-D-aldosterone from its binding sites. It can be seen that when the tissue is incubated with ^{3}H-D-aldosterone alone, a peak of radioactivity is present which is associated with the protein peak and exists in a bound (macromolecular) form. When the incubation is performed in the presence of excess deoxycorticosterone, much of the ^{3}H-D-aldosterone is displaced and little radioactivity is found associated with protein (After Alberti and Sharp[68]; courtesy Biochim. biophys. Acta)

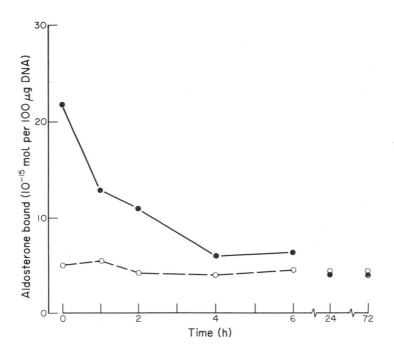

Figure 5.20.   Rate of breakdown of the bound complex of ^{3}H-D-aldosterone and the effect of excess (displacing) deoxycorticosterone. Paired tissues were incubated with 10^{-8} M ^{3}H-D-aldosterone alone (continuous line), or 10^{-8} M ^{3}H-D-aldosterone + 10^{-5} M deoxycorticosterone (broken line). The amount of ^{3}H-D-aldosterone bound was determined by Sephadex G-25 chromatography at zero time of the supernatant from sonicated nuclei. Aliquots of the supernatant were subjected to Sephadex G-25 chromatography at 1, 2, 4, 6, 24, and 72 hours after preparation. It can be seen that the non-displaced bound complex breaks down rapidly with a half-life of approximately 2 hours. A stable component, which is not displaced by deoxycorticosterone, remains intact for as long as 72 hours. During this time the supernatant was kept at 0–4 °C. Thus the displaceable material, which must contain the mineralocorticoid receptor sites for aldosterone, is extremely unstable, while a non-specific stable component also exists in the nuclei (After Alberti and Sharp[68]*; courtesy Biochim. biophys. Acta)*

179

In experiments to determine the site of the aldosterone in toad bladder, 'displaceable' aldosterone binding was not found in the mitochondria or microsomes, but was detected in purified nuclei[70]. 6.5×10^{-14} mol ^{3}H-D-aldosterone per 100 µg DNA were calculated to be bound in the nuclei. The specificity of this binding was shown by the fact that bound ^{3}H-D-aldosterone could be displaced by 9α-fluorocortisol, cortisol and deoxycorticosterone in addition to unlabelled aldosterone, and that all these steroids stimulate sodium transport in the toad bladder. Spironolactone (SC 14266), which is a competitive structural antagonist of aldosterone, also displaced ^{3}H-D-aldosterone from the nuclei. Testosterone, inactive on sodium transport, had no effect upon binding.

Further studies showed that the bound aldosterone could be extracted from the nuclei by sonication, and then exists in the supernatant of the sonicated material as macromolecular complexes. This can be detected by separation of bound from free aldosterone on Sephadex columns (see Figure 5.20). The macromolecular complexes were judged to be protein by enzyme studies, and at least three sets of sites were characterised by their rates of dissociation, by displacement studies, and by agarose gel chromatography[68]. A rapidly dissociating fraction and a very stable fraction can be detected by time-course studies (Figure 5.20), while the stable fraction can be separated into two components by agarose column chromatography (Figure 5.21). These stable components were not affected by an excess of unlabelled steroids, and consist of either two nonspecific steroid binding proteins or two large pools of low affinity sites. Much of the bound aldosterone could be displaced by unlabelled mineralocorticoids. A hundredfold excess of testosterone did not displace ^{3}H-D-aldosterone in experiments using the purified nuclear pellet, a fact that indicates the specificity of the interaction. The mineralocorticoid binding sites were attributable, on the basis of displacement studies, to the rapidly dissociating complex.

In a separate study, the binding of ^{3}H-cortisol was compared with that of ^{3}H-D-aldosterone. As aldosterone is thought to stimulate sodium transport maximally in toad bladder at 10^{-7} M, while cortisol is approximately 50 times less active, cortisol would be expected to occupy only a small percentage of the sites at 10^{-7} M. The results showed that little ^{3}H-cortisol was bound to the rapidly dissociating complex, but did show binding to the non-displaceable non-mineralocorticoid sites. ^{3}H-D-aldosterone, on the other hand, was bound to both the stable and unstable components, thus adding further evidence that the rapidly dissociating complex is relevant to the mineralocorticoid effect.

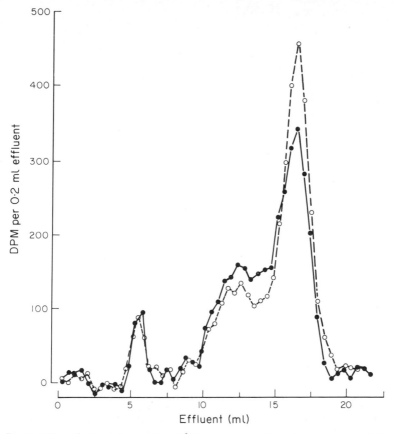

Figure 5.21.   Elution pattern of bound ^{3}H-D-*aldosterone from the supernatant of the sonicated crude nuclear fraction on agarose. Bladders were incubated in* 10^{-7} *M* ^{3}H-D-*aldosterone before separation of nuclear fraction and sonication. The supernatant was passed down a G-25 Sephadex column and the separated bound material divided into two portions. One was placed on an agarose column immediately, while the other was stored on ice and placed on the agarose column two hours later. The first elution pattern (continuous line) shows three peaks, the first material with a molecular weight greater than* 10^6*, the second approximately* 10^5*, while the third peak is free aldosterone. This last peak must be derived from the bound aldosterone which dissociated during the 2-hour period required for complete elution of the column, because only bound hormone was put on the agarose column. The second elution pattern, from the sample put on at 2 hours (broken line), shows the first peak is unchanged, the second slightly diminished, and the third slightly increased. When displacement experiments are performed all the displacement occurs in the third, free aldosterone peak. Thus the first two peaks represent the stable, non-dissociable bound* ^{3}H-D-*aldosterone detected previously by time course studies on G-25 Sephadex. The free steroid in the third peak is derived from the displaceable, rapidly dissociable, mineralocorticoid receptor sites (After Alberti and Sharp*[68]*; courtesy* Biochim. biophys. Acta)

THE ROLE OF RNA METABOLISM

The possibility that aldosterone acted via messenger RNA synthesis was suggested by Edelman et al.,[65] who found that actinomycin D blocked the stimulation of sodium transport by aldosterone in toad bladder, and by Williamson[71], who demonstrated inhibition by actinomycin D of the effect of aldosterone in adrenalectomised rats. The incorporation of ^{3}H-uridine into total tissue water-soluble RNA, as judged by release of counts with ribonuclease, was reported to be greater in aldosterone-treated toad hemibladders than in control hemibladders, and it has been reported by Rousseau and Crabbe[72] that aldosterone can increase ^{3}H-uridine incorporation into RNA extracted by the method of Kirby[73].

The difficulty in relating these effects upon RNA metabolism to the stimulation of sodium transport by aldosterone lies in the fact that cortisol-like steroids may have marked effects upon the incorporation of uridine into RNA. Furthermore, they can enhance precursor pool labelling such that increased RNA labelling occurs without necessarily increasing RNA synthesis. In most studies of aldosterone on precursor incorporation into RNA, the dose or concentration of aldosterone used has been maximal or supramaximal. For a conclusive demonstration of an action of aldosterone on RNA metabolism, which is related to the stimulation of sodium transport, it is necessary to use doses of the hormone within the dose-response range. Preferably, the incorporation experiments should show a graded response within this range. Furthermore, increased RNA synthesis should be demonstrated during the latent period of the hormonal effect before the metabolism of the tissue is changed by the increased work of transport. These requirements may prove to be too severe, in that modern techniques for the study of RNA metabolism may not be sufficiently sensitive to detect a very small change relevant to the stimulation of sodium transport by aldosterone.

THE ROLE OF PROTEIN SYNTHESIS

A considerable body of evidence has been accumulated that aldosterone acts via protein synthesis[65, 74c]; for instance, (1) puromycin abolishes the effect of aldosterone on sodium transport, without disturbing the basal rate of sodium transport or the subsequent response to vasopressin. Puromycin is known to interfere with protein synthesis by preventing the assembly of polypeptide chains in ribosomes. (2) Cycloheximide, similarly, abolished the stimulation of transport by aldosterone without an effect upon basal transport. Cycloheximide inhibits amino acid transfer from amino acyl RNA to the polypeptide chains. (3) Fluorophenylalanine abolished the response to aldosterone (and also accelerated the rate of decay of the basal sodium transport in control tissues), while addition of excess phenylalanine completely reversed the effect of

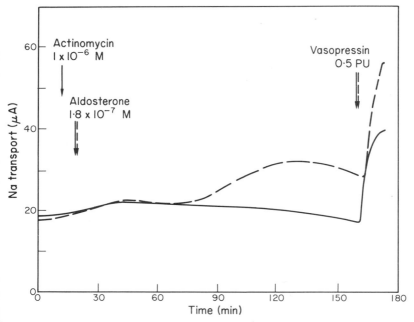

Figure 5.22. The effect of actinomycin D (1 × 10⁻⁶ M) on the stimulation of sodium transport by aldosterone. Actinomycin D can be seen to block the stimulation of sodium transport by aldosterone. The specificity of this inhibitory effect is illustrated by the lack of effect of actinomycin D either on basal sodium transport or on the stimulation of transport by anti-diuretic hormone (After Sharp and Leaf [8]; courtesy Recent Prog. Horm. Res.)

fluorophenylalanine. (4) 1-canavanine, an arginine analogue, had a similar effect to fluorophenylalanine. In contrast, the serine analogues, α-methylserine and β-thienylserine, had no effect upon the aldosterone response. These results are in accord with the ability of fluorophenylalanine to inhibit phenylalanine incorporation into proteins by 61%, whereas β-thienylserine failed to inhibit serine incorporation into proteins. Thus, the ability of inhibitors of protein synthesis and amino acid analogues to inhibit amino acid incorporation into proteins correlates well with their ability to inhibit the effect of aldosterone.

It seems unlikely that puromycin and actinomycin D are exerting non-specific inhibitory effects in blocking the action of aldosterone on sodium transport, because these agents fail to interfere with the basal level of sodium transport in this tissue or the response to

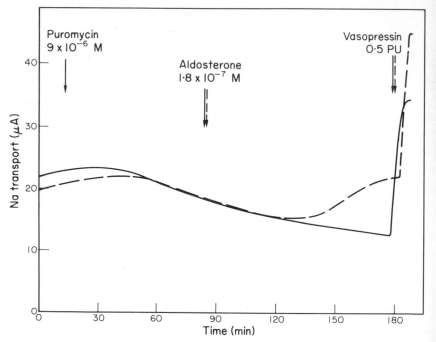

Figure 5.23. The effect of puromycin (9×10^{-6} M) on the stimulation of sodium transport by aldosterone. The stimulation due to aldosterone is completely blocked, while no effect of puromycin is seen on basal sodium transport or on the stimulation by anti-diuretic hormone (After Sharp and Leaf [8]; courtesy Recent Prog. Horm. Res.)

vasopressin. This is shown in Figures 5.22 and 5.23 for actinomycin D and puromycin, respectively. In these experiments, the bladder was exposed to the inhibitor before addition of aldosterone. The basal rate of transport appears unaffected, and a good response to vasopressin is elicited, although no effect from the aldosterone is seen.

The time course of the action of these inhibitors is of interest in telling us the rates of synthesis and inactivation of the messenger

RNA and protein involved. Figures 5.22 and 5.23 document that actinomycin D (1.2 μg ml^{-1}), and puromycin (5 μg ml^{-1}), given before the aldosterone, produces total blockade of the aldosterone response. The same concentration of puromycin, added at the moment of onset of the stimulation of sodium transport by aldosterone, inhibits this response in 10–12 minutes. Actinomycin D, added 10 minutes prior to aldosterone, completely blocks the hormonal effect. When added 20 minutes after the aldosterone, its inhibitory effect is then manifest in approximately 75 minutes. Thus, the latent period for the onset of the inhibitory effect of actinomycin D when added after aldosterone must be at least as long as that for the onset of the hormonal stimulation of sodium transport. The half-lives of the protein and messenger RNA, calculated from a series of experiments such as these, were estimated to be 40 minutes and 1–2 hours, respectively[74b]. When the bladders are incubated for longer periods, e.g. 12–15 hours before administration of inhibitors, the half-life of the protein is prolonged.

THE ROLE OF INDUCED PROTEINS

If aldosterone stimulates DNA-dependent RNA synthesis with the formation of new protein, the role of these proteins in transepithelial sodium transport remains to be elucidated. Two different views have been adopted regarding the answer to these important questions. It has been suggested that the primary action of aldosterone involves NADH-linked electron transport, coupled to phosphorylation of ATP. Increased sodium transport is thus a consequence of increased production of ATP[75, 76]. However, a recently reported demonstration of an aldosterone effect on sodium transport under completely anaerobic conditions[55] is difficult to reconcile with a system presumed dependent on oxidative metabolism. Furthermore, according to this view, increased levels of ATP and creatine phosphate would be expected in the presence of aldosterone, and such increases have not been found. In fact, Handler et al.[55] demonstrated a significant decrease in creatine phosphate in bladders exposed to aldosterone. This finding is not compatible with the suggestion that aldosterone stimulates sodium transport primarily by stimulating the production of energy, and in fact suggests a primary stimulation of sodium transport per se.

The enhancement of metabolism by sodium ions within the tissue implies a stimulation of some step in metabolism, i.e. an enzymatic reaction, directly by sodium. The sodium–potassium-dependant adenosine triphosphatase (ATPase) system of this tissue might seem an obvious candidate for such a sodium-dependant enzymatic reaction, since the resulting increase in ADP may be expected to stimulate metabolism. However, an effect of aldosterone on activity of ATPase in the toad bladder has not been demonstrated[77].

Another possibility is that aldosterone induces the synthesis of a membrane protein which facilitates the entry of sodium into the responsive cells of the epithelium. The evidence in support of this view is as follows:

1. *Tissue labelling.* After exposure to radioactive sodium in the mucosal medium, the radioactivity within the epithelium of paired portions of the same hemibladder, with and without aldosterone, has been assayed. The hormone-induced increase in sodium transport is associated with a slight but significant, increase in radioactivity within the tissue[2]. This finding implies an action of aldosterone to increase the permeability of the apical (mucosal) surface of the cells to sodium. However, these experiments are open to serious criticism, because they fail to localise the radioactive sodium within the transporting cells.

2. *Substrate utilisation in response to aldosterone.* The increased sodium transport, induced by aldosterone, is associated with an increased rate of oxygen, glucose, pyruvate, and acetoacetate utilisation, in response to the extra work performed. If sodium is omitted from the mucosal medium, obviously no stimulation of transport can occur after administration of hormone. As the hormone can prepare the tissue for increased transport even in the absence of sodium, the rates of substrate utilisation under such conditions were studied. No increased metabolism was detected in the aldosterone-treated tissue in the absence of sodium, so that the increased substrate utilisation after aldosterone depends upon the presence of mucosal sodium[69]. Thus, a primary increase in metabolism cannot be demonstrated, a finding consistent with the view that the action of the hormone is to facilitate entry of sodium across the apical membranes of the epithelial cells, making more sodium available for transport. Again, the evidence does not provide proof of this hypothesis, because a very small increase in metabolism might not be detectable by the techniques used.

3. *Substrate utilisation after amphotericin B.* If the metabolic effects of aldosterone are secondary to increased entry of sodium into the cells, then any other means of increasing sodium entry into these same cells should reproduce similar metabolic effects to those

seen with aldosterone. Amphotericin B does have the ability to increase sodium transport, by causing increased sodium entry across the apical membrane. After amphotericin B, increased sodium transport is associated not only with increased oxygen consumption, but with increased utilisation of precisely those substrates which sustain the increased sodium transport induced by aldosterone (Figure 5.24). As the effects of amphotericin B on utilisation of substrates are secondary to sodium entry, so, by analogy, the substrate effects induced by aldosterone may be similarly mediated.

4. *Electrical resistance changes in the response to aldosterone.* In recent studies, Hoffman and Civan[78] have examined the transepithelial electrical resistance of the toad bladder in response to various treatments. A fall in resistance, averaging 18%, regularly accompanied the increase in sodium transport stimulated by aldosterone. On the other hand, the increase in sodium transport that follows addition of pyruvate to the medium of aldosterone-treated, but substrate-depleted bladders, was not accompanied by a detectable change in electrical resistance of the tissue. These measurements of electrical resistance are most readily explained by an action of aldosterone to increase the rate of sodium movement across the apical surface of the epithelial cells.

5. *Actions of aldosterone and vasopressin.* Very good evidence is available to indicate that vasopressin increases the permeability of a barrier to sodium at the apical or mucosal surface of the epithelium[29]. Since one possible action of aldosterone, as described above, is to increase the permeability of this same barrier, a comparison of the actions of vasopressin on sodium transport, in the presence and absence of aldosterone, is of importance to our understanding of the transport processes.

It has been shown that the increment in sodium transport induced by vasopressin is the same whether the vasopressin is added to tissue that has been exposed to aldosterone, or to tissue without this hormone (Figure 5.25). It should be mentioned that this result is seen only when vasopressin is added on top of an initial response by fresh tissue to aldosterone. Tissue that has been exposed for some hours to aldosterone will show a greater effect of the vasopressin on sodium transport than the control tissue, an effect that may be related to a non-mineralocorticoid effect of aldosterone.

This equality of the increment of sodium transport after vasopressin in the presence of aldosterone seems difficult to reconcile with an action of aldosterone, that stimulates the active transport

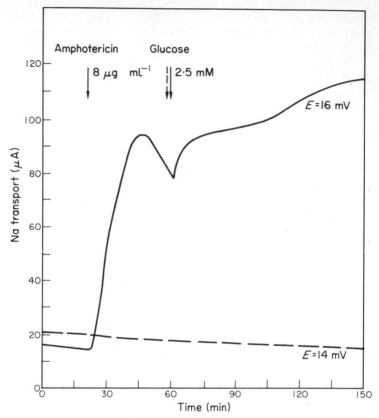

Figure 5.24. The effect of amphotericin B and glucose on sodium transport in the absence of aldosterone. The paired quarter-bladders were mounted in substrate-free Ringer's solution 12 hours before zero time, on the figure, to partially deplete the tissues of endogenous substrate. Amphotericin B, added to the mucosal bathing solution, caused a pronounced rise in sodium transport which reached peak values at about 15 minutes and then declined (presumably due to lack of endogenous substrate). The addition of glucose to the serosal bathing medium induced a sustained increase in sodium transport. No effect of glucose was seen in the control tissue not treated with amphotericin B. In other experiments, it was shown that the toad bladder, after treatment with amphotericin B, responds to the substrates that are effective after treatment with aldosterone. Since amphotericin B admits increased amounts of sodium into the tissue across the apical surface and induces the same susceptibility to substrates as aldosterone, it suggests that aldosterone, too, accomplishes its effect secondarily to facilitating the entry of sodium across the apical permeability barrier of the mucosal layer of cells

mechanism at the basal or serosal surface of the mucosal cells, either directly or by providing an increased supply of high-energy fuel for the active transport system. The provision by vasopressin of an increased amount of sodium to such a primed serosal transport

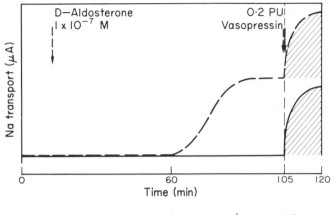

Sodium transport (μequiv mg^{-1} per 15 min)

	Aldosterone		$\Delta \pm$ s.e.m.	p
	Absent	Present		
Without vasopressin	0·076	0·194	0·118±0·016	<0·001
Increment induced by vassopressin	0·132	0·128	0·004±0·013	<0·8

Figure 5.25. The effect of vasopressin on sodium transport in the presence and absence of aldosterone. Twelve paired experiments were performed with, and without, D-aldo-sterone (1 × 10⁻⁷ M) on fresh tissues. A large increase in sodium transport occurred in the aldosterone treated tissues. Despite this, the increment in sodium transport induced by vasopressin was the same whether aldosterone was present or not. This observation of the equality of vasopressin responses in the presence or absence of aldosterone is explicable if aldosterone increases the entry of sodium through pathways in the mucosal membrane that are parallel to, and unaffected by, those on which vasopressin exerts its effects (After Sharp and Leaf[8]; courtesy Recent Prog. Horm. Res.)

system would be expected to cause some enhancement of the vasopressin response in the presence of aldosterone. On the other hand, if the aldosterone functionally duplicates the action of vaso-pressin on the mucosal or apical surface, the expected response is not so easily predicted, as it will depend on the kinetics of the serosal 'pump'. With the 'pump' mechanism able to accommodate an increase of sodium in the intracellular transport pool by a linear increase in the rate of transport, results like those observed would

189

be expected. However, if the transport mechanism was already partially or completely saturated by the maximal aldosterone effect, then addition of vasopressin would be expected to give a lesser response in the presence of aldosterone than in its absence.

The observations are readily explained if aldosterone increases the entry of sodium through pathways in the mucosal surface, which are unaffected by vasopressin, and which are parallel to those on which vasopressin exerts its effects. Indeed, recent findings suggest that the channels, which are affected by aldosterone, are parallel and separate, not only at the mucosal surface, but throughout their extent, from those affected by vasopressin. Thus, there could be two separate pathways for transport of sodium across the bladder: a basal, and an aldosterone-dependent transport system. The evidence suggesting two separate and parallel pathways for sodium transport across the tissue may be summarised as follows. (a) Puromycin and actinomycin D block only the aldosterone-stimulated transport of sodium, without interfering with the basal rate of transport and with little, if any, effect on the response to vasopressin. (b) Ouabain at 10^{-5} M will block partially the aldosterone-stimulated transport of sodium, but only at 10^{-4} M is the basal rate depressed. (c) There is an equality of response to vasopressin in the presence or absence of aldosterone as documented above. (d) Specific substrates, i.e. pyruvate and acetoacetate, will stimulate sodium transport in the bladder exposed to aldosterone, whereas they have essentially no effect in the absence of this hormone. All this evidence taken together gives rise to the possibility that the pathways for the basal and aldosterone-dependent transport of sodium are parallel and separate. Such parallel pathways could exist in separate cell types or side by side within the same cells. It should be stressed, however, that the existence of such parallel pathways is by no means certain.

It is clear from this general account of toad bladder that only a little is known about the transport of Na^+ and water and the regulatory process controlling these functions. Hormone actions by aldosterone at the nuclear level and by ADH at the membrane level modify Na^+ and water reabsorption. More subtle modification by concentration of different electrolytes, catecholamines, prostaglandins, and probably other agents also occurs. Much remains to be learned from amphibian epithelial tissues.

REFERENCES
1. DUBOIS-REYMOND, E., *Untersuchungen über tierische Elektrizität*, Berlin (1848)
2. GALEOTI, G., *Z. phys. Chem.* (Leipzig) **49**, 542 (1904)
3. USSING, H. H., and ZERAHN, K., *Acta physiol. scand.*, **23**, 110 (1951)

4. BENTLEY, P. J., *Biol. Rev.*, **41**, 275 (1966)
5. USSING, H. H., in *The Alkali Metal Ions in Biology*, Springer, Berlin, 49 (1960)
6. USSING, H. H., *The Harvey Lectures*, **59**, 1 (1963)
7. LEAF, A., *Ergebn. Physiol.*, **56**, 216 (1965)
8. SHARP, G. W. G., and LEAF, A., *Recent Prog. Horm. Res.*, **22**, 431 (1966)
9. SHARP, G. W. G., and LEAF, A., *Handbook of Physiology*, Vol. 6, sect. 6, Livingstone, London (1971)
10. CHOI, J. K., *J. cell Biol.*, **16**, 53 (1963)
11. KELLER, A. R., A thesis. Department of Biochemical Sciences, Harvard College, Cambridge, Mass. (1960)
12. PAK POY, R. F. K., and BENTLEY, P. J., *Expl. Cell. Res.*, **20**, 235 (1960)
13. PEACHEY, L. D., and RASMUSSEN, H., *J. biophys. biochem. Cytol.*, **10**, 529 (1961)
14. DIBONA, D. R., CIVAN, M. M., and LEAF, A., *J. Cell Biol.*, **40**, 1 (1969)
15. HAYS, R. M., SINGER, B., and MALAMED, S., *J. Cell Biol.*, **25**, 195 (1965)
16. LOEWENSTEIN, W. R., SOCOLAR, S. J., HIGASHINO, S., KANNO, Y., and DAVIDSON, N., *Science*, **149**, 295 (1965)
17. EWER, R. F., *J. exp. Biol.*, **29**, 173 (1952)
18. BENTLEY, P. J., *J. Endocr.*, **17**, 201 (1958)
19. HAYS, R. M., and LEAF, A., *J. gen. Physiol.*, **45**, 905 (1962)
20. DIBONA, D. R., CIVAN, M. M., and LEAF, A., *J. Membrane Biol.*, **1**, 79 (1969)
21. SHARP, G. W. G., and LEAF, A., *Nature, Lond.*, **202**, 1185 (1964)
22. LEAF, A., *J. gen. Physiol.*, **43**, 175 (1960)
23. FRAZIER, H. S., *J. gen. Physiol.*, **45**, 515 (1962)
24. FRAZIER, H. S., and LEAF, A., *Medicine*, **43**, 281 (1964)
25. FRAZIER, H. S., DEMPSEY, E. F., and LEAF, A., *J. gen. Physiol.*, **45**, 529 (1962)
26. LATIMER, W. M., *J. chem. Phys.*, **23**, 90 (1955)
27. FRAZIER, H. S., *J. clin. Invest.*, **43**, 1265 (1964)
28. DIAMOND, J. M., and WRIGHT, E. M., *A. Rev. Physiol.*, **31**, 581 (1969)
29. CIVAN, M. M., and FRAZIER, H. S., *J. gen. Physiol.*, **51**, 589 (1968)
30. BENTLEY, P. J., *J. Physiol.*, **195**, 317 (1968)
31. EHRLICH, E. N., and CRABBE, J., *Pflügers Archiv. ges Physiol.*, **302**, 79 (1968)
32. KELLER, A. R., *Anat. Rec.*, **147**, 367 (1963)
33. FARQUHAR, M. G., and PALADE, G. E., *J. Cell Biol.*, **17**, 375 (1963)
34. SKOU, J. C., *Physiol. Rev.*, **45**, 596 (1965)
35. POST, R. L., MERRITT, C. R., KINSOLVING, C. R., and ALBRIGHT, C. D., *J. biol. Chem.*, **235**, 1796 (1960)
36. KATZ, A. I., and EPSTEIN, F. H., *New Engl. J. Med.*, **278**, 253 (1968)
37. MAFFLY, R. H., HAYS, R. M., LAMDIN, E., and LEAF, A., *J. clin. Invest.*, **39**, 630 (1960)
38. LEAF, A., and HAYS, R. M., *J. gen. Physiol.*, **45**, 921 (1962)
39. ANDERSEN, B., and USSING, H. H., *Acta physiol. scand.*, **39**, 228 (1957)
40. LICHTENSTEIN, N. S., and LEAF, A., *J. clin. Invest.*, **44**, 1328 (1965)
41. LICHTENSTEIN, N. S., and LEAF, A., *Ann. N. Y. Acad. Sci.*, **137**, 556 (1966)
42a. GINETZINSKY, A. G., *Nature, Lond.*, **189**, 236 (1961)
42b. SCHWARTZ, I. L., RASMUSSEN, H., SCHUESSLER, M. A., SILVER, L., and FONG, C. T. O., *Proc. Natn. Acad. Sci., U.S.A.*, **46**, 1288 (1960)
43. SCHWARTZ, I. L., RASMUSSEN, H., and RUDINGER, J., *Proc. Natn. Acad. Sci., U.S.A.*, **52**, 1044 (1964)
44. ORLOFF, J., and HANDLER, J. S., *Am. J. Med.*, **42**, 745 (1967)
45. SUTHERLAND, E. W., OYE, I., and BUTCHER, R. W., *Recent Prog. Horm. Res.*, **21**, 623 (1965)
46. ORLOFF, J., and HANDLER, J. S., *Biochem. biophys. Res. Commun.*, **5**, 63 (1961)
47. ORLOFF, J., and HANDLER, J. S., *J. clin. Invest.*, **41**, 702 (1962)
48. HANDLER, J. S., BUTCHER, R. W., SUTHERLAND, E. W., and ORLOFF, J., *J. biol. Chem.*, **240**, 4524 (1965)

49. BROWN, E., CLARKE, D. L., ROUX, V., and SHERMAN, G. H., *J. biol. Chem.*, **238**, 852 (1963)
50. PETERSEN, M. J., and EDELMAN, I. S., *J. clin. Invest.*, **43**, 583 (1964)
51. LIPSON, L. C., and SHARP, G. W. G., *Am. J. Physiol.*, **220**, 1046 (1971)
52. ORLOFF, J., HANDLER, J. S., and BERGSTROM, S., *Nature, Lond.*, **205**, 397 (1965)
53. KRISHNA, G., WEISS, B., and BRODIE, B. B., *J. Pharmac. exp. Ther.*, **163**, 379 (1968)
54. HYNIE, S., and SHARP, G. W. G., *Biochim. biophys. Acta*, **230**, 40 (1971)
55. HANDLER, J. S., PRESTON, A. S., and ORLOFF, J., *J. biol. Chem.*, **244**, 3194 (1969)
56. BISHOP, W. R., MUMBACH, M. W., and ACHEER, B. T., *Am. J. Physiol.*, **200**, 451 (1961)
57. MAETZ, J., JARD, S., and MOREL, F., *C. r. hebd. Séanc. Acad. Sci., Paris*, **247**, 516 (1958)
58. MAETZ, J., 'The method of isotopic tracers applied to active ion transport'. *The Saclay Conference*, Pergamon, New York, 185 (1959)
59. MCAFEE, R. D., and LOCKE, W., *Am. J. Physiol.*, **200**, 797 (1961)
60. TAUBENHAUS, M., and MORTON, J. V., *Proc. Soc. Biol.*, **98**, 162 (1958)
61. SAWYER, W. H., *Mem. Soc. Endocr.*, **5**, 44 (1956)
62. CRABBE, J., *Endocrinology*, **69**, 673 (1961)
63. CRABBE, J., *J. clin. Invest.*, **40**, 2103 (1961a)
64 PORTER, G. A., and EDELMAN, I. S., *J. clin. Invest.*, **43**, 611 (1964)
65. EDELMAN, I. S., BOGOROCH, R., and PORTER, G. A., *Proc. natn. Acad. Sci., U.S.A.*, **50**, 1169 (1963)
66. SHARP, G. W. G., and LEAF, A., *J. biol. Chem.*, **240**, 4816 (1965)
67. SHARP, G. W. G., LICHTENSTEIN, N. S., and LEAF, A., *Biochim. biophys. Acta*, **111**, 329 (1965)
68. ALBERTI, K. G. M. M., and SHARP, G. W. G., *Biochim. biophys. Acta*, **192**, 335 (1969)
69. SHARP, G. W. G., KOMACK, C. L., and LEAF, A., *J. clin. Invest.*, **45**, 450 (1966)
70. AUSIELLO, D. A., and SHARP, G. W. G., *Endocrinology*, **82**, 1163 (1968)
71. WILLIAMSON, H. E., *Biochem. Pharmac.*, **12**, 1449 (1963)
72. ROUSSEAU, G., and CRABBE, J., *Biochem. biophys. Acta*, **157**, 25–32, 1968 (1968)
73. KIRBY, K. S., *Biochem. J.*, **64**, 406 (1956)
74a. FANESTIL, D. D., and EDELMAN, I. S., *Fedn Proc. Fed. Am. Socs exp. Biol.*, **25**, 912 (1966)
74b. SHARP, G. W. G., and LEAF, A., *Physiol. Rev.*, **46**, 593 (1966)
75. FIMOGNARI, G. M., PORTER, G. A., and EDELMAN, I. S., *Biochem. biophys. Acta*, **135**, 89 (1967)
76. EDELMAN, I. S., and FIMOGNARI, G. M., *J. biol. Chem.*, **243**, 3849 (1968)
77. HOFFMAN, R. E., and CIVAN, M. M., *Am. J. Physiol.*, **220**, 324 (1971)

6

ACTIVE TRANSPORT AND IONIC
CONCENTRATION GRADIENTS IN MUSCLE

R. P. KERNAN

Department of Physiology, University College, Dublin

INTRODUCTION

Most ions found in the living organism are unevenly distributed between the cell interior and the external bathing fluid. For example, in frog skeletal muscle (Table 6.1) the concentration of potassium ions in the fibre water appears to be about 50 times greater than that in the external fluid or plasma, while the sodium ions are mainly confined to the external medium. Permeability studies[2, 3] revealed that while the muscle fibre membrane is virtually impermeable to sodium ions, potassium ions may pass freely into the cell fluid. It has been shown by these and other workers that the virtual exclusion of sodium from the fibre water, coupled with the presence within the cells of indiffusible, negatively charged molecules, such as nucleoproteins, has imposed a pattern of selectivity on movement of permeable ions through the cell membrane and so modified their distribution between the inside and outside of the cell. In the near-absence of sodium and hydrogen ions from the fibre water, potassium ions must enter the cell in quantity to neutralise the negative charge on proteins. Chloride ions tend to be excluded by the internal negativity of the cell. On the outside of the muscle fibres chloride is the major counter-ion for sodium ions in the interstitial fluid. The distribution of the permeable ions potassium and chloride across the cell membrane under these circumstances is then described by the Donnan equilibrium as follows:

$$[K]_i \times [Cl]_i = [K]_o \times [Cl]_o \qquad (6.1)$$

or

$$[K]_i/[K]_o = [Cl]_o/[Cl]_i \qquad (6.2)$$

H

where $[K]_i$, $[Cl]_i$ and $[K]_o$, $[Cl]_o$ are the concentrations of potassium and chloride ions in muscle fibre water and external fluid, respectively. The resting membrane potential, that is, the electrical potential difference between the inside and outside of the muscle fibre, measured by the microcapillary electrode[4], has been shown to depend on the concentrations of permeable ions inside and outside

Table 6.1

	Muscle	Plasma		Concn. in fibre water
Potassium	83·8*	2·15	0·3	124
Sodium	23·9*	103·8	21·5	3·6
Calcium	4·0*	2·0	0·3	4·9
Magnesium	9·6*	1·2	0·2	14·0
Chloride	10·7	74·3	9·7	1·5
HCO_3 (incl. Ba-sol. CO_2)	11·6	25·4	3·3	12·4
Phosphate	5·3	3·1	0·4	7·3
Sulphate	0·3	1·9	0·3	0·4
Phosphocreatine	23·7			35·2
Carnosine	11·0			14·7
Amino acids	6·8	6·9	0·9	8·8
Creatine	5·3	2·1	0·3	7·4
Lactate	3·1	3·3	0·5	3·9
Adenosine triphosphate	2·7			4·0
Hexosemonophosphate	1·7			2·5
Glucose	0·5	3·9	0·5	
Protein	1·5	0·6	0·1	2·1
Urea	1·6	2·0	0·25	2·0
Water (g kg^{-1})	800	954		
Interspace water (g kg^{-1})	127			

Allowance for interspace (including external fiber sodium) shown in second column of plasma. After Conway[1].

the skeletal muscle fibre *in vitro*[5-7a]. So, it may be said that confirmation has been obtained for the membrane theory of Bernstein[8], according to which the magnitude and polarity of the membrane potential is determined by the relative permeabilities and ionic gradients of the principal monovalent inorganic ions inside and outside the cell. If the muscle fibre membrane at rest is impermeable to all but the potassium and chloride ions then the relation between membrane potential, E_m, and the distribution of these ions between the inside and outside of the fibre should, according to the membrane theory, be defined by the Nernst equation,

$$E_m = - RT/F \ln [K]_i/[K]_o = RT/F \ln [Cl]_i/[Cl]_o \qquad (6.3)$$

where R and F are the gas constant and Faraday constant and T is the absolute temperature.

When E_m was measured in isolated frog skeletal muscles immersed in Ringer's fluid in which $[K]_o$ was varied over a wide range it was found that at concentrations greater than 25 mM K[5] or even 10 mM[9] the Nernst equation was valid for muscle or, in other words, that the membrane behaved as a K-electrode. When the Ringer's fluid contained the normal concentration of potassium, namely 2·5 mM, however, E_m was less negative than the calculated potential by about 8 mV. In these experiments, $[K]_o$ was increased in some cases by substitution for Na in presence of an impermeant anion, while in other cases KCl was added to normal Ringer's and time was allowed for restoration of a Donnan equilibrium before potentials were measured. When muscles were immersed in plasma of the respective animal, instead of Ringer's, during membrane potential measurement, it was found[9, 10] both in frog sartorius and in rat toe muscle, that the Nernst equation applied even at low external potassium concentrations comparable to those found *in vivo*.

Hodgkin and Horowicz[7a] maintained $[K]_o$ constant and varied both $[Cl]_i$ and $[Cl]_o$, and they found that the resulting changes in E_m were consistent with the view that the fibre membrane behaved as a Cl-electrode. Therefore, it might be said that skeletal muscle fibres appeared to exhibit a high permeability for both potassium and chloride ions and, as these were in equilibrium with the membrane potential, a Donnan equilibrium existed with respect to these ions.[1] The ions under these circumstances are at the same electrochemical potential inside and outside the fibre, and therefore the maintenance of the existing concentration gradient does not call for a direct expenditure of energy.

In rabbit atria, Vaughan-Williams[11] made a study of the effect of changing $[K]_o$ on the resting potential of isolated rabbit atria, taking special precautions to avoid the artifact of tip potential[5] in the microelectrodes he used. It was found that at 31 °C the regression line for E_m change at K_o values from 4–11 mM had a slope of $-56·03 \pm 1·77$ mV for a tenfold change in K_o. There was a significant difference ($P = 0·05$) between this value and the slope for the theoretical K-equilibrium potential, which was 59·5 mV. In spite of this divergence, which was probably due to leakage of sodium ions into the muscle fibres, it was evident that in this tissue also potassium ions were close to electrochemical equilibrium in the resting state. Active transport, which is the carriage of a substance across the cell membrane against an electrochemical gradient, need not be considered as responsible directly for striking concentration gradients for potassium and chloride ions across the muscle fibre membrane in amphibia and mammals (Tables 6.1, 6.2). However, the same cannot be said for other ions such as sodium and calcium, which were far removed from equilibrium with the resting membrane potential.

Table 6.2

	Mammalian skeletal muscle		Mammalian plasma	
	mmol kg^{-1}	mmol l^{-1} fibre water	mmol kg^{-1}	mmol l^{-1} plasma water
Potassium	101 ± 1·9	152	5·9 ± 0·2	6·4
Sodium	27·1 ± 1·1	16	138 ± 1·8	150
Calcium	1·6 ± 0·04	(1·9)	3·1 ± 0·1	3·4
Magnesium	11·0 ± 0·3	(16·1)	1·5 ± 0·1	1·6
Chloride	16·2 ± 1·5	5·0	110 ± 0·9	119
Bicarbonate	2·5 ± 0·6	1·2	22·4 ± 0·5	24·3
Total phosphorus	82·9 ± 0·9	—	—	—
Total ac.sol.P.	61·1 ± 0·8	—	—	—
Phosphocreatine	24·4 ± 0·4	36·9	—	—
Adenosinetriphosphate	6·5 ± 0·8	9·8	—	—
Total hexose monophosphate	10·0 ± 0·9	15·1	—	—
Triosephosphate and phospho- glycerate	3·5	5·3	—	—
Fructose diphosphate	0·2 ± 0·06	0·3	—	—
Remaining acid-soluble phos- phate (incl. inorganic P)	(3·5)	5·3	2·2 ± 0·5	2·3
Phospholipids	(10·0)	(15·0)	—	—
Carnosine	1·4 ± 0·1	2·1	—	—
Anserine	20·3 ± 4·9	30·7	—	—
Urea	5·2 ± 0·6	7·0	6·7 ± 0·8	7·0
Amino acids	(20)	(30)	3	3
Lactate	(2)	(3)	1	1
Glucose	—	—	5	5
Acid-labile CO_2 (Ba-soluble)	(10)	(15)	—	—
Protein	3	4	1	1
Water (g kg^{-1})	768	—	923	—

After Conway[12].

The resting potential (the most negative potential during the contraction–relaxation cycle) was measured in visceral smooth muscle with microelectrodes[13, 14] and was found to be as low as − 50 to − 54 mV. It also differed from that of skeletal and cardiac muscles, in that it changed by only 35 mV for a tenfold change in $[K]_o$, even at high potassium concentrations.

Casteels[15] measured the ionic gradients of potassium, sodium, and chloride between cell fibre water and external fluid in guinea pig taenia coli, and also their unidirectional fluxes across the cell membrane by means of radioisotopes. He then used the Goldman constant-field equation to evaluate the contribution of passive ion movements to the membrane potential of this tissue. In this equation, conditions of zero net flux of ions are assumed to exist and the influence of a significant membrane permeability to sodium ions on the potential is determined from the following equation:

$$E_m = RT/F \ln \frac{P_K[K]_o + P_{Na}[Na]_o + P_{Cl}[Cl]_i}{P_K[K]_i + P_{Na}[Na]_i + P_{Cl}(Cl]_o} \qquad (6.4)$$

where P_K, P_{Na}, and P_{Cl} are permeability coefficients. Adrian[5] and Hodgkin and Horowicz[7a], for example, have used this equation to explain the discrepancy between E_K and measured potential in frog sartorii in normal Ringer's solution. They considered that they were dealing with steady-state conditions with the Na–K pump operating in an electrically neutral manner and preventing the net entrance of sodium ions into the muscle fibres. They obtained good agreement between E_m and that calculated from the Goldman equation, when they made P_{Na}/P_K equal to 0·01. As mentioned above, steady-state conditions do not apply in fact and it is possible to demonstrate net uptake of Na by isolated sartorii when bathed in normal Ringer's solution.

Casteels found the permeability constants for taenia coli to have the following values; $P_K = 11 \times 10^{-8}$ cm s^{-1}, $P_{Cl} = 6·7 \times 10^{-8}$ cm s^{-1}, and $P_{Na} = 1·8 \times 10^{-8}$ cm s^{-1}. When these values were used in the Goldman equation, the calculated membrane potential was $- 37$ mV, a value significantly less than that measured. He suggested that the difference here might be due to electrogenic pumping of sodium ions in the tissue. While the value of P_{Na} found by him was similar to that found by Adrian in skeletal muscle, his values for P_K and P_{Cl} were much lower. This would indicate that the greater P_{Na}/P_K ratio in smooth muscle was due not to high sodium permeability but rather to low potassium permeability. Low chloride permeability was also indicated by the fact that this anion was more concentrated in the cells than was compatible with a Donnan distribution, and it appeared to be actively transported into the cells[16].

CHLORIDE PERMEABILITY OF MUSCLE

The relative membrane permeabilities of chloride and potassium ions have been determined by Hutter and Noble[17] by measurement of the changes in resting membrane conductance resulting from replacement of external chloride ions by impermeant anions such as methyl sulphate and pyroglutamate. Their results showed that in resting muscle, chloride accounted for about 68 % of conductance and that this value varied with E_m because of changes in K conductance in presence of a relatively constant Cl permeability. One effect of removing Cl_o here was a slow repolarisation after an action potential[18], showing that net influx of this anion normally contributed to the repolarising current. Several other factors, including the pH of the external medium and the presence of foreign anions and cations in the bathing fluid, were capable of modifying considerably the mobility of chloride ions through the muscle fibre membrane and therefore the membrane resistance. The membrane conductance measured by the microelectrode in frog sartorius and toe muscles was found[19a] to decrease markedly in acid solution and to increase in alkaline solution when the external pH was varied over the range of pH 5–9.8. The fact that replacement of external chloride by methylsulphate virtually removed the pH sensitivity suggested that it was chloride permeability, rather than potassium permeability, that was modified by hydrogen ions here. When conductance change was plotted against pH, a sigmoid curve was obtained that had its steepest part about pH 7, so the pH effect might be significant in vivo. The participation of chloride ions in the pH-sensitive component of membrane conductance was also confirmed in the following experiment. Hutter and Warner[19b] immersed isolated frog muscles in Ringer's with 100 mM K and 216 mM Cl labelled with ^{36}Cl and ^{42}K, so that the fibres accumulated these ions. The efflux of these isotopes into inactive Ringer's of the same composition was then followed at external pH values ranging from 5 to 9.8. The efflux of ^{36}Cl from the fibres into acid solution was so slow that it could easily be distinguished from the efflux from the extracellular space, but when the pH of the medium was increased into the alkaline range ^{36}Cl efflux was raised by a factor of 4.5–6.7 and this increase was reversible. The efflux of ^{42}K from the muscles, on the other hand, was only slightly increased in response to increase of pH of the medium. The buffers acetylglycine and tris maleate were used for adjusting pH.

Carmeliet and Bosteels[20] found that the Cl^- efflux from cardiac Purkinje fibres was sensitive to changes in external pH, but only when the bathing fluid contained sodium. They concluded that the primary effect of pH was on Na^+ movement, which was increased by about 50 % when the pH of the medium was raised from 5.0 to 9.0.

The nature of the influence of pH on anion permeability here is not quite clear, as it seems to be contrary to that found elsewhere[21, 22] and to that which might be expected from theoretical consideration of fixed membrane charge. If pH change modifies net charge on membrane pores, then it might be expected that decrease in hydrogen ion concentration of the external medium might increase their negativity or decrease positivity, thereby making the membrane channels less receptive to chloride ions.

Such was found to be the case in muscle fibres of the giant barnacle (*Balanus nubilis*), when either the external or the internal pH was changed over the range of pH 3·5 to 10·0. A sharp increase in Cl^- conductance was observed[23] when pH was lowered to values below pH 5. Relative membrane conductances contributed by Cl and K ions were 1/6, respectively, at pH 7·7, but changed to 9/1 at pH 4. The fact that increase in external pH actually increased chloride permeability made it necessary for Hutter and Warner to put forward a modified scheme for alteration of membrane fixed charge by hydrogen ions. It was proposed that H^+ competed with Ca^{2+} for monovalent anionic sites on the membrane. In this way, increased alkalinity of the bathing fluid might increase the positivity of the membrane charge by the following reactions:

$$RH \rightleftharpoons R^- + H^+$$
$$R^- + Ca^{2+} \rightleftharpoons (RCa^+)$$

However, cations other than hydrogen ions were found[24] to depress ^{36}Cl efflux from frog muscles that had been equilibrated with KCl fluid at pH 7. For example, when heavy metals, including Cu^{2+}, Zn^{2+}, or UO_2^{2+}, at a concentration of 10^{-4} M, were added to inactive bathing fluid, the ^{36}Cl efflux from muscle was markedly inhibited. Since this observation did not conform with the scheme for mediation of fixed Ca^{2+} in production of membrane charge as outlined above, it seemed that the mechanism involved in regulating chloride permeability was somewhat more complicated. Before such a mechanism is considered, it should be mentioned that certain foreign anions also interfere with the chloride permeability of the muscle fibre membrane and this must also be taken into account.

A fall of membrane conductance was noted[25, 26] when Cl^- in the bathing fluid was replaced by nitrate and by other foreign anions. While replacement of external chloride fully or in part by NO_3, I^-, or CNS^- did not have an immediate effect on the turnover of labelled Na+ or K^+ in frog muscles[27], an effect of such replacement on Cl^- fluxes was evident. For example, Harris[28], using muscles that had been labelled in either normal or high-KCl fluid, measured their rate of Cl^- output during reimmersion in inactive Ringer's, and he found that Cl^- movement across the muscle fibre membrane was slowed in the presence of foreign anions, including Br^-, NO_3^-, I^-, ClO_4^-, and CNS^-, in order of increasing effect. The fact that non-permeant anions were no more effective, and in some cases less effective, in bringing about inhibition of chloride movement than the slightly permeant anions mentioned indicated that the rate of Cl^- efflux here was not solely determined by the rate at which it could be replaced by the foreign external anion. The Cl^- output from the muscles appeared to be the sum of two processes, exchange for permeable anion and net output accompanied by cation; the latter predominated in muscles given a preliminary Cl loading in KCl fluid. The rapid onset of Cl inhibition induced by the foreign anions indicated that their action was at the membrane surface, possibly through adsorption on specific anionic sites. Since the order of their relative effectiveness in modifying chloride permeability corresponded to the lyotropic series, involvement of distribution of the foreign anions at a protein–water interface seemed likely.

Hutter and Warner[29] also distinguished between anions too large to penetrate the membrane, and whose presence in the external bathing fluid did not normally interfere with Cl^- movement, and anions that penetrated the muscle fibre membrane slowly and in so doing inhibited Cl movements. The former included sulphate and methylsulphate, and the latter those anions already mentioned. Spurway[30, 31] compared the washout rates for ^{82}Br and ^{36}Cl from muscles into various inactive fluids, containing 50 mM K, 30 mm permeant anion, and Na^+ and sulphate making up the osmotic deficit. The mean half-times for washout of ^{36}Cl were 6.7 ± 0.5 min into Cl^-, about 12 min into Br^-, and 22 ± 2.0 min into NO_3^-. The washout values for ^{82}Br from the muscles into these fluids gave comparable results, so it seemed that the external anion, and not the internal one, was important in determining anion efflux rate.

In another series of experiments, Spurway measured the relative rates of net anion influx into muscles soaked in high-K media in the presence of different internal anions. These were determined by following rates of muscle swelling (cf. reference 3). He concluded from these experiments that the rates of anion uptake were not influenced by the nature of the internal anion. He proposed a model to account for the effect of both external pH and external anion on chloride permeability. According to this, specific positive binding sites for anions might be located near to the chloride channels in the membrane, leaving the cation-binding sites remote. The binding of H^+ with the negative sites, he believed, might increase the affinity of the positive sites for the otherwise loosely held Cl^-, thereby impeding chloride movement through the channels. When foreign, slightly permeable, anions were present, these would have a greater affinity than did Cl^- for the anion-binding sites and so could influence Cl^- movement through the channels. The binding the anions by proteins in solution has been demonstrated by Carr[32] and the affinity for anions increased with protonation, so thus far the model seems feasible. The absence of any immediate effect on cation permeability in the experiments discussed may be due to the fact that chloride channels seem, from conductance measurements, to be located at the fibre surface, while K^+ channels appear to be more concentrated in the transverse tubular system of the frog skeletal muscle[33].

MEASUREMENT OF INTRACELLULAR SODIUM CONCENTRATION

At one time, it was thought that sodium ions were completely excluded from muscle fibres because they could not penetrate the lipoprotein membrane enveloping the sarcoplasm. However, this view changed when it was found that muscles tended to accumulate this ion *in vivo* during fatigue[34] and when rats were put on a diet deficient in potassium[35]. The determination of the actual concentration of sodium and of other ions within the muscle fibre water presents a number of difficulties that might be discussed at this stage. The concentration of ions in terms of milliequivalents per unit weight of tissue may as a rule be readily available following wet or dry ashing of the tissue and analysis by means of a flame photometer[36]. However, these ions within the muscle are dissolved in the aqueous phase (making up about 80% of the muscle weight) and this water is distributed between the intracellular compartment and the interstitial space. In the latter space, concentrations of ions are similar to those found in plasma (Tables 6.1 and 6.2), so it is important that the distribution of muscle water between interstitial fluid and sarcoplasm be known if intracellular concentrations are to be calculated. The intracellular concentration of potassium ions in frog muscle (Table 6.1), for example, was calculated by the following equation:

$$[K]_i = \frac{[K] - [K]_o \times IF}{TW - IF} \qquad (6.5)$$

where $[K]$ was the potassium concentration expressed in terms of unit weight of wet tissue, $[K]_i$ and $[K]_o$ were potassium concentrations in sarcoplasm and interstitial fluid expressed per unit volume of water in these compartments, and TW and IF were total muscle water and interstitial or extracellular water volumes expressed as fractions of total muscle weight. Taking total water and extracellular water volumes to be 80% and 13% of total muscle weight, respectively, the values in Table 6.1 for intracellular potassium concentration were calculated as follows,

$$[K]_i = \frac{83\cdot8 - 0\cdot13 \times 2\cdot15}{0\cdot80 - 0\cdot13} = 124 \qquad (6.6)$$

The total water of muscle may be conveniently measured by drying the tissue to constant weight in an oven at about 110 °C and expressing the loss of weight as a fraction of the original muscle weight. Determination of interstitial fluid volume has been approached in a number of ways and with a variety of results.

EXTRACELLULAR SPACE MEASUREMENT

Attempts have been made to determine the extracellular fluid volume in tissues by measuring the concentration within them of a substance which seems to be confined to that region. For example, it was believed at one time that under *in-vivo* and *in-vitro* conditions chloride ions were completely extracellular. If the chloride concentration in this space is the same as that of plasma, namely 74·3 mequiv l^{-1} (Table 6.1), then a muscle containing 10·7 mequiv Cl per kg wet weight will have an interstitial volume of 0·14–0·15 ml g^{-1} of muscle. Since the space occupied by the impermeable polysaccharide inulin in muscle under equilibrium conditions was usually significantly less than the chloride space[37], it appeared that chloride might enter the fibres to a significant extent, particularly when a muscle was isolated and soaked in Ringer's fluid for more than 15 min[38]. These workers found the mean inulin space in frog muscles soaked in inulin–Ringer's to be 0·096 ml g^{-1}, compared with mean chloride and sodium space of about 0·13 ml g^{-1}. The latter were calculated from the kinetics of washout of sodium and chloride into isotonic glucose solution. Cotlove[39] measured the rate and extent of distribution of inulin and sucrose in the chloride space of rat muscle *in vivo*. Here, plasma concentration of saccharide was maintained constantly by infusion. It was found that there were two distinct phases of entry of the saccharide. During the first phase, which occupied 1–2 h, the saccharide space was about 80% of the chloride space, which was roughly comparable to the results obtained in frog muscle *in vitro*[38]. The second phase lasted for about 15 h, during which 93–98% of the Cl space became occupied by

202

saccharide. This experiment illustrated the maintenance of intra-cellular chloride at a low concentration *in vivo*, in contrast to the situation of isolated muscles in Ringer's, since in the latter case chloride entry had the effect of lowering the fractional chloride space occupied by inulin, Harris[40] measured the loss of ^{36}Cl during 1 min from sodium-enriched frog skeletal muscles into inactive Ringer's to determine the chloride space and he found a mean value of 0.129 ± 0.02 ml g^{-1}. The sodium space in the same muscles measured by a 1 min washout of this ion into sodium-free medium was 0.117 ± 0.025 ml g^{-1}. The 1 min chloride space of muscles given a preliminary soaking in 25 mM K Ringer's increased to 0.169 ± 0.039 ml g^{-1} but the extra chloride here must have been of intracellular origin.

Although most of the values for extracellular space quoted above are grouped around a value of 0.13 for isolated frog skeletal muscle in Ringer's, and this is a value which has been widely used in the determination of intracellular potassium concentrations in muscle fibre water[41, 5, 1, 9], a range of values as wide as 8–40% has been reported for a variety of methods of measurement[42].

At the lower extreme, we have a value of 0.08–0.09 ml g^{-1} for 'true' extracellular space reported by Ling and Kromash[43] for a variety of frog skeletal muscles. In their work, they employed radioactive isotopes of inulin, sorbitol, sucrose, and poly-*l*-gluta-mate as probe materials. Particular attention was paid in this extensive work to the question of equilibria with reference to relative diffusion coefficients of probing substances used and to mean diffusion path lengths. It was concluded that inulin reached 99% of equilibrium distribution with the true extracellular compartment within 4 h and that it then continued to enter a fibre space for up to 18 h. The space entered by ^{14}C-labelled inulin after 60 h at 0 °C during this slower phase of penetration was 0.16–0.19 ml g^{-1} of muscle. As the space occupied during the slow phase here, namely, $0.19 - 0.09$ ml g^{-1}, was in excess of the 0.005 ml g^{-1} that might be attributed to the transverse tubules of the sarcotubular system that Ling and Kromash considered to be the only part of that system confluent with extracellular space, they suggested that the inulin was being absorbed or was entering into compartments other than extracellular space. Harris[40], on the other hand, argued that frog skeletal muscle had three compartments. The first, occupying 0.13 ml g^{-1}, was that which contains the chloride fraction that exchanged within 1 min and might be considered as the space between the individual fibres. The second, termed a 'non-specific' fibre region, was supposed to be occupied normally by sodium and to be penetrated only slowly by the usual probing molecules such as inulin. It was considered by Harris to occupy a further 0·10 ml g^{-1} of the muscle, so giving a total sodium space of 0·23 ml g^{-1} of muscle. Harris arrived at this conclusion through a study of the K^+, Na^+, and Cl^- contents of muscles after exposure to a variety of media. After subtracting the ions and water associated with the 'absolute' extracellular space of 0·13 ml g^{-1}, he discussed the possible distribution of the remaining ions between two cellular compartments, sodium ions being present in only one of these. He concluded that this arrangement gave better agreement with the Donnan relation than found on the basis of one cellular compartment[2]. In favour of Harris's view are (1) the phenomenon of exchange-diffusion of sodium in muscle, to be discussed later, and (2) the electronmicroscopic studies of Birks and Davey, which indicated that the sarcoplasmic reticulum in skeletal muscle must also be considered as an extracellular region. The capacity of the reticular system would add as much as another 13% of the cell volume to the non-specific space measured by many workers[38, 41].

Birks and Davey[44] found that changes in the volume of the sarcoplasmic reticulum (SR) induced by soaking muscles in hypotonic and hypertonic fluids were always the opposite of those occurring within the sarcoplasm (Figure 6.1). This striking demonstration of the extracellular nature of this region confirmed earlier observations by Dydynska and Wilkie[45] of anomalous osmotic behaviour of the muscle fibre spaces.

Peachey[46] estimated from electronmicrographs that the total volume of SR and transverse tubules of frog sartorius (Figure 6.2) was about 13% of the fibre volume. As the solid matter in these spaces was about 3% of the fibre volume, leaving 10% of water, this, expressed as a percentage of the total muscle volume, was about 9%. This would yield a total of about 0·22 ml g^{-1} for the total extracellular water of this muscle, and leave a volume of 0·58 ml g^{-1} for the intracellular water. If this value is used in equations (6.5) for the calculation of $[K]_i$ this will be found to increase from 124 to about 143 mequiv l^{-1} f.w., but it will only increase E_K (equation 6.3) by about 3 mV.

Yamada[47] made a histological study of frog sartorii following their immersion for up to four days in normal or K-free Ringer's at 0°C. These studies revealed that even after one day there was an increase in the interfibre space attributable to a loosening of the normally tightly packed fibres. This first appeared in the outermost fibres, but spread progressively through the tissue over the total period examined and the author remarked on the parallel between these histological observations and the steady increase of extracellular space with prolonged soaking observed by Tasker and his

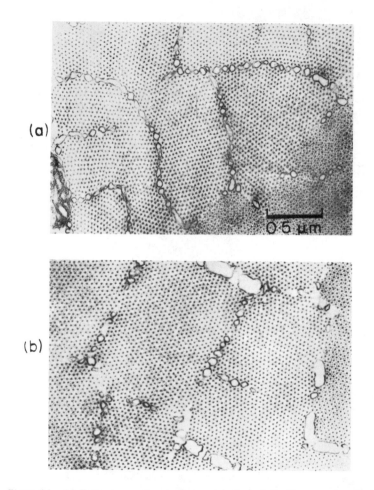

Figure 6.1. (a) *Transverse section passing mainly through the H-zone region of a frog sartorius muscle, showing cross-sections of the H-zone sacs of the sarcoplasmic reticulum. The muscle was soaked for 10 min in normal Ringer's (N Ringer's) before fixation in acrolein. All figures at same magnification. (b) Similar view to (a) from a muscle soaked for 10 min in hypertonic Ringer's (1·5 × N) before fixation in acrolein. Note dilated H-zone sacs (From Birks and Davey*[44]; *courtesy* J. Physiol)

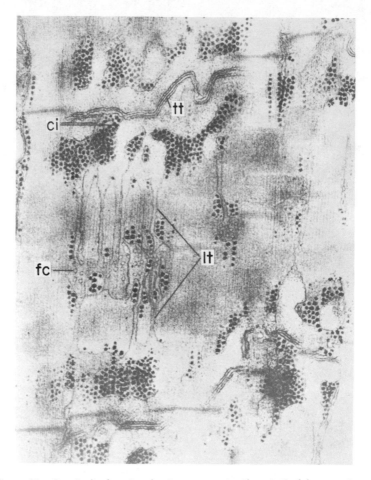

Figure 6.2. Longitudinal section showing an extensive 'face view' of the sarcoplasmic reticulum. Near the top of the figure, the transverse tubule (tt) can be followed for almost 2 μm as a continuous structure. A discontinuity in a terminal cisterna (ci) is indicated. Note the serrate appearance of the 200 Å spaces between the transverse tubules, and the terminal cisternae. This appearance results from the presence of dense lines running between these structures and spaced about 200 Å apart. Other elements of the sarcoplasmic reticulum are also visible and marked for identification: lt, longitudinal tubules; fc, fenestrated collar (From Peachey[46]; courtesy J. Cell. Biol.)

associates[42]. In the light of this finding, one wonders about the significance of the slow phase of inulin entry.

SODIUM DISTRIBUTION IN FROG SKELETAL MUSCLE

The sodium concentration in frog sartorius muscle is about 24 mequiv kg^{-1} wet weight of tissue (Table 6.1). Using the lower extracellular space volume of 0.13 ml g^{-1} and an extracellular sodium concentration of 104 Na, the total sodium in this space would be 13.5 mequiv kg^{-1}, leaving almost half the sodium to be accounted for within the fibres. If, as suggested by Harris[40], the space occupied by sodium is as much as 0.23 ml g^{-1} then practically all the sodium is extracellular with only about 1 mequiv l^{-1} muscle fibre water. Lev[48] measured the activities of sodium and potassium ions in frog sartorius muscles by means of microcapillary electrodes made from ion-selective glass developed by Eisenman and his colleagues[49]. They obtained a range of 0.0053 to 0.0055 for the sodium ion corresponding to a minimal intracellular concentration of about 7 mequiv l^{-1} f.w. for maximum possible activity coefficient of 0.770. A value of 0.20 ml g^{-1} for the volume of the extracellular space[50] would give a similar intracellular sodium concentration. A similar concentration for intracellular sodium was also found by Keynes and Swan[51] from that fraction of muscle sodium which exchanged with a time constant of 1 h.

Conway[1] considered the sodium within frog skeletal muscle to be divided into three fractions as follows:

interstitial space Na 13.5 mequiv kg^{-1}
'external fibre Na' 7.8 mequiv kg^{-1}
internal fibre Na 2.6 mequiv kg^{-1}

The external fibre sodium, also referred to as that of a 'special region', possibly the sarcolemma, was determined on the basis of kinetic measurements of radioisotopic fluxes. It exceeded the chloride present in the muscle fibre space and it was thought that it might be adsorbed on fixed charges at the external surface of the fibres. When the Cl^- in the interstitial space, 9.7 mequiv kg^{-1}, was subtracted from the total muscle chloride, only 0.8 mequiv Cl per kg muscle remained in the fibre, compared with 10.4 mequiv Na per kg.

The ion exchange properties of the 'external fibre' sodium were suggested by the following experiments of Carey and Conway[52]. When freshly isolated, frog sartorii were immersed in isotonic glucose; the sodium in the interstitial space was rapidly lost, but the remaining sodium moved much more slowly (Figure 6.4), only

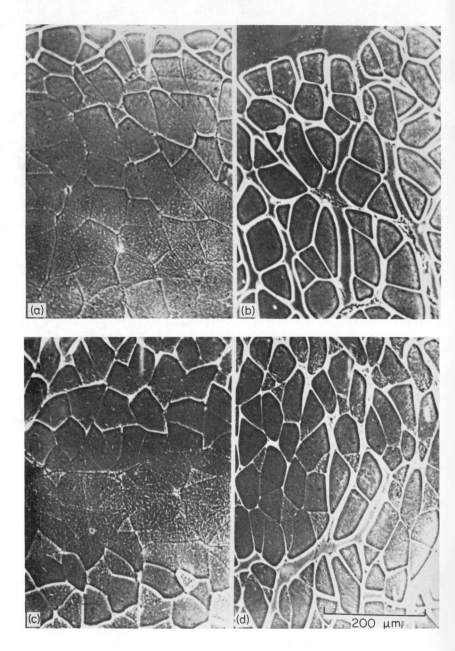

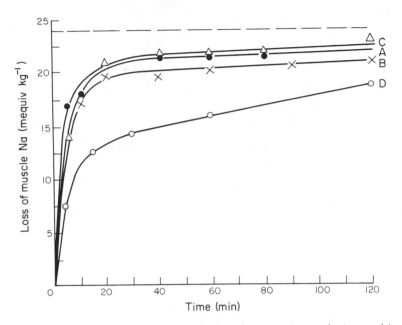

Figure 6.4. *The curves show the loss of sodium from sartorius muscles immersed in glucose solutions with and without potassium ions.* A (●) *in 3·2% glucose solution with 100 mmol KCl per litre.* B (×). *in 2·8% glucose solution with 5 mmol* K_2SO_4 *per litre.* C (△), *same external fluid as in* B, *but the* Na^+ *in the muscle was labelled by injecting labelled* Na^+ *into the frog the previous day. Concentrations of Na determined from the relative counts.* D (○), *in 3·2% glucose solution (After Carey and Conway[52]; courtesy J. Physiol.)*

Figure 6.3 (left). *Phase contrast microphotographs of the transverse sections of two pairs of sartorii fixed with gluteraldehyde in veronal buffer:* (a, c) *control muscles;* (b) *companion muscle of* (a). *which has been stored in the cold low-K soaking in solution for two days;* (d) *companion muscle of* (c), *which has been stored in the cold low-K soaking-in solution for four days. The epimysium on the lower surface of the muscle, which is thinner than that on the upper surface and is usually chosen for microelectrode penetration, is at the top of each section (From Yamada[4]; courtesy* Kumamoto Med. J.)

60% being lost in 3–4 h. When 5 mM K_2SO_4 was added to the glucose, a much greater fraction of the sodium was in the early fast phase, and this apparently exchanged for potassium. When frog sartorii were labelled *in vivo* by injecting frogs with ^{24}Na-labelled saline 24 h before the experiment, it was found that 19·7 of the 23·9 mequiv Na per kg muscle had mixed with the isotope, leaving

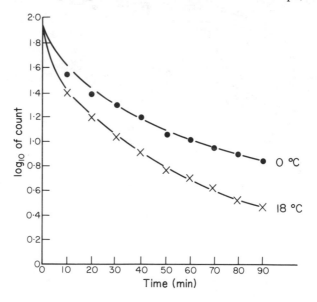

Figure 6.5. Effect of temperature on the loss of ^{24}Na, from sartorii of frogs labelled in vivo by injecting ^{24}Na the previous day. The logarithm of the count under the conditions described in the text is plotted against time (min); at zero time the count is taken as 100. Companion muscles were used for the observations at 0 °C (●) and 18 °C (×) and each ● × pair represents the mean obtained in nine experiments (After Carey, Conway, and Kernan[53]; courtesy J. Physiol.)

about 4·2 mequiv unmixed. The difference was variable but highly significant. When muscles labelled in this way were then reimmersed in isotonic glucose with 10 mM K as much as 90% of the ^{24}Na was lost from the muscles within 20–30 min and this phase of exit was indistinguishable from efflux from interstitial space. There followed a slow ^{24}Na efflux in which 2·4 mequiv Na per kg was lost over 2 h. The temperature coefficient of the fast exit phase was low, Q_{10} of 1·39 suggesting a physical process (Figure 6.5). When frog muscles were labelled with ^{24}Na, under conditions where an appreciable quantity of this ion entered the fibre water, a completely different picture emerged. Sodium enrichment of the fibre water was carried out by soaking the isolated muscles overnight in K-free Ringer's containing labelled sodium. After 90 min in inactive Ringer's, at 0 °C, as much as 12–13 mequiv Na remained in these muscles, compared with about 1·5 mequiv Na per kg in muscles which had been labelled *in vivo*. The temperature coefficient for the

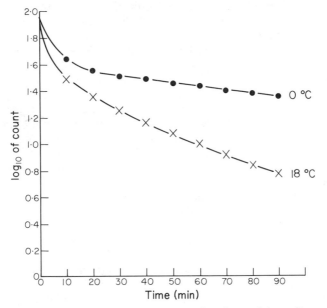

Figure 6.6. As Figure 6.5, but using sartorii made sodium-rich by soaking overnight in K-free Ringer's fluid labelled with ^{24}*Na (After Carey, Conway, and Kernan*[53]*; courtesy* J. Physiol)

exit of the ion remaining in the sodium-enriched muscles here into inactive Ringer's with 10 mM K was high, Q_{10} being over 2·33 (Figure 6.6). The same difference in temperature coefficient was also described by Harris[54].

As will be seen later, the presence of an appreciable fraction of sodium in a special-region, whether it be sarcolemma, connective tissue, or sarcoplasmic reticulum, introduces problems for the interpretation of radioisotopic studies as related to net ion movements across the muscle fibre membrane. It might therefore be profitable to consider firstly experiments in which the operation of the sodium pump in muscle was examined through measurement of net changes in the sodium and potassium concentrations of whole muscle and in muscle fibre water.

NET SODIUM MOVEMENTS IN ISOLATED MUSCLES

In spite of the fact that electrical and chemical energy gradients favour sodium entrance into muscle fibres, this process in isolated amphibian muscles immersed in normal Ringer's fluid can be fairly slow. For example, Draper and his colleagues[55], in 200 experiments carried out over a period of three years, found that the resting membrane potential and muscle fibre potassium concentration of isolated frog sartorii immersed in Ringer's fluid did not change significantly for up to 4 h after isolation. Intracellular sodium concentration increased by about 20 mequiv l^{-1} fibre water over this period, but part of this increase was due to water loss from the fibres. The overshoot of the action potential, which is directly related to a concentration gradient of Na ions across the fibre membrane[56], fell from $42 \cdot 7 \pm 0 \cdot 6$ to $29 \cdot 4 \pm 1 \cdot 4$ mV, confirming the entrance of this cation into the cell fluid. Carey and Conway[52] found that if frog plasma was used instead of Ringer's as a bathing fluid for the isolated sartorii, the net uptake of sodium ions by the muscles was greatly reduced. Creese and Northover[57], confirming this protective action of plasma against sodium entrance, found that serum which had been dialysed against Ringer prevented sodium accumulation in isolated rat diaphragm maintained at body temperature for 2 h. Even when the insulin within the serum was inactivated by pretreatment with cysteine–HCl, the serum was still effective in maintaining a low intracellular sodium concentration in the muscles. In an important development of this work, Creese[58] sought to determine whether the serum here acted by decreasing sodium influx or by increasing its efflux by measuring unidirectional ^{24}Na fluxes across the muscle fibre membrane in the presence and absence of serum. It was found that the sodium influx rate was significantly less in serum than in Ringer's. These were 10 pmol cm^{-2} s^{-1} and 15 pmol cm^{-2} s^{-1}, respectively, and the sodium concentrations in the muscle fibre water at the end of the immersion period were 9 and 13 mequiv l^{-1} fibre water, respectively.

While it is possible that the sodium permeability of the muscle fibre membrane at rest may be very low, so that the maintenance of the existing electrochemical gradients may not require an appreciable expenditure of metabolic energy, there can be little doubt that a significant net uptake of sodium into the fibre water takes place during muscle activity. Hodgkin and Horowicz[7b] measured the ^{24}Na and ^{42}K fluxes in isolated single fibres from the frog semitendinosis both at rest and during activity. The mean diameter of these fibres was 100 µm and their average length $1 \cdot 6$ cm. They found the following average movement of ions to be associated with conduction of *one* impulse at 21 °C.

^{24}Na: influx 19·6 ± 1·5, efflux 3·8 ± 1·0;

net entry 15·6 ± 1·8 pmol cm^{-2}

^{42}K: influx 1·8 ± 0·3, efflux 11·4 ± 0·6;

net exit 9·6 ± 0·7 pmol cm^{-2}

The average ^{24}Na influx in resting muscle was 3·5 pmol cm^{-2} sec^{-1}. Net chloride movement probably accounted for the difference between sodium and potassium exchange revealed by these flux studies. Earlier studies[59], in which rat skeletal muscles were stimulated *in situ* at a rate of 400 impulses per minute for half an hour, revealed an average gain of 8·3 mM Na and 2·8 mM Cl with a loss of 6·1 mM K on the basis of 1 kg dry weight of tissue.

The maintenance of the low intracellular sodium concentration of muscle *in vivo* in the face of net uptake associated, for example, with conduction of action potentials is the special function of the sodium pump. This must excrete the cation from the fibre water into the external bathing fluid against an electrochemical potential gradient and therefore requires the use of energy derived from metabolism.

SODIUM PUMP IN VIVO

The active excretion of sodium from muscles *in vivo* was first investigated in rats that had been put on a K$^+$-deficient diet for a week or more, so that their tissues gained sodium in exchange for potassium lost[35]. When rats thus treated were then put on a K-rich diet or injected intraperitoneally with 8–9 KCl, and their muscles analysed after a further week, it was found[60, 61] that the muscle electrolyte concentrations had returned almost to normal. During the period of K depletion, 3 K$^+$ were lost for every 2 Na$^+$ taken up, so that alkalosis developed[62, 63]. Part of the excess K loss may have been the result of readjustment of the Donnan equilibrium as the product $[K]_i \times [Cl]_i$ exceeded $[K]_o \times [Cl]_o$, due to both fall in $[K]_o$ and increase of $[Cl]_i$[61]. There was also an uptake of Mg^{2+} during K depletion[64]. When the rats were returned to a high-K diet, the plasma K$^+$ concentration was restored to normal within a day and $[K]$ within the muscles reached normality quite rapidly, but muscle sodium did not fall significantly within two days and even after five days was still slightly above the normal level. The extremely slow rate of sodium excretion *in vivo*, with a half-period of about three days, has made it difficult to study the process and much use has therefore been made of a technique *in vitro* devised by Desmedt[41] .

SODIUM EXCRETION IN VITRO

Skeletal muscle
When Desmedt soaked isolated frog sartorii in cold K-free Boyle–Conway Ringer's[2] containing 120 mM Na for upwards of 68 h, the intracellular sodium concentration of the muscles increased to

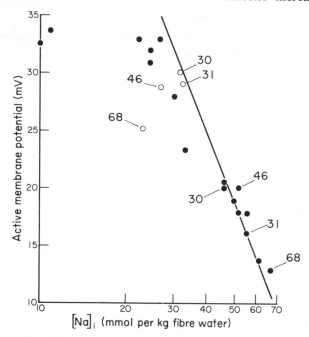

Figure 6.7. The relation between active membrane potential and logarithm of intracellular sodium concentration. Each point represents a separate experiment. O: *long-soaked muscles after recovery in high potassium. Straight line drawn with a slope of 58 mV for a tenfold change. The figures refer to the number of hours spent in low potassium. External sodium concentration: 104 mM in all experiments (From Desmedt[41]; courtesy J. Physiol.)*

nearly 70 mequiv per litre f.w. and $[K]_i$ fell to about 77 mequiv per litre f.w. Increase in $[Na]_i$ was reflected in a fall of the positive phase of the action potential from $+35$ mV to $+15$ mV. The relationship found by Desmedt between the two is shown in Figure 6.7. He examined the recovery of these sodium-enriched muscles by re-

214

immersing them for 3 h in fluid containing 10 mM K at 20 °C with $[Na]_o$ reduced from 120 to 104 mM Na. The intracellular sodium concentration decreased with a half-time of 30–40 min to an apparent plateau of 19 mequiv per kg f.w., while $[K]_i$ increased a corresponding amount. The fall of $[Na]_i$ produced a corresponding increase in the magnitude of the action potential. Under similar experimental conditions, Carey, Conway, and Kernan[53] found a level of $59 \cdot 0 \pm 0 \cdot 7$ mequiv Na per kg in 166 muscles after sodium enrichment with a final concentration of $39 \cdot 4 \pm 0 \cdot 3$ mequiv per kg for 390 muscles after 2 h in recovery fluid at 20 °C, an average loss of $19 \cdot 6 \pm 0 \cdot 8$ m mequiv or, allowing for loss from the interstitial space, an active excretion of $17 \cdot 6$ mequiv per kg. The lowest $[Na]_i$ value found after recovery in Conway's lab under these conditions was $37 \cdot 3 \pm 1 \cdot 4$ mequiv per litre f.w.[65] for *Rana temporaria* and $20 \cdot 8 \pm 1 \cdot 5$ mequiv per litre f.w.[66] in *Rana pipiens*. The difference in recovery between various species of frog under similar experimental conditions is worth noting, also differences in the same species obtained from different sources. For example, Desmedt[41] and Frazier and Keynes[67], all working at Cambridge, found comparable low sodium levels in muscles after recovery, and the latter workers were unable to find evidence of inhibition of the net transport of sodium by cyanide. In Conway's laboratory, however, with the same species of frog and under identical experimental conditions, there was a higher concentration of sodium in the muscles after recovery and also a 60% inhibition of the sodium pump by cyanide and by anoxia[53].

The quantity of sodium excreted by the sodium-rich muscles during recovery was found to depend on the external concentrations of both sodium and potassium. When, for example, the reimmersion fluid contained 10 mM K and 120 mM Na, instead of 104 mM Na, there was no apparent excretion of sodium from the muscles. In order to examine this effect further, experiments were conducted in which active transport by Na-rich muscles was measured in various bathing fluids based on the Boyle–Conway formula containing 10 mM K, but with concentrations of sodium ranging from 80 to 120 mM. Osmotic equilibrium was maintained here by replacement of sodium chloride by glucose in the Ringer's and vice versa. The results obtained in these experiments of Conway and his associates[63] are shown in Table 6.3. Here, it can be seen that active transport of sodium took place when $[Na]_o$ was less than about 115 mM, but above this value there was no significant excretion. However, in the presence of 120 mM Na, if $[K]_o$ was increased from 10 to 30 mM, so as to reduce the membrane potential, net sodium transport was restored.

THE CRITICAL ENERGY BARRIER CONCEPT

The explanation put forward for this effect was that there was a critical energy barrier to sodium transport in muscle fibres, above which the sodium pump was unable to function. The total energy requirement for the net excretion of sodium may be expressed as the sum of the chemical and electrical energy gradients in the following equation.

$$\mathrm{d}G/\mathrm{d}n = RT \ln [\mathrm{Na}]_o/[\mathrm{Na}]_i - E_m F \qquad (6.7)$$

where $\mathrm{d}G/\mathrm{d}n$ is the free energy change per equivalent of sodium transported, E_m the membrane potential of the fibre measured

Table 6.3

Procedure	Membrane potential E_m (mV)	$E_m F$ (cal mequiv^{-1} Na)	$RT \ln [\mathrm{Na}_o]/[\mathrm{Na}_i]$	Total cal mequiv^{-1}	Average Na excretion (mequiv kg^{-1})
120,0/120,10	67·3 ± 1·5	1·55	0·44	1·99	1·4 ± 2·4
120,0/115,10	(66·0)	1·52	0·41	1·93	3·7 ± 2·3
120,0/110,10	(64·7)	1·49	0·38	1·87	18·2 ± 1·7
120,0/104,10	63·4 ± 0·7	1·46	0·35	1·81	19·6 ± 1·5
120,0/80,10	56·8 ± 0·8	1·32	0·22	1·54	19·8 ± 1·8
120,0/120,30	48·9 ± 0·7	1·12	0·44	1·56	14·6 ± 1·8
120,0/120,60	41·6 ± 1·0	0·96	0·44	1·40	15·6 ± 2·9
104,0/104,10	71·7 ± 0·9	1·65	0·51	2·16	1·0 ± 0·9
104,0/80,10	54·8 ± 0·7	1·27	0·39	1·66	12·6 ± 2·2
104,0/104,30	46·6 ± 2·2	1·07	0·51	1·58	9·0 ± 1·7

The brackets used with some membrane potentials indicate interpolated values. membrane potential (E_m) is taken for convenience as positive (outside minus inside).
After Conway, Kernan and Zadunaisky[18]

during sodium transport, and R, F, and T the gas constant, Faraday constant, and absolute temperature, respectively. When the total energy gradient was calculated for the conditions described above it was found that if this exceeded about 2 cal mequiv^{-1} Na, no active transport occurred, but if it was less, sodium was excreted by the muscles (Table 6.3). The critical energy barrier calculated from equation (6.7) at the end of the two-hour recovery period where a net transport of sodium had taken place also had a value of about 2 cal mequiv^{-1} Na.

The concept of the critical energy barrier was not without its critics. Steinbach[50], using *Rana pipiens* enriched with sodium up to levels as high as 130 mequiv per kg wet weight of tissue, and then

216

reimmersed for 2 h in recovery fluid with 10 mM K and 130 mM Na, found a loss of about 72 mequiv Na per kg from the tissue to a final recovery level of 58 mequiv per kg muscle. On the basis of this and similar experiments, he suggested that it was unnecessary to reduce the sodium concentration of recovery fluid to bring about active transport, and therefore it was implied that the concept of a critical energy barrier to sodium transport in muscle was superfluous. A study of Steinbach's data revealed that the reason for the apparent insensitivity of the sodium pump to $[Na]_o$ was the unusually high value of $[Na]_i$ obtained in the preliminary soaking in K-free fluid.

Table 6.4

Procedure	Energy barriers at beginning and end of 'recovery' phase (cal per mmol Na)	
	Beginning	End
115,0/115,10	1·42	2·07
125,0/125,10	1·45	2·17
110,0/130,10		
A	1·56	2·47
B	1·52	2·25
C	0·52	2·11
Means ± standard error	1·29	2·21 ± 0·1

After Conway, Harrington, and Mullaney[68].
In the calculations the interfibre space was taken as 20%, in accordance with Steinbach's measurements. The temperature was taken as 20 °C.
The energy barriers to Na secretion were calculated from the equation[69]:

$$dG/dn = RT \ln [Na]_o/[Na]_i + E_m F$$

E_m being measured from without inwards. The E_m value was obtained from the relation $E_m = RT \ln ([K]_i/[K]_o)$ where $[K]_i$ and $[K]_o$ are the concentrations in the fibre water and in the external fluid [1, 7a, 70].

Steinbach's results were used to calculate energy barriers at the beginning and end of the period of sodium excretion[68] and the results are shown in Table 6.4. Even allowing for the fact that E_m measured during Na excretion is usually about 10 mV more negative than E_K used here to calculate the electrical energy gradient, the mean energy gradient to active transport at the time of reimmersion would still be 1·52 cal mequiv^{-1} compared with the critical energy barrier of about 2 cal mequiv^{-1} obtained in Conway's lab, so there would seem to be no reason why active transport should not take place, even without a reduction of Na_o to 104 mM. Furthermore, it might be noted that the critical energy barrier in *Rana pipiens*[66] of

2·4 cal mequiv^{-1} was also greater than that found in *Rana temporaria*.

Mullins and Awad[71] criticised the critical energy barrier concept, mainly on theoretical grounds. They suggested that net loss of sodium from Na-rich muscles during recovery was dependent on its efflux via the pump being greater than its passive entrance, and they believed the absence of net excretion in the presence of high $[Na]_o$ might be due to increased Na influx exceeding the capacity of the sodium pump. Two pieces of evidence might be quoted in reply to this suggestion. First, Na-rich muscles were reimmersed for 2 h at room temperature in a fluid containing 10 mM K and 120 mM Na, where no excretion takes place, and then compared with their companions that had been immersed in a similar fluid to which sodium pump inhibitors ouabain and iodoacetate had been added. There was no significant difference in the sodium concentrations in these two sets of muscles after the period of reimmersion[65]. This clearly suggested that the sodium pump was not functioning in the presence of 120 mM Na. Another relevent observation was that when oxygen consumption by the Na-rich companion muscles was compared during a 2 h period of reimmersion with one set of muscles in 10 mM K and 120 mM Na and the other set excreting sodium in fluid containing 10 mM K and 104 mM Na, the oxygen consumption was much greater in those latter excreting sodium. There appeared to be a direct relationship between the quantity of sodium excreted and the extra sodium oxygen consumed. It might be argued, however, that the extra oxygen consumption during sodium excretion reflected a metabolic change not related to the operation of the sodium pump but dependent rather on the ionic composition of the fibre water. However, this objection is answered by the finding[65] that there was no significant correlation between oxygen consumption and sodium concentration in the fibre water.

Smooth muscle

Daniel and Robinson[72, 73] applied a procedure similar to that described above to study the net transport of sodium in isolated segments of uterus as representing smooth muscle. They used uterine tissue from rabbits treated with oestrogens, and in some cases with progesterone, and also similar tissue obtained from pregnant cats. These were made sodium-rich, as already described, and were then reimmersed in recovery fluid containing 5 mM K for about 2 h at 37 °C. During the first immersion 40–50 mequiv K per kg muscle was lost with a corresponding gain of sodium. In recovery fluid, about 19 mM of sodium was excreted in exchange for 25 mM K taken up. The extra K^+ was balanced by 8 mM of Cl^- uptake. When the myometrium and endometrium were examined separately, it was found that, in the former, exchange of ions during recovery was on a one-for-one basis. Anoxia and most inhibitors of oxidative metabolism, including cyanide and antimycin, had relatively little effect on ion transport, while inhibitors of anaerobic glycolysis, including iodoacetate and fluoride, inhibited active transport strongly. The possible significance of these and other observations of the effects of metabolic inhibitor on sodium transport in muscle will be considered later.

Cardiac muscle

In papillary muscle of cat heart[74], enriched with Na^+ by cooling to 2–3 °C for 90 min. $[K]_i$ fell from 160 to about 120 mequiv per kg cell fluid, while $[Na]_i$ increased correspondingly to about 80 mequiv per kg cell water. When these muscles were rewarmed to about 27·5 °C for a further hour in the same bicarbonate-buffered fluid with 5 mM K, $[Na]_i$ fell to about 45 mequiv and $[K]_i$ increased again to about 170 mequiv per kg cell fluid, owing to the activity of the sodium pump. Active transport was dependent on $[K]_o$ and was inhibited when this was reduced to 1 mM or less. The degree of inhibition of active transport by the cardiac glycoside ouabain was also dependent on $[K]_o$. Inhibition by 10^{-6} M ouabain was almost completely reversed when $[K]_o$ was increased from 2·5 to 25 mM. This was interpreted as indicating that these two substances were competing for the same sites on the ion carrier molecule in the membrane. Before discussing further the effect of various factors on active transport in muscle, and their possible relationship to the mechanism of sodium excretion, we return now to a consideration of the radioisotopic flux studies in isolated frog skeletal muscles and their relevance to active transport of sodium.

MEASUREMENT OF FLUXES OF RADIOACTIVE ISOTOPES OF SODIUM AND POTASSIUM IN ISOLATED FROG MUSCLES

The use of radioactive isotopes in the study of sodium transport by muscle has given rise to some bizarre effects, by virtue of the fact that two completely separate processes may be contributing to the loss of labelled sodium from the tissue. One of these is the net efflux of sodium from the muscle fibre water. The measurement of this process should ideally be conducted under conditions where the specific activity of the sodium within the muscle fibre, that is, the ratio of labelled to total sodium present, remains constant and loss of radioactivity from tissue to inactive bathing fluid is proportional to the net transport of this cation from the cell. However, in practice, this situation is not achieved when there is unlabelled sodium in the external bathing fluid because of a process of exchange diffusion[75] by which labelled sodium ions in the muscle exchange for un-labelled sodium in the bathing fluid, without net loss of cation from the tissue taking place. In this way, the specific activity of sodium within the muscle may decrease continually so that a simultaneous net efflux of sodium will no longer be proportional to loss of radio-activity from the tissue and will therefore be difficult to determine. The very extensive investigations of isotopic fluxes in muscles carried out by Keynes and Swan[51], Harris[40, 76], Mullins and Fru-mento[77], Keynes[78], and Keynes and Steinhardt[79], and also others, with a great variety of bathing fluids and in some instances with the use of metabolic inhibitors have shown how these two processes may be studied separately and characterised. The need for brevity here makes it difficult to deal adequately with finer points of pro-cedure in their experiments.

Generally, labelling of isolated muscles was carried out by immersion in Ringer's fluid containing either ^{22}Na or ^{24}Na. The advantage of using the former is that its long half-life of 2·6 years, compared with 14·9 h in the latter, makes it unnecessary to apply a correction for decay of the isotope. The muscles used were either freshly dissected, in which case the intracellular sodium concentra-tion was very low, or they were given a preliminary treatment that increased $[Na]_i$ substantially. In some cases, this was done by soaking the isolated muscles for 12 h in normal or K-free Ringer's at a low temperature. Labelling of sodium-rich muscles was brought about either by incorporating the radioisotope in the fluid used for sodium enrichment or by reimmersion of the sodium-rich muscles for a further hour or two in a similar fluid containing the labelled sodium. This second reimmersion had to be of sufficiently long duration to allow the specific activity of the sodium in the muscle fibres to reach the same level as that in the bathing fluid.

Keynes and Swan[51] reimmersed ^{24}Na-labelled, freshly dissected frog leg muscles in inactive Ringer's containing sodium and measured the ^{24}Na loss from the tissue over a period of an hour. They then replaced the bathing fluid with sodium-free fluid, lithium or choline replacing this cation, and found that the rate of ^{24}Na efflux was reduced reversibly by half (Figure 6.8a). They concluded that the fraction of ^{24}Na efflux that was removed in the absence of external sodium was that associated with exchange diffusion. When this experiment was repeated on sodium-rich muscles, there was no reduction of ^{24}Na efflux rate on removing external sodium, but the efflux then became more dependent on the presence of external

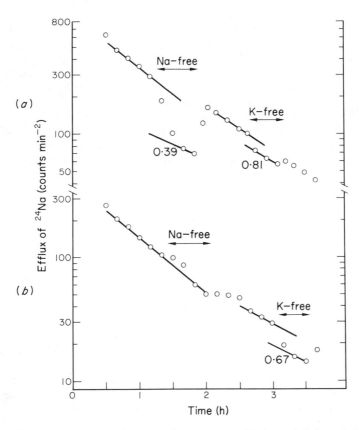

Figure 6.8. Reciprocal changes in the responses to Na-free and K-free solutions for paired muscles stored in normal Ringer's solution for different periods. Ordinate (a, b), efflux of ^{24}Na (counts min^{-2}); abscissa (a), time in hours, zero 3·5 h after dissecting muscle; abscissa (b) time in hours, zero 23·5 h after dissecting muscle. Sartorii, weight 44.7 mg (a), 44·8 mg (b). Temperature 19·7 °C (a), 19·5 °C (b). Semilog scale (After Keynes and Swan[51]*; courtesy J. Physiol.)*

potassium, as it decreased when external potassium was removed (Figure 6.8b). It seemed then that exchange diffusion, as opposed to net sodium efflux, was more predominant in freshly dissected muscles, but was reduced and replaced by Na : K exchange as $[Na]_i$ was increased.

The efflux of ^{24}Na from muscles was not linearly related to $[Na]_o$, but was about twice as great at low $[Na]_o$ as that expected from direct proportionality. Here, it seemed that exchange diffusion alone occurred at high $[Na]_o$ but was augmented by net sodium efflux as the energy barrier to this process was reduced by lowering of $[Na]_o$. Another observation of Keynes and Swan that has prompted further investigation and speculation was that the rate of ^{24}Na efflux from freshly dissected muscles was proportional to the cube of $[Na]_i$, after these muscles had been soaked in lithium Ringer's for a long time. Mullins and Frumento[77], in an experiment of a similar type, but using tris or choline Ringer's instead of lithium Ringer's, found that if they plotted $[Na]_i$ calculated by a back-addition of efflux counts against ^{24}Na efflux on a log–log scale the initial slope at $[Na]_i$ values of 3–8 mequiv per litre f.w. ranged from 2 to 3·5, but as $[Na]_i$ increased further the slope decreased and the 'cubic law' no longer applied. They interpreted this finding as indicating the presence of three specific binding sites on the carrier molecule in the membrane responsible for active transport of sodium. Harris[76] confirmed that the 'cubic law' applied over the $[Na]_i$ range of 1–3 mM per kg tissue for ^{24}Na-labelled frog sartorii immersed in potassium methylsulphate solution, and he suggested that the greatest dependence of efflux on $[Na]_i$ might be found under conditions favouring onset of active exchange of muscle sodium for external potassium. Keynes[78] found that this high dependence of sodium efflux on its internal concentration was absent when he used Ringer's buffered with a high concentration of bicarbonate instead of the usual phosphate buffer. He believed the use of this anion might lead to a fall of intracellular pH, so that internal hydrogen ions would then compete for sites on the carrier normally occupied by sodium ions. The use of bicarbonate, in any event, resulted in the efflux rate for ^{22}Na being proportional to the first or second power rather than to the third power of $[Na]_i$. In testing the cubic relationship in ^{22}Na-labelled frog sartorii, Keynes[78] varied $[Na]_i$ reversibly by adding 200 mM dextrose to the external bathing fluid, which resulted in a loss of water from the muscles which was half-complete within 3 min. During a 16 min collecting period, after addition of dextrose, ^{22}Na efflux rate increased by a factor of 3·5.

Active transport of sodium by muscles was most conveniently studied by labelling sodium-enriched sartorii and reimmersing them at room temperature in sodium-free media, so that exchange diffusion was abolished. Active transport was favoured under these conditions by reduction of the electrochemical potential gradient against which the sodium was moved. Although the resulting energy gradient was negligible, the term 'active' transport was still applied, as the carriage of sodium from the fibres seemed to be oubain-sensitive[79] and also dependent on external potassium. Harris[76]

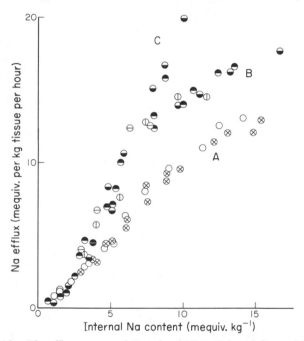

Figure 6.9. The efflux-content relations found (A) *in K-free choline or lithium mixtures,* (B) *with 0·5 mM K plus lithium or choline or 3 mM K plus lithium and also 3 mM K plus choline in one experiment.* (C) *with 3 mM K plus choline in two further experiments* (*From Harris*[76]*; courtesy* J. Physiol.)

plotted the rate of ^{24}Na efflux against $[Na]_i$ in Na-rich muscles reimmersed in various Na-free media, and found the highest efflux into inactive Ringer's containing mainly potassium methylsulphate and bicarbonate. Where the bathing fluid contained magnesium acetate instead of a sodium salt, with 3 mM K as lactate, the ^{24}Na efflux was considerably less, but where even as little as 0·5 mM K was present in the medium the sodium efflux was significantly greater than in the absence of potassium for a given $[Na]_i$ (Figure 6.9). Keynes and Swan[51] and Kernan[80] also found labelled sodium efflux from Na-rich muscles into choline and lithium Ringer's to be greatly reduced in the absence of external potassium. The dependence of active sodium excretion on $[K]_o$ here might seem to indicate that a coupled exchange of these cations was involved in the process; however, when muscles were analysed after a number of hours in sodium-free fluid containing K^+, it was found[77, 76] that sodium loss from the muscles greatly exceeded potassium uptake, and there appeared to be a cation deficit within the fibres. Between 15 min and 4 h after reimmersion in choline or tris fluid with 2·5 mmol K, Na-rich muscles lost 26 mequiv Na per kg and gained only 5 mequiv per kg muscle[77], the difference apparently being made up by entrance of choline or tris into the fibres. The fact that increase of $[K]_o$ stimulated sodium efflux from these muscles, without entering the fibres in quantity, would seem to raise the possibility that $[K]_o$ may of itself influence membrane permeability to choline, lithium, and tris, by allowing them to enter the fibres more freely in exchange for sodium excreted, thereby facilitating the maintenance of electrical equilibrium within the fibres. Beaugé and Sjodin[81] obtained evidence of a ouabain-sensitive ion-for-ion exchange of sodium for lithium in sodium-enriched muscles placed in K-free Ringer's containing 60 mM LiCl and 50 mM NaCl for 2 h at 20 °C. It seemed then that lithium ions themselves could exchange for sodium on the pump, thereby activating it. By the use of the inhibitor 10^{-5} M strophanthidin, it was estimated that under these conditions 23% of the total ^{22}Na efflux was by exchange diffusion, the remaining 67% being actual ouabain-sensitive net efflux. There remain several problems of interpretation in relation to the results of Beaugé and Sjodin on the one hand and those of Harris[76] and others on the other. If it is true that external lithium can stimulate the sodium pump in muscle, while at the same time exchanging on the carrier, as suggested by the strophanthidin sensitivity, one wonders at the striking augmentation of Na efflux brought about by addition or increase of external potassium in a situation where $[K]_o$ must be competing with $[Li]_o$ for external carrier sites. This situation is probably greatly complicated by simultaneous exchange of internal potassium for external lithium.

Keynes and Steinhardt[79] also examined the ouabain sensitivity of the efflux of labelled sodium from frog muscle and found that the fraction of total ^{22}Na loss blocked by this specific inhibitor of the sodium pump varied considerably depending on the experimental conditions. For example, in the case of freshly dissected labelled muscles immersed in normal Ringer's with 2·5 mM K, about 37%

of the efflux was blocked by 10^{-4} M ouabain, compared with 75% of the efflux of muscles that had been previously enriched with sodium. In Na-rich frog sartorii, it was found that 10^{-3} M ouabain produced about 80% inhibition of net sodium excretion[53], which is in close agreement with this observation. The effects of adding K^+

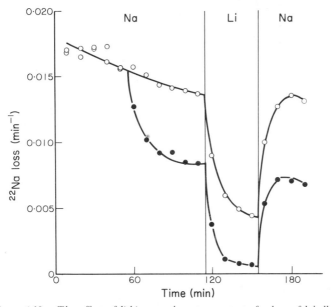

Figure 6.10. The effect of lithium on the rate constants for loss of labelled sodium from a pair of freshly dissected sartorius muscles. The muscles were loaded with $22Na$ for 3 h, and washed in several changes of inactive Ringer's containing 2·5 mmol K for 30 min before starting measurements. The washing-out solutions all contained 10 mmol K and 111 mmol Na or Li. ○ control muscle; ● 10^{-4} mol ouabain added. Temperature 21·5 °C. Experiment of 13 January 1966 (After Keynes and Steinhardt[79]; courtesy J. Physiol.)

and then replacing Na^+ by Li^+ in the bathing fluid in the presence or absence of the inhibitor ouabain in freshly dissected (Figure 6.10) and sodium-rich muscles (Figure 6.11) illustrate the strikingly different characteristics of exchange diffusion and net sodium transport.

Sodium efflux from ouabain-treated freshly dissected muscles remained unaltered when $[K]_o$ was increased from 0 to 10 mM, while in an untreated companion muscle it more than doubled. The ouabain-sensitive component of freshly dissected muscles differed

225

I

from that of sodium-enriched muscles, in that it seemed to be insensitive to replacement of external sodium by lithium. While the ouabain-sensitive sodium efflux was always dependent on $[K]_o$, both the influx and efflux of sodium were reduced by about 20% by addition of ouabain in absence of K_o^+ and it seemed likely that this inhibitor-sensitive Na : Na interchange may have been similar in nature to that found by Garrahan and Glynn[82] in erythrocytes.

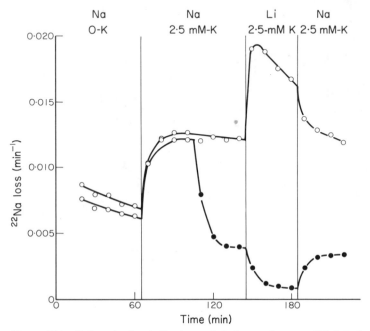

Figure 6.11.   Before the first collecting period, the muscles spent $22\frac{1}{2}$ h in inactive K-free Ringer's at $2°C$, then 4h in K-free ^{22}Na Ringer's at room temperature. Solutions were then changed as shown. (O control muscle: ● muscle in 10^{-4} mM ouabain.) Temperature $21\cdot3$ °C. Experiment of 25 June 1965 (After Keynes and Steinhardt[79]; courtesy J. Physiol.)

Harris[76], from studies of the effect of $[K]_o$ on the kinetics of Na-efflux, concluded that increase of $[K]_o$ raised the saturation level of the sodium efflux and Keynes[78] agreed with this view. The value of $[Na]_i$ at which Na efflux was half-saturated seemed, however, to be relatively independent of changes in $[K]_o$.

While the sodium efflux that is ouabain- and potassium-sensitive is most likely to be from the sarcoplasm and via the pump, the location of that fraction of the efflux that is dependent on the presence of external sodium is less clear. This is perhaps Conway's sodium of the 'special region', which he thought might be located within the sarcolemma. Zadunaisky[83] has, with the electron microscope, found some evidence for its presence there. But there is a fraction of nonexchangeable sodium in muscle, varying from about 2 mequiv in fresh muscle to 5 mequiv per kg in sodium-enriched muscle, that is probably bound to some structure such as sarcolemma or possibly in the mitochondria. There remains, however, the possibility that the muscle sodium involved in exchange

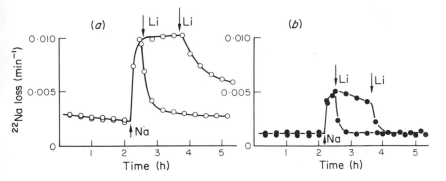

Figure 6.12. The effect on the final efflux in Ringer's of varying the duration of an interpolated period in Na Ringer's. Counts taken during a preliminary 3 h period in Li Ringer's are not shown. In each experiment, both muscles were transferred from Li to Na Ringer's at the first arrow; one muscle was then left in Na Ringer's for only 20 min (until the second arrow), while the other remained for altogether 90 min. A: muscles not treated with ouabain. B: the washing out solutions contained 10^{-4} mol ouabain. Temperature 21 °C (After Keynes and Steinhardt[79]; courtesy J. Physiol.)

diffusion may be in the sarcoplasmic reticulum or in some part of the sarcotubular system. While Keynes and Swan[51] considered that a two-compartment system in the muscle fibre might explain the nonlinearity of the plot of ^{24}Na efflux against $[Na]_i$ they rejected this view in favour of a one-compartment system with cube-law kinetics on the grounds that it did not explain certain experimental findings. For example, when muscles were placed in lithium Ringer's for a long time so that Na efflux rate fell to a low level, and the muscles were then exposed to Ringer's containing sodium, the efflux of the labelled sodium increased substantially because of Na : Na interchange. Horowicz, Taylor, and Waggoner[84] obtained a similar result but part of the increased Na efflux on returning the muscles to Na Ringer was strophanthidin-sensitive. On returning the muscles once more to lithium Ringer's, it was expected that the Na efflux would fall once more to the low level reached before addition of the external sodium, as the Na efflux in K-free fluid should be proportional to the radioactivity in the intracellular compartment, which by now was at a very low value. However, this was found not to be the case, for when the muscles were returned to Na-free ringer for a second time sodium efflux was found to have increased. This observation seemed difficult to explain in terms of two fibre compartments in parallel, and so this scheme was dropped, until it was realised that provided that there could be free interchange of ions between these fibre compartments the scheme was applicable. Such a system could account for the recharging of the intracellular space with labelled sodium during immersion of the muscles in Na Ringer's. In confirmation of their revised scheme of parallel fibre compartments exchanging ions with one another and with the external fluid, Keynes and Steinhardt[79] showed (Figure 6.12) that the longer the muscles were retained in sodium Ringer's between immersions in lithium fluid the higher was the level of ^{22}Na efflux in the final immersion. It was suggested that ^{22}Na efflux into Na-free fluid took place from the sarcoplasm and that ^{22}Na loss in exchange for unlabelled ions was mainly from the sarcoplasmic reticulum. This view is in agreement with the suggestion of Harris[76] and of Birks and Davey[44] that the bulk of the fibre sodium in the freshly dissected muscle was in this region.

COUPLING OF SODIUM–POTASSIUM EXCHANGE DURING ACTIVE TRANSPORT

The manner of coupling of sodium and potassium exchange during active transport in muscle has been of interest mainly because of the light it may shed on the mechanism of carriage of ions across the fibre membrane. Formerly, this coupling was considered to be mainly chemical, with one potassium ion being transported into the fibre for each sodium ion excreted during one cycle of operation of the pump[85]. More recently, however, it has seemed likely that electrical coupling may play a significant part in the accumulation of potassium by the cell and that this cation may enter passively under the influence of an electrical potential difference created by net excretion of sodium. When the mean membrane potential of Na-rich frog sartorii was measured by the microelectrode 10–20 min after immersion at room temperature in recovery fluid containing 10 mM K, and this was compared with the potassium equilibrium potential calculated from the Nernst equation following chemical analysis of the muscles and bathing fluid, it was found[86] that E_m was more negative than E_K by 11 mV and the difference was highly significant ($P < 0.01$). At the end of the recovery period, E_m and E_K were practically the same. Here, the distribution of potassium ions across the fibre membrane was far removed from equilibrium, with the electrochemical potential difference favouring the passive movement of potassium into the fibres. Since the difference between E_m and E_K was also present in chloride-free fluid, and increased to about 33 mV when sodium excretion was enhanced by immersion of the muscles in sodium-free recovery fluid, it was suggested that the membrane hyperpolarisation might be due to the active transport of sodium ions from the cell. Subsequently, other workers[87, 71, 88–91] confirmed that the sodium pump in muscle was indeed electrogenic in its operation and they extended the investigations exploring other possible explanations for the membrane hyperpolarisation. Perhaps the most likely of these was that during active transport of ions on an electrically neutral pump, the concentration of potassium in external fluid close to the muscle fibre membrane might fall significantly below that in the bulk of the bathing medium; then E_K calculated on the basis of the latter value would have been underestimated and E_m and E_K might in fact be identical. Ritchie and Straub[92] used this argument to explain post-tetanic hyperpolarisation in nerve fibres.

Harris and Ochs[90] found that addition either of cocaine or of the antihistamic mepyramine to sodium-rich muscles in recovery fluid increased the electrical resistance of the fibre membrane from about 12 kΩ cm^{-2} to as much as 58 kΩ cm^{-2}, while decreasing net uptake of K^+ to about a fifth of the quantity of sodium excreted. At the same time, E_m increased by about 10 mV over that measured in the absence of the drug. Adrian and Slayman[91] made a similar observation with cocaine. The increase in E_m here, in spite of inhibited K^+ uptake, indicated that sodium extrusion rather than depletion of $[K]_o$ was responsible for membrane hyperpolarisation.

Adrian and Slayman also examined the effect of using 10 mM Rb instead of 10 mM K in recovery fluid on the membrane potential of sodium-rich muscles during active transport. The lower permeability of isolated frog skeletal muscles to Rb^+ as compared with K^+ had been established by Conway and Moore[3] by osmotic studies, and was also evident in the fact that about twice as much Rb^+ as K^+ had to be added to Ringer's fluid to produce an equal membrane depolarisation[93, 94]. Therefore, if the coupling of sodium excretion with inward movement of cation was electrical, the sodium pump would need to generate a greater potential to move rubidium into the fibre than in accumulating potassium at a similar

229

rate. Movement of K or Rb into the muscle fibre would tend to reduce the potential generated by the sodium pump. Adrian and Slayman maintained their sodium-enriched frog sartorii at 1–3 °C for 2 h to equilibrate before the temperature of the bathing fluid was raised to activate the sodium-pump. In the presence of 10 mM K, E_m was found to increase from about -50 mV at low temperature to a maximum of -85 mV after 30 min in recovery fluid at 20 °C; whereas, with muscles in 10 mM Rb, E_m increased to about -100 mV. In freshly dissected muscles, E_m at a particular $[Rb]_o$ was never more negative than 7 mV with respect to a companion muscle at a similar $[K]_o$, which is far less than the difference observed in the sodium-rich muscles during active transport. It should be mentioned that if membrane hyperpolarisation here was due to depletion of the external exchangeable cation, K^+ or Rb^+, then as the latter was accumulated more slowly than the former, the increase in E_m might be expected to be somewhat greater in K^+ than in Rb^+. Although sodium was readily excreted in both recovery fluids, $[Rb]_i$ increased at a rate of 50 mequiv per kg f.w. per hour compared with 70 mequiv per kg f.w. per hour for K^+ over the first half-hour of sodium excretion. These workers concluded that the sodium pump here was electrogenic, and further calculations satisfied them that the potential generated by the pump was sufficiently large to account for K^+ uptake by a passive movement. However, in the case of rubidium, when they used the Goldman flux equations[95] to calculate theoretical influx rates corresponding to the membrane potentials measured in the presence and absence (ouabain inhibition) of active transport, taking advantage of the fact that Rb^+ permeability is independent of E_m, they obtained an influx ratio M_a/M_r of about 2. Here, the equation used to calculate the ratio of influx during active transport, M_a, to influx in presence of ouabain, M_r, had the following form:

$$\frac{M_a}{M_r} = \frac{E_a e^{E_r F/RT} - 1}{E_r e^{E_a F/RT} - 1} \tag{6.8}$$

where E_a and E_r were the membrane potentials measured during sodium extrusion and in absence of active transport, respectively. When the actual influx ratio was measured here, it was found to be 20, that is, about 10 times greater than the calculated value. From this it was concluded that only about 10% of Rb^+ uptake could be accounted for by passive influx. If 90% of its accumulation here was in fact by an active process, it seems reasonable to suppose that the carrier sites occupied by Rb^+ here may normally be used by K^+

230

and therefore that there is provision for active carriage of K^+ into the fibres, even though passive uptake through permeability channels may be an alternative means of accumulation during membrane hyperpolarisation. In support of this view, it might be said that Cross, Keynes, and Rybova[87] found E_m to be less negative than E_K during active transport of sodium in some muscles of frogs that had been stored in the cold, and under these conditions K^+ uptake would be against an energy gradient and therefore active.

Several investigators have tried to determine the possible contribution of electrical, as opposed to chemical, coupling in potassium accumulation during active transport by sodium-rich muscles. For example, some have estimated the membrane current that might result from the application of the potential difference $E_m - E_K$ across the membrane resistance, and have compared this current, presumably carried by K^+, under Cl-free conditions with the measured net influx of cation. Harris and Ochs[90] found the average K^+ accumulation during reimmersion in recovery fluid was 20 mequiv $kg^{-1} h^{-1}$, which corresponded to a membrane current of $1.3 \, \mu A \, cm^{-2}$. To drive this current through the membrane resistance of $12 \, k\Omega \, cm^{-2}$ would require a voltage of $15 \, mV$, which was somewhat less than the hyperpolarisation measured in Cl-free recovery fluid. So it seemed likely that K^+ uptake could be completely coupled electrically to sodium extrusion from the fibre. Cross, Keynes, and Rybova[87] using a K^+ conductance value of $480 \, \mu mho \, cm^{-2, 7a}$ calculated that the inward K^+ current that would be produced by a $10 \, mV$ increase in E_m was $4.8 \, \mu A \, cm^{-2}$, corresponding to $50 \, pmol \, cm^{-2}$. This was within the range of measured K^+ uptake derived from the data of Desmedt[41], so it was concluded that K uptake could be accounted for as a passive response to the membrane hyperpolarisation $E_m - E_K$ generated by the sodium pump, provided that K conductance, g_K, was similar in sodium-rich and freshly dissected muscles. However, the fact that the specific resistance of muscle fibres increased to twice its normal value when K^+ was replaced by Na^+ within the fibre[96] raises some doubts on this point.

Geduldig[97] found that the membrane conductance of Na-rich frog sartorii increased sharply when the rate of sodium pumping was reduced by addition of ouabain. Although it was uncertain to what extent K ions contributed to such conductance, it was suggested that the G_K values determined by Harris and Ochs[90] and by Adrian and Slayman[91] might be much higher than those actually existing when pumping of sodium was most rapid. If such was the case, then passive movement of K^+ into the cell through permeability channels would be insufficient to account for its accumulation.

The finding[90, 91, 98] that the temperature coefficient of E_m during active transport in frog and rat skeletal muscles was greater than that of the resting potential itself was taken as evidence of the direct participation of metabolism and of enzymatic processes, as distinct from physical ones, in its generation. It is evident from the Nernst equation that the resting membrane potential should be proportional to the absolute temperature, provided temperature change does not alter the relative permeability of sodium and

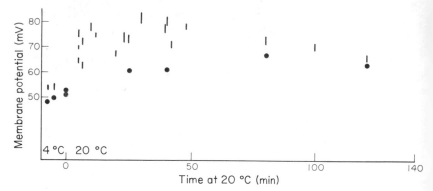

Figure 6.13.   Membrane potentials of initially K^+-depleted muscles in a solution having 10 mequiv K per litre during active ion movement. The points at the left of $t = 0$ were measured at 3–5 °C after at least 15 min equilibrium; active movement is slight at this temperature. On warming to between 18 and 20 °C, the measured potential rises and then gradually falls again. The vertical lines are twice the s.e. in length. The potassium ratios found from terminal analysis of muscles either taken directly from the cold solution or after various times in the warm have been used to obtain the ideal K^+ electrode potentials shown as blobs (After Harris and Ochs[90])

potassium ions. The measured Q_{10} of E_m in resting frog muscle of 1·033[99] showed conformity with the equation. On this basis, an increase in temperature of bathing fluid from 2 °C to 21 °C might be expected to increase E_m by about 5 mV. Harris and Ochs, however, found that E_m increased by almost 25 mV when bath temperature was raised by about 20 °C in the case of Na-rich frog muscle reimmersed in recovery fluid containing mainly choline methansulphonate with 10 mM K. The effect of temperature here, and the relationship between E_m and E_K, are illustrated in Figure 6.13.

Another approach to the investigation of the coupling ratio of sodium–potassium exchange on the pump[78, 100] has been to compare the ouabain-sensitive sodium efflux and potassium influx.

In freshly dissected muscles, Keynes found that the fraction of ^{22}Na efflux that was blocked in the presence of 10^{-4} M ouabain was about five times as great as ouabain-sensitive ^{42}K influx, indicating a coupling ratio of 5 : 1 in favour of sodium, which supported the view that the sodium pump was electrogenic. In sodium-rich muscles, the coupling ratio tended to approach unity. Sjodin and Beaugé, on the other hand, found the ouabain-sensitive fluxes to be 0·7 pmol cm^{-2} s^{-1} for K and 0·9 pmol cm^{-2} s^{-1} for Na in low-Na muscles, suggesting neutral coupling. In muscles with higher $[\text{Na}]_i$ and with 5 mM K outside, K influx was 5 pmol cm^{-2} s^{-1} and the sodium efflux was 'considerably greater'.

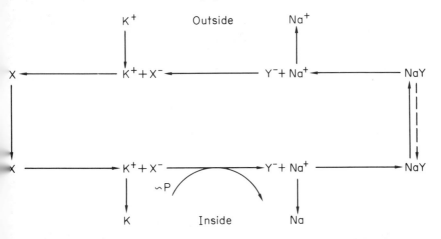

Figure 6.14.   Scheme of a circulating carrier molecule for the active coupled exchange of sodium and potassium ions across the cell membrane in tight linkage. Y^- is the phosphorylated form of the carrier that specifically binds sodium ions within the cell. At the outer face of the membrane this carrier is transformed into X^-, which has specificity for the potassium ions

The coupling ratio of sodium and potassium transport on the pump is of some importance in relation to present theoretical models for carrier mechanisms. These are predominantly based on the assumption that a phosphorylated intermediate compound, possibly a protein, within the membrane serves as a carrier molecule, shuttling backwards and forwards from one surface to the other and that a membrane Na–K-activated adenosine triphosphatase system is involved in this mechanism. As the ATPase theory of ion transport is discussed at some length in another chapter, it will be dealt with only briefly here.

233

I*

This enzyme system, as found in erythrocytes, exhibits an assymetrical activation[101, 102] requiring sodium ions at the inner surface of the membrane and potassium outside for optimal rate of ATP hydrolysis. Since enzyme activation must necessarily involve some form of chemical combination between activating ion and enzyme, it was tempting to attribute to the adenosine triphosphatase system the role of ion carrier, especially as it was also concerned with energy transfer. The first step in active excretion of sodium

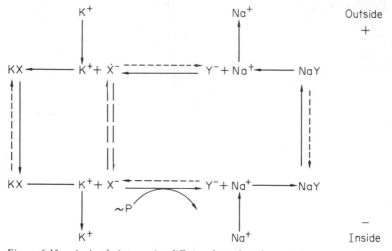

Figure 6.15. A circulating carrier differing from that shown in Figure 6.14 in that form X⁻ is capable of crossing the membrane without bound potassium. This would make it possible for the pump to become at least partly electrogenic (After Cross, Keynes and Rybova[87]; courtesy J. Physiol.)

must be its specific binding to anionic sites on a carrier system at the inner surface of the cell membrane (Figure 6.14), forming an electrically neutral complex. In this form, the carrier with sodium attached may move to the outer surface of the membrane where, due to activation by external potassium, energy-rich phosphate bonds are broken, yielding energy for release of sodium ions into the external bathing fluid. It has been suggested that at the hydrolysis of the energy-rich bond the specificity of the carrier molecule changes towards potassium in preference to sodium, so that on its return to the inner membrane surface the carrier now transports an equivalent amount of the K^+ to the sarcoplasm. The chemical coupling of the sodium and potassium movements here favours the view that the pump acts in an electrically neutral manner, taking up one potassium ion for each sodium ion excreted. However, as it appears that the pump may be electrogenic in its operation, it has been necessary to modify this scheme somewhat to account for this behaviour[103, 87] and the revised scheme is shown in Figure 6.15. This allows for a less tightly coupled exchange of the cations, with some degree of electrical coupling of ion movement.

METABOLIC ASPECTS OF ACTIVE TRANSPORT IN MUSCLE

INHIBITION OF ACTIVE TRANSPORT

Information concerning the mechanism and direct energy source for active transport in muscle has been sought through observation of the effect of various metabolic inhibitors, hormones, and metabolites on this process. Unfortunately, the effectiveness of this approach has been hampered by the difficulty of distinguishing between primary and secondary effects. For example, according to Conway's redox pump theory of active transport, which is probably more directly applicable to hydrogen ion transport in gastric secretion[13] and mitochondria[104], respiratory enzymes are assumed to act as ion carriers during their oxidation-reduction cycle, the direct energy source being electron energy. It might be considered that acceptance of this theory is favoured by evidence that sodium transport is inhibited by substances that block electron transfer at the cytochrome level, including cyanide and antimycin A. This is not the case, however, as it could be argued that inhibition of cellular respiration might simply deprive the cell of ATP required to fuel a Na–K-activated ATPase pump. In skeletal muscles, as in smooth muscle[72], inhibitors of anaerobic glycolysis such as 2×10^{-3} M iodoacetate[53] and 7×10^{-4} M p-chloromercuribenzoate[105] fully blocked net excretion of sodium. As these substances should effectively inhibit ATP synthesis, this information does not go very far towards helping one to decide between a redox or an ATPase mechanism. The cyanide inhibition of sodium transport found in Conway's laboratory[53] was not confirmed by Frazier and Keynes[67]. Caldwell[103] has suggested that this discrepancy might have been due to differences in the metabolic state of the muscles used, and he drew attention to the fact that the mean concentration of energy-rich phosphate ($\sim$P) of Na-rich muscles in Conway's experiments was only 9·8 mM kg^{-1}, which is lower than that found elsewhere. Dydynska and Harris[132], for example, found a mean phosphocreatine concentration of $15·5 \pm 0·88$ and ATP concentration of $2·6 \pm 0·05$ mequiv kg^{-1} in their Na-rich muscles. While this might explain the difference in cyanide sensitivity of the muscles here, and might also explain the lower $[Na]_i$ found by Frazier and Keynes at the end of the recovery period, it should be pointed out that these workers did not provide any information as to the $\sim$ P in their muscles.

More satisfactory results have been obtained with other inhibitors of the respiratory chain, particularly with those blocking at the level of cytochromes b and c. Antimycin A has been reported to block electron transfer at this point[106-108]. At a concentration of 10^{-5} M, this substance was found to inhibit net sodium transport in frog muscle by about 44%[11]. Bittar[109] also found a similar inhibition in *Maia* muscle fibres at 10^{-4} M concentration. Finally, Creese and his colleagues[110] found that the ^{24}Na efflux from denervated rat diaphragm was significantly depressed by 10^{-5} M antimycin A. Another inhibitor of electron transfer at this level, namely, 2.3-dimercaptopropanol (BAL)[111], also inhibited net sodium transport by 89% at a concentration of 1 mM[105]. In view of this evidence for participation of cytochromes in active transport, it might be asked whether blocking of electron flow here might interfere with sodium excretion by preventing synthesis of ATP. As suggested by Caldwell[103], such interference with the cytochrome chain should not interfere with synthesis of ATP at the substrate level, so it seems unlikely that this is the cause of the inhibition of the sodium pump here. In smooth muscle, 2,4-dinitrophenol, which appears to uncouple oxidation from phosphorylation[112] produced a 97% decrease in phosphocreatine and a 77% fall in ATP content of frog muscle[113] and caused a marked inhibition of K uptake. At a concentration of 10^{-3} M, Daniel and Robinson[72] found that DNP completely blocked reaccumulation of K by Na-rich segments of rabbit uterus while reducing Na excretion by about 50%. In incubated single muscle fibres from dog, DNP produced a profound and rapid loss of K^+[114] that was even more marked than the loss that took place in presence of 10^{-3} M ouabain or 10^{-3} M monoiodoacetate. As will be seen later, in skeletal muscle this substance at a similar concentration may even stimulate sodium transport.

STIMULATION OF ION TRANSPORT IN MUSCLE

The active transport by sodium-enriched muscles has been found to be augmented in the presence of a number of substances, including insulin. Manery, Gourley, and Fisher[115] found that addition of 5 mM lactate and of insulin (5×10^{-2} units ml^{-1}) to isolated frog

236

leg muscles bathed in normal Ringer's, caused net uptake of K^+ that was usually but not invariably associated with increased oxygen consumption. In an extension of this work, Smillie and Manery[116] found that insulin was more effective than lactate in promoting K^+ accumulation by the tissue, but the effect of lactate in increasing respiration was more consistent than that of insulin. As an example of the *in vivo* action of this hormone on ion movement, it might be mentioned that Regan and his associates[117] found that injection of insulin into the coronary artery of dogs increased both the uptake of K^+ and lactate from the blood while raising the respiratory quotient towards unity.

Zierler[118, 119] observed a hyperpolarisation of skeletal muscles in the rat when insulin $(0.1\text{--}0.3$ units $ml^{-1})$ was added to the Ringer's fluid they were immersed in and this was accompanied by K^+ uptake. Since the increase in $[K]_i$ brought about by the insulin here was not sufficient to account for the increase in E_m, he suggested that the K^+ accumulation might be a response to 'a change in the fixed electrical charge within the membrane produced by association with insulin'[120]. The probable nature of the hyperpolarisation became evident with the finding that the sodium excretion by muscles could be electrogenic.

Sodium uptake by isolated rat diaphragm immersed in Ringer's was prevented when insulin was present in the bathing fluid[57] and an augmentation of net transport of ions from sodium-rich sartorii by this hormone was also demonstrated[121]. When the latter muscles were reimmersed at 20 °C in Ringer's fluid containing 2·5 mM K and 120 mM Na for 2 h, no excretion took place, as the energy gradient for sodium excretion here was in excess of the critical energy barrier of 2 cal mequiv^{-1} Na. When insulin$(0.03\text{--}0.10$ units $ml^{-1})$ and lactate (1–5 mM,) separately or together, were added to this reimmersion fluid, sodium was excreted in amounts ranging from 10 to 22 mequiv per kg muscle. When the recovery fluid contained 2·5 mM K and 104 mM Na, muscle sodium was reduced from 59.4 ± 2.4 to 25.1 ± 1.0 mequiv per kg muscle, which was not significantly different from the sodium concentration found in freshly dissected muscles. The increased critical energy barrier to sodium excretion that could be surmounted might be equated with the amount of additional energy available to the sodium pump. The charge was from 2·0 cal in the absence of insulin to 3·74 cal per mequiv Na when 0·1 units ml^{-1} of insulin was added to the recovery fluid.

When sodium was being actively extruded into Ringer's containing 2·5 mM K and 120 mM Na in the presence of insulin and lactate the addition of orthophenanthroline, a Zn chelator and inhibitor of lactate dehydrogenase[122] reduced the transport by about 80%, an effect which could be reversed by subsequent addition of 1 mM $ZnSO_4$. This was interpreted as indicating that lactic dehydrogenase was involved in providing the extra energy required here for sodium excretion. It should also be mentioned that the effect of insulin on sodium transport here was not dependent on the presence of glucose in the ringer. Caldwell[103] has suggested that the stimulation of metabolism by insulin or by the oxidation of lactate may increase the $\sim P$ within the muscle fibres, making more energy available to an ATPase pump. As will be seen below, there are reasons for doubting this explanation of the phenomenon.

Further evidence of a direct action of insulin on the sodium pump has been obtained by Creese[58], by Herrera and his colleagues[123, 124], and by Crabbe[125]. Creese found that insulin (0·02 unit ml^{-1}) increased the rate constant for ^{24}Na efflux from rat diaphragm by about 50% and reduced the final concentration of sodium in the muscle fibres. The other workers found that the hormone increased the short-circuit current in toad bladders[124], isolated frog skin[123] and in toad colon[125] and this current has been shown to be directly related to net sodium transport across the epithelium[126]. It seems likely, therefore, as suggested by the author[121], that insulin hyperpolarised the muscle fibre membrane by stimulating an electrogenic sodium pump and that potassium uptake was a passive response to the increase in membrane potential here.

Augmentation of active transport of sodium and potassium ions associated with membrane depolarisation has been found in rat skeletal muscles[127]. When the procedure of Desmedt[41] was applied to the study of active transport in rat toe muscles, it was found that during a 2 h period of sodium enrichment in cold K-free modified Krebs fluid containing 160 mM Na, muscles lost about 76·5 mequiv K in exchange for 57·5 mequiv Na per litre fibre water. When these muscles were reimmersed for a further 2 h at 37 °C in oxygenated Ringer's with 10 mM K and 137 mM Na, they excreted 20 mequiv Na in exchange for only 5 mequiv K taken up and the muscles gained water to the extent of 9·6% of their weight over the 4 h period. When muscles were left innervated during immersion in K-free fluid, they lost about 10 mequiv less K per litre f.w., while sodium uptake was about the same as before. When innervated muscles were reimmersed in recovery fluid, they excreted an average of 38·5 mequiv Na, in exchange for 37·3 mequiv K per litre fibre water during the 2 h recovery period[127].

In some cases, these muscles were separated from their nerves just before reimmersion, and here it was found that only 22·9 mequiv Na was excreted in exchange for 20·5 mequiv K per litre f.w. When

E_m was measured in both the innervated and recently denervated companion muscles 10–20 min after reimmersion in recovery fluid, it was found to be about 4·8 mV more negative than E_K in the latter muscles, the difference being highly significant ($P < 0.01$), but in the innervated muscles E_m and E_K were almost identical. It seemed, therefore, that muscles that were innervated during reimmersion in recovery fluid excreted twice as much sodium as their completely isolated companions, while at the same time they pumped sodium in an electrically neutral manner. Several possible explanations for the effect of nerve on active transport come to mind. First, the nerve might influence active excretion of sodium through its depolarising action on the muscle membrane. This would have the effect of reducing the electrical energy gradient to sodium excretion. Horowicz and Gerber[128], for example, found that addition of 5 mM sodium azide to Ringer's fluid bathing frog striated muscles increased the ouabain-sensitive ^{24}Na efflux rate. This effect was attributed to the depolarising action of this substance on the membrane. This effect may also be related to the increased rate of sodium excretion from sodium-rich muscles found[53] when 2.7×10^{-5} M DNP was incorporated in recovery fluid. Koketsu[129] confirmed this observation and found ^{24}Na efflux associated with membrane depolarisation of about 25 mV in the presence of this substance. It is interesting to note that Koketsu found K^+ influx depressed here. In model systems composed of phospholipid bimolecular membranes, it has been found[130] that increased electrical conductance through the membranes occurs in presence of DNP, and Mitchell[131] has suggested that the uncoupling of oxidative phosphorylation by this substance is not brought about by inhibition of a specific enzymatic reaction *per se* but through this substance acting as a lipid soluble proton donor–acceptor system across the membrane.

While it is possible that membrane depolarisation early in the recovery period, when E_m is usually more negative than E_K, could arise through an increase in K permeability and could produce a transient increase in sodium efflux in exchange for potassium, it seems unlikely that this would influence the total quantity of ion transported, which is probably largely determined by the critical energy barrier to sodium excretion. At the end of the recovery period, E_m was equal to E_K and the latter was dependent on the final concentrations of sodium and potassium in the muscle fibres. Dydynska and Harris[132] have raised the question of whether a downhill movement of potassium across a favourable electrochemical potential gradient, such as that generated by an electrogenic sodium pump, might assist the latter process 'by feeding energy in'. While this should influence the kinetics of the process, it should not change the critical energy barrier at the end of the recovery period as defined by the equation

$$dG/dn = RT \ln [Na]_o/[Na]_i + RT \ln [K]_i/[K]_o \qquad (6.9)$$

In innervated rat muscles after 2 h in recovery fluid, E_m was $- 63\cdot5$ mV, $[Na]_i$ was $29\cdot4 \pm 1\cdot3$ mequiv per litre f.w., and the critical energy barrier was $2\cdot41$ cal per mequiv Na. In muscles that had been completely isolated before reimmersion, these values were $- 58$ mV, $45\cdot1 \pm 1\cdot6$ mequiv per litre f.w., and $2\cdot02$ cal per mequiv Na, respectively, at the end of the 2 h recovery. The ability of the innervated muscles to excrete sodium against a higher electrochemical gradient suggests a direct effect of nerve on the pump. Another significant observation here was that while 10^{-6} M ouabain was without effect on active transport in completely isolated muscles, it completely blocked the extra ion transport in the innervated muscles[133].

EFFECTS OF DENERVATION ON ELECTROLYTES AND ON SOME METABOLITES IN MUSCLE

Denervation of skeletal muscle is followed by changes in membrane properties. These are accompanied by a spreading of the acetyl choline-sensitive area of the membrane to the whole fibre surface from the normal confines of the endplate region[134, 135]. In both amphibian and mammalian skeletal muscles, there has been evidence of a fall in membrane permeability to K^+ following neurotomy. There was an increase in membrane resistance[136, 137] that could be attributed to fall in K conductance. Hubbard[137] found K conductance to decrease from 52 to 33 μmho cm^{-2} at 37 °C following denervation, while Cl conductance increased slightly. Ion fluxes also reflected this change. Harris and Nicholls[138] found that 22 days after denervation of frog muscle, the uptake rate for ^{42}K was consistently less in the denervated than in control muscles. However, as the K^+ content of the muscles seemed unaffected, K efflux rates from the denervated preparation must have been similarly reduced. Klaus and his colleagues[139] also confirmed that the unidirectional fluxes of ^{42}K were depressed by about a third following denervation of rat diaphragm, while at the same time E_m fell from $- 88$ mV to $- 79$ mV, also indicating a fall in the P_K/P_{Na} ratio for the tissue. In considering the effects of neurotomy on skeletal muscle, a distinction must be made between changes induced in red tonic muscle fibres on the one hand and in the whiter phasic fibres on the other. There appears to be little or no parallel between the effects produced by denervation on the electrical, mechanical, and metabolic properties of these muscle types[140], and therefore it seems unlikely that the electrolyte movements in these phasic and tonic fibres will respond in the same way to section of the motor neuron. Using gastrocnemius and soleus muscles of rat as representing phasic and tonic muscles, respectively, Kovacs and his colleagues[141] examined the effect of denervation on Na^+, K^+, and water content of the muscle fibres for up to 21 days post-operatively. They found that in phasic muscles, $[Na]_i$ increased from about 14·5 to 23·8 mequiv per litre fibre water over this period, while $[Na]_i$ of the tonic muscles decreased slightly but not significantly. Denervated muscles also showed a significant loss of K^+ and of water, particularly in the phasic type. An attempt was also made in these experiments to determine how Na^+ permeability might be effected by denervation, and to this end Li movement into the fibres was measured as this cation has the same permeability properties as sodium but is not effectively excreted by the fibres[142]. When Li was introduced intravenously into the rat's circulation, it was found to enter denervated phasic muscle steadily over a 21-day period, while its concentration in control muscles did not change significantly. In tonic muscles, however, Li tended to enter fibres of both control and denervated muscles and a significant increase of cell Li in favour of denervated muscles did not develop until the third week after neurotomy. These results were interpreted as showing increased permeability to Li and possibly to Na in phasic muscles following denervation, with little change in tonic muscle except perhaps a delayed stimulation of the sodium pump.

As already mentioned, insulin has been found to have a stimulating effect on the active transport of sodium from skeletal muscles[121, 58]. This hormone was also found to increase the rate of entrance of nonutilisable amino acids into isolated rat diaphragm[143]. Harris and Manchester[144] found in the same preparation that denervation of the diaphragm made it markedly less responsive to insulin with respect to transport of amino acids into the muscle fibres, but at the same time increased both accumulation of nonutilisable amino acids into the cells and also incorporation of normal ones into the cellular proteins. While the effect on protein synthesis is probably related to the hypertrophy found in denervated diaphragm, and may be due to more general disturbance of cellular metabolism, the effects on amino acid transport may give a lead as to the relationship between movement of electrolyte and metabolites across the cell membrane. If, as indicated above, denervation leads to increased turnover of sodium and stimulation of its excretion at the membrane,

241

and there is a correlation between sodium and amino acid transport as found in other tissues[145, 146], then there may be a complimentary stimulation of amino acid transport by insulin on the one hand and after denervation on the other, mediated through an effect on sodium transport through the membrane.

EFFECT OF NERVE STIMULATION AND OF ACETYLCHOLINE ON THE PERMEABILITY OF THE MUSCLE FIBRE MEMBRANE

At the neuromuscular junction in skeletal muscle, the transmission of excitatory impulses from nerve to muscle seems to depend on the release of acetylcholine from nerve endings, its diffusion to the muscle endplate and on its depolarising action there. The application to minute amounts of this transmitter substance at the endplate region by means of micropipettes[147] was found to depolarise the membrane, apparently through increased membrane permeability to all ions but particularly to sodium ions. The evidence for a non-specific increase in membrane permeability here was that there was no reversal of membrane potential at the endplate as found during the action potential. Furthermore, when the NaCl of the bathing fluid was completely replaced by K_2SO_4 and the membrane repolarised artificially by application of electric current from a micro-electrode, stimulation of the nerve to the muscle tended to depolarise the membrane and in the absence of external Na^+ it seemed that the only way this could happen was by increased entrance of K^+ into the fibres. The electrical and chemical changes that occur at the neuromuscular junction have recently been reviewed and commented on by a prominent worker in the field[148] and will not be referred to further here. However, the actions of acetylcholine and of vagal stimulation on the excitability of the cardiac auricular and pacemaker cells will now be considered, since here the effect of acetylcholine is to decrease rather than increase the excitability of the muscle fibre membrane.

Hoffman and Suckling[149], by means of microelectrodes, examined the effects produced on dog heart *in situ* and in isolation, by stimulation of the vagus and by application of acetyl choline to the auricles. They found that both produced an increase in the rate of repolarisation of the action potential without any decrease in either resting or action potentials, and they suggested that increased permeability to both Na^+ and K^+ might be responsible for this effect. In the pacemaker cells of the sinus of spontaneously beating turtle hearts, Hutter and Trautwein[150] were able to demonstrate both an increase in the diastolic membrane potential from -55 mV to -65 mV and a suppression of the pacemaker potential during vagal stimulation. That these effects were due to increased membrane permeability to potassium was suggested by the work of Harris and Hutter[151], who found that addition of acetyl choline at a concentration of 10^{-7} to 2×10^{-6} M induced a two- to threefold increase in the rate constant for ^{42}K efflux from labelled sinus venosus of tortoise and frog hearts immersed in inactive Ringer's with or without K^+. A four- to sevenfold increase in ^{42}K efflux rate was induced by addition of acetyl choline when $[K]_o$ was increased above 9 mM. The rate of K^+ influx was also increased by addition of acetyl choline and by electrical stimulation of the vagus. Raynor and Weatherall[152] confirmed the acceleration by cholinergic agents of the unidirectional K^+ fluxes in spontaneously beating and electrically driven rabbit auricles, and as the effect on efflux tended to predominate there was a net loss of K^+ from the muscles. If the effect of acetyl choline here was to increase K permeability relative to sodium permeability, then it might be expected that the diastolic membrane potential should approach the E_K value, which in rabbit auricles may be as high as -93 mV[12], while at the same time there would be a net loss of this cation, tending to repolarise the tissue. This should account for the observed increase in diastolic membrane potential and increase rate of repolarisation and in part at least for the change in the slope of the prepotential induced by acetyl choline and by vagal stimulation.

SOURCE OF ENERGY FOR THE SODIUM PUMP: OXYGEN CONSUMPTION

An increase in oxygen consumption associated with transport of sodium has been observed in muscles[153] and in nerve[222]. When oxygen consumption by sodium-rich muscles was compared during reimmersion in Ringer's with 10 mM K and 120 mM Na where no net excretion occurred, and in Ringer with 10 mM K and 104 mM Na where up to 22 mequiv Na per litre f.w. was excreted[65], mean oxygen consumption in the former over a 2 h period was 10.5 ± 0.8 mmol kg^{-1}, compared with 16.3 ± 1.0 mmol kg^{-1} in the presence of sodium transport. If the former value is taken as the basic oxygen uptake, this is increased by about 56% in excreting about 22 mequiv Na and ther ratio of sodium ions transported to oxygen molecules consumed was about 3.8. This is close to the value of 4 that would be expected if this cation were transported by a redox pump[13]. In recovery fluid with 10 mM K and 104 mM Na, there appeared to be a direct relationship between the quantity of sodium excreted and the extra oxygen consumption. A correlation of 0.68 ± 0.09 was found between the two quantities, and the regression line relating net sodium loss to oxygen consumption by the sartorii during the 2 h recovery was

$$O_2 \text{ uptake (mmol kg}^{-1}) = 0.26 \times \text{Na loss (mequiv kg}^{-1}) + 6.04$$

from which it can be seen that 3.85 mequiv sodium excretion was associated with 1 mmol oxygen uptake, and mean oxygen consumption in the absence of sodium transport was 6.04 mmol kg^{-1}.

TURNOVER OF ENERGY-RICH PHOSPHATE DURING ACTIVE TRANSPORT OF SODIUM

The use of $\sim$ P in active transport by muscles has been examined quantitatively by Dydynska and Harris[132]. They measured the changes in the concentrations of creatine phosphate (CrP), of adenosine triphosphate (ATP), of inorganic phosphate (P$_i$), and of lactic acid (LA) in muscles under anaerobic conditions in the presence of active transport brought about by addition of 10 mmol K to the immersion fluid. They also used the metabolic inhibitors, iodoacetate, oligomycin, and 0.38 mM dinitrofluorobenzene (DNFB) in some of their experiments.

Under aerobic conditions they found no significant change in [CrP] to be associated with net extrusion of 27.4 mequiv Na per kg muscle, but there was a slight decrease in [LA] and [P$_i$] and increase of [ATP]. These results were in agreement with the changes from state 3 to state 4 in the respiratory chain[105], indicative of an increase in phosphate potential[154], when active transport was induced in Na-rich muscles in oxygenated Ringer's with 10 mM K Ringer's by reduction of [Na]$_o$. Dydynska and Harris found that net sodium excretion in an atmosphere of nitrogen was accompanied by a marked fall in [CrP] with no appreciable change in [ATP], while [P$_i$] and [LA] increased. In the presence of DNFB which blocks ATP-Cr transferase (the Lohmann reaction), [CrP] did not change during anoxic sodium extrusion but [ATP] was significantly decreased, suggesting that the latter was the more immediate energy source for the pump. When oligomycin, an inhibitor of mitochondrial oxidative phosphorylation, was added to the muscles excreting sodium aerobically, the ion movement was accompanied by a fall in [CrP] and the formation of [P$_i$] and [LA], as in anaerobic conditions. This would seem to indicate that the main function of the aerobic respiration was to replenish energy-rich stores. When

0.4–0.6 mM iodoacetate was used to inhibit glycolysis, ion movements were reduced but the breakdown of CrP and ATP per equivalent of cation moved increased considerably. Much of the $\sim$ P here was used to form fructose 1–6 diphosphate, so it was necessary to make allowance for this fact. It was also considered desirable to follow the changes in adenosine diphosphate concentration [ADP], as well as [ATP], as much of the former was converted into AMP which was then deaminated. Harris[155] therefore extended this work, making allowance for these sources of error, and came to the conclusion that 3 Na^+ were extruded for each $\sim$ P utilised during active transport in presence of iodoacetate and DNFB. Active transport here in absence of oxygen would appear to be more efficient than that studied under aerobic conditions, where about four Na^+ were transported for each molecule of oxygen consumed[65]. Assuming a P : O ratio of three in aerobic oxidative phosphorylation, two oxygen atoms would be equivalent to 6 $\sim$ P, giving a Na^+ : $\sim$ P ratio of only 2 : 3. On the other hand, a procedure in which active transport was induced by reduction of $[Na]_o$ from 120 to 104 mM would seem to be less subject to extraneous metabolic changes than one in which it was induced by increasing $[K]_o$ from 0 to 10 mM.

THE CALCIUM PUMP IN MUSCLE

Calcium ions have an important role to play in muscle contraction. Increase in the concentration of ionised calcium within the sarcoplasm, following influx across the fibre membrane and release from membrane binding sites, is believed to be directly responsible for the onset of contraction in muscle. This calcium combines as cofactor with myofibrillar ATPase, bringing about activation of the contractile protein, breakdown of ATP, and shortening of the sarcomeres. An essential feature of the relaxation process, on the other hand, appears to be removal of calcium ions from the sarcoplasm once more by sequestration in internal membranes and by storage in microsomal vesicles of the sarcoplasmic reticulum[156]. The uptake of calcium ions by mitochondria[157] and its binding by soluble relaxing factors[158, 159] may also contribute to the rapid removal of sarcoplasmic and myofibrillar calcium that terminates the active state. Before discussing the investigations undertaken to study the calcium sequestration in microsomal vesicles of the sarcoplasmic reticulum, it might be worth mentioning some of the difficulties that arise in relation to the study of calcium movement between muscle fibre and the external environment.

DISTRIBUTION OF CALCIUM IN MUSCLE

CALCIUM CONTENT OF MUSCLE

This seems to vary considerably, particularly in relation to season. In frog sartorius, for example, mean concentrations range from 0.74 mM kg^{-1} in spring to 1.79 mM kg^{-1} in autumn[160]. Bozler[161] found the total calcium in frog stomach to be about 0.35 mM kg^{-1} higher in November and December than during the remainder of the year.

The total calcium in muscle is distributed in the following regions: (1) interstitial fluid, (2) connective tissue, (3) unit membrane and sarcoplasmic reticulum, and (4) sarcoplasm. Estimates of the calcium concentration in the sarcoplasm of the resting muscle fibre have been made on the basis of the maximum concentration consistent with relaxation, which appears to be about 10^{-7} M[160, 162]. Activation of the contractile proteins in muscle by the direct action of calcium ions has been demonstrated by their microinjection into the cell[163-165] and by direct application to the myofibrils in skinned fibres[166], and from such experiments the $[Ca^+]_i$ needed to bring about shortening of the sarcomere has been estimated. When the sarcoplasmic calcium and that of the interstitial fluid have been subtracted from the total calcium of resting muscle, it becomes evident that a substantial fraction of the total must reside in other regions.

Gilbert and Fenn[167] found that when isolated frog muscles were soaked for 5 h in Ca-free Ringer's, their calcium concentration decreased by only 0.5 mM to about 1.12 mM per kg tissue and even treatment with the chelating agent, EDTA in Ca-free fluid, for a similar period did not reduce [Ca] below 0.9 mM per kg. On the basis of ^{45}Ca exchange experiments, they concluded that 10% of muscle Ca^{2+} was in a surface phase exchanging within 10 min, 12% was in the interstitial fluid and 17% in connective tissue being replaced by tracer in 30 min, 24% in the intracellular space exchanging in 5 h, the remaining 37% being nonexchangeable. Harris[168] confirmed that an appreciable fraction of calcium in muscle was bound apparently in a sparingly soluble complex. He found that after an initial rapid phase, ^{45}Ca uptake rate decreased with time to such an extent that ^{45}Ca in the muscle was about the same after 16 h and even 20 h from time of immersion. While the quantity of ^{45}Ca taken up was similar in amount to that which could be removed again in Ca-free Ringer's, there was some incorporation in a bound fraction from which it was not removed even in the presence of EDTA. Harris felt there was no justification for attempting to

246

resolve calcium movements into a number of first-order processes reflecting compartmentalisation since, with the extreme variability found here, these phases might just as easily have been due to changes in the strength of binding of the ion to the membrane.

Table 6.5 CALCIUM CONTENTS OF MUSCLE AND PLASMA

Tissue	Total (μmol g^{-1})	Rapidly exchangeable (μmol g^{-1})	Slowly exchangeable (μmol g^{-1})	Nonexchangeable (μmol h^{-1})	Ca_o (mM)
Sartorius (Bianchi[160])	1·81	0·45	1·0	0·36	1·36
Turtle ventricle (Henrotte, Cosmos and Fenn[172])	1·19				
Guinea pig atrium	1·7	1·01	0·2	0·44	2·25
Rat uterus (van Breemen, and Daniel[173])	2·23	0·8	1·0	0·45	1·5
Frog stomach (Bozler[61])	1·54	1·38	0·12	0·4	
Frog muscle (Gilbert and Fenn[67])	1·62	0·63	0·39	0·6	

Shanes[169] has stressed the importance of surface calcification in maintaining the normal resting membrane permeability of monovalent cations and he summarised the evidence that excess $[Ca]_o$ decreased P_K and P_{Na}, while reduction of $[Ca]_o$ had a contrary effect. The terms P_K and P_{Na} are permeability coefficients in the Goldman constant-field equation, which defines the membrane potential in terms of concentration gradient and relative permeabilities of diffusible ions across the membrane as shown in equation 6.4 on p. 197.

Because of the large amount of bound calcium in skeletal muscle (only about 20% being freely exchangeable), the use of ^{45}Ca flux measurements in the study of net calcium movement across the cell membrane appeared to be of limited value. Indeed, the work of Cosmos and Harris[170] illustrated the importance of supplementing tracer studies with determination of the actual concentrations of this cation in the tissue because of the uncertainty as to the relative contributions of net movement and exchange of labelled for unlabelled cation to fluxes observed under a particular set of conditions. This is reminiscent of the exchange diffusion already discussed in relation to sodium fluxes. The exchange process predominates *in vivo*, where labelling of muscles with injected ^{45}Ca is observed[171] in the absence of net influx. Cosmos and Harris[170], with isolated frog sartorii, found that the rate of Ca^{2+} uptake increased significantly under conditions where the muscles were losing K^+ in exchange for Na^+. These conditions included the addition of 1 µg ml^{-1} of strophanthidin to the normal bathing fluid containing 3 mM K and 1 mM Ca and also the omission of K^+ from the Ringer's. Here, there was a high correlation between net uptake of sodium and of Ca^{2+} by the tissue. The tracer studies also revealed an increase in calcium uptake (measured by ^{89}Sr) when K^+ was absent from the bathing fluid. For example, after 2 h in K-free Ringer muscles had gained 1·94 µmol Ca per g, compared with 1·44 µmol g^{-1} in control muscles in normal Ringer's. Net Ca^{2+} uptake observed in these experiments was reversible, as its concentration in muscle returned to normal when the muscles had excreted sodium once more during immersion for a number of hours in recovery fluid with K^+. This was the first demonstration of net active transport of Ca^{2+} from muscles.

EFFECT OF $[Na]_o$ ON Ca INFLUX

In the experiments described the measured uptake of labelled ion was found to be about twice as great as the net accumulation of Ca^{2+}, a result illustrating how misleading flux studies can be. A good example of increased exchange which did not necessarily involve net uptake was that which followed an increase in the $[Ca^{2+}]_o : [Na^+]_o$ ratio brought about by replacement of sodium in the Ringer's by sucrose. Niedergerke and Harris[174] found that when $[Na]_o$ was reduced in this way there was increased labelling of ventricle strips with ^{45}Ca. The essential difference is that when $[Na^+]_i$ rises on account of K^+ loss, both Na^+ and Ca^{2+} move in together. When $[Na^+]_i$ is reduced in response to lowered $[Na^+]_o$ then Ca^{2+} enters to replace some of the lost Na^+. Formally, the system behaves as if the internal Na^+ and Ca^{2+} share and compete for occupation of a space from which they are both excluded by an active process linked to K^+ retention, so its capacity rises when $[K^+]_i$ falls, and conversely. This effect was therefore re-examined by Cosmos and Harris in frog skeletal muscle by comparing calcium exchange measured with ^{89}Sr in 1 mM Ca Ringer's containing 120 and 30 mM Na (replacement of NaCl by sucrose). It was found that in this tissue also the rate of tracer uptake increased as $[Na]_o$ was reduced, but a comparison of net Ca^{2+} accumulation with total labelling of the tissue under these conditions revealed that it was mainly the exchange component of tracer uptake that was increased by raising the $[Ca^{2+}]_o : [Na^+]_o$ ratio. In the rat uterus, van Breemen, Daniel, and van Breemen[175] failed to find evidence of sodium competing with calcium for inward transport where $[Na]_o$ was varied.

MAGNESIUM MOVEMENT

A relationship between the net movement of monovalent cations and of another divalent cation, namely magnesium, has been described. A net uptake of magnesium has been noted[64, 176] in potassium deficient tissues, including muscle, and restoration of the diminished $[K^+]_i$ also required the presence of $[Mg^{2+}]_o$[177, 178].

COMPARATIVE EXCHANGEABILITY OF MUSCLE CALCIUM

Some comparisons of exchangeability of Ca^{2+} in various types of muscle are worth mentioning in so far as they may relate to differences in structural features such as fibre diameter and the degree of organisation of the sarcoplasmic reticulum, which is believed to be an important storage depot for that fraction of fibre Ca^{2+} concerned with muscle contraction. For figures of relative exchangeability of calcium between muscles, see Table 6.5.

Henrotte, Cosmos, and Fenn[172] found that the Ca^{2+} exchangeability measured with ^{45}Ca was much greater in turtle ventricles than in frog skeletal muscles, and they suggested that this fact might explain the more immediate failure of contractility in the former muscle, following immersion in Ca-deficient media. Bozler[161] examined the distribution and exchange of Ca^{2+} in smooth muscle, making allowance for the large quantity of the cation contained in tendon and fascia. He found that exchange of Ca^{2+} in frog stomach was much greater than that in skeletal muscles. More than 50% of the fibre Ca^{2+} was lost to a Ca^{2+}-free solution within 10 min. Calcium within the connective tissue exchanged more slowly than that in the fibres.

CALCIUM FLUXES DURING EXCITATION

Stimulation of most excitable cells including muscle is accompanied by increase in Ca^{2+} exchange. In frog sartorii, Bianchi and Shanes[179] found that ^{45}Ca entry into muscle, which was about $28\cdot2$ pmol $g^{-1}s^{-1}$ in 1 mM Ca, doubled during electrical stimulation of the tissue. When $[Cl]_o$ was fully or partially replaced by nitrate or by thiocyanate, the resting influx remained practically unchanged, but influx during excitation increased to as much as 10 times the resting value, and this increase paralleled a potentiation of contraction tension. As a general rule, while stimulation increased Ca^{2+} influx it also raised the efflux rate, so that there appeared to be a dissociation and increased turnover of bound calcium of the membrane without a measurable increase in $[Ca^{2+}]_i$. The membrane depolarisation and contracture produced by exposure of muscles to high-K Ringer's was also found to lead to release of Ca^{2+} from membrane sites into muscle fibres in amounts of up to 26 nmol g^{-1} [179]. When muscles were exposed to ^{45}Ca and high-K simultaneously, no measurable uptake of tracer occurred, but when the muscles were exposed to ^{45}Ca first to allow time for the tracer to enter the interstitial space and exchange for bound Ca^{2+} in the membrane, then subsequent addition of high-K evoked a transient three- to five-fold increase in influx. The failure of muscles to sustain increased Ca influx and contracture tension in presence of high-K was probably due to depletion of membrane Ca^{2+} through competition with K_o. Frank[180] demonstrated the importance of Ca^{2+} influx in the development of K-contracture tension when he showed that soaking muscles for up to 1 min in Ca-free Ringer before adding high-K eliminated the mechanical response to depolarisation. In rat uterus[173], high-K actually slowed ^{45}Ca influx and caused a net Ca^{2+} efflux; however, the latter did not appear to be from the sarcoplasm but probably originated in a membrane-bound fraction. In rat uterus depolarised by exposure to isotonic K_2SO_4 Ringer's, addition of $[Ca^{2+}]_o$ was found to produce graded and reversible contraction[181], which was antagonised by Mg^{2+} and abolished by DNP. In this depolarised preparation, after eliciting the maximal contraction by addition of calcium, increase of tension by as much as 90% could be induced by addition of acetyl choline. The latter contracture in depolarised smooth muscle of cat intestine[182] could take place even when over two thirds of the muscle Ca^{2+} had been removed. In Ca-free media it was possible to demonstrate contracture in response to addition of acetylcholine, even when K contracture was absent, so the two may be independent.

While depolarisation may serve as a trigger for the release of membrane calcium into the sarcoplasm in bringing about contracture, other substances, including caffeine[183, 184], ryanodine[185], and quinine[186], can produce contracture without any apparent depolarisation and may therefore act directly on the membrane-bound calcium.

ISOTOPE STUDIES IN FROG HEART

Niedergerke[187] used ^{45}Ca to examine the uptake and release of Ca^{2+} in isolated frog ventricles at rest and during contracture induced by exposure to high-K and to Na-deficient fluids. The mean resting influx in normal Ringer's with 1 mM Ca was about 0·009 pmol $cm^{-2} s^{-1}$. This increased to about 0·11 pmol $cm^{-2} s^{-1}$ when contracture was induced by addition of Ringer's with 100 mM K or of Ringer's in which 50% of Na^+ was replaced by choline, lithium, or sucrose. These fluids contained 1 mM Ca. When sodium was entirely omitted from the Ringer's further three-fold increase in Ca^{2+} uptake was noted. The efflux of Ca^{2+} also increased in response to application of these contracture fluids. It was estimated, however, that the exchangeable muscle Ca^{2+} increased from about 0·49 mmol per litre fibre during rest to about 2·09 mmol per litre during activity. A demonstration of net efflux of ^{45}Ca was found possible by transferring muscles from Na-free Ringer's to normal Ringer's. (This again raises the paradox of the apparent correlation of Na^+ and Ca^{2+} influx[170] and the apparent competition between these cations for uptake.) The movement of Ca^{2+} in beating ventricles has also been examined by Niedergerke[188], in relation to changes in $[Na]_o$. Both ^{45}Ca uptake and release from the muscles increased during activity, the amount of cellular Ca involved in the exchange being 0·6 mmol per litre fibre. The extra Ca^{2+} influx per beat through the average heart fibre at 7 °C was about 0·15 pmol cm^{-2} in 1 mM Ca Ringer's, increasing to 0·23 pmol in 2 mM Ca. The response of Ca^{2+} influx in active cardiac muscle to increase of $[Ca^{2+}]_o$ was quite unlike that of frog skeletal muscle, where increase of $[Ca]_o$ from 1 to 3 mM led to a three-fold increase in resting uptake without changing its uptake during stimulation[160]. Replacement of 50% of $[Na]_o$ by choline here, in the presence of as little as 0·5 Ca, increased the Ca^{2+} influx during activity to 0·25 pmol $cm^{-2} s^{-1}$. The apparent competition between Na^+ and Ca^{2+} for uptake during excitation was interpreted as indicating that both ions combine with specific molecules within the membrane for transfer into the muscle fibre. This common mechanism was also indicated by experiments in which Ca efflux from auricular muscle was found to decrease when these muscles were transferred from Ca-free Ringer's to a medium deficient in both Ca^{2+} and Na^{+} [189].

Finally, to obtain a comparison of ^{45}Ca uptake by quiescent and active frog ventricles, Niedergerke, Page, and Talbot[190] treated some hearts with tetrodotoxin, which blocks action potentials, and then perfused these and untreated hearts with ^{45}Ca-labelled Ringer's with 1·5 mM Ca. The quantity of ^{45}Ca taken up or released by these hearts in successive perfusion periods were then calculated from the measured concentrations and volumes of perfusate. In the resting hearts there was an initial rapid phase of ^{45}Ca influx into the ventricles amounting to about 0·2 mmol per litre heart cells, followed by a slower phase saturating at about 0·6 mmol per litre. In the untreated muscles, stimulation was commenced after 40–60 min in ^{45}Ca fluid, at which time fast phase of entry was completed and the slow influx rate had been established. There was a very marked increase in the rate of ^{45}Ca uptake with onset of activity from about $4\cdot1 \times 10^{-15}$ to 130×10^{-15} mol cm^{-2} s^{-1} per heart beat. Voltage-clamp experiments[191] in ventricular myocardial fibres from dog have provided evidence for the existence of a slow inward ionic current sensitive to variations in Ca^{2+} when the early rapid sodium current had been eliminated by previous depolarisation. This current, which would occur during the plateau of the action potential, was insensitive to tetrodotoxin (which abolishes the depolarising sodium current and spike) and so may be due to Ca^{2+} influx. Depolarising calcium currents have also been reported in smooth muscle[192] and in the giant barnacle muscle fibre[23].

THE SARCOPLASMIC RETICULUM AND CALCIUM UPTAKE AND STORAGE

The internal calcium required for the activation of the contractile protein seems to reside mainly within the sarcoplasmic reticulum, especially in skeletal muscle. It has been calculated that a calcium concentration of 2×10^{-6} M or more would be needed to activate the actomyosin[103], on the assumption that each myosin molecule binds 1–2 Ca^{2+}. The Ca pump located apparently within the vesicles of the sarcoplasmic reticulum seems to be mainly responsible for removing Ca^{2+} from the sarcoplasm to restore the fibres to the resting state.

The preparation of the sarcoplasmic reticulum (SR) which has been extensively used in the study of the calcium transport here has been a particulate fraction of homogenised tissue, rich in vesicles (Figure 6.16), obtained by differential ultra-centrifugation, followed in some cases by purification on a sucrose density gradient. Palmer and Posey[194] obtained a rather high yield of vesicles, amounting to 10 mg per g muscle, but comparative studies of binding capacity of Ca^{2+} suggested incomplete purification here. Van der Kloot[195] estimated 2 mg per g as the normal yield of vesicles from skeletal muscle, and this in the presence of ATP and Mg^{2+} could complex 0.2 μmol Ca^{2+}, which was about twice the amount of Ca^{2+} that would combine with 100 mg of myofibrillar proteins in 1 g muscle. Weber and her associates[157] calculated that about 1 μmol of Ca^{2+} required to be removed from 1 g of actomysin to bring about relaxation, and this is of the same order of magnitude.

An important preliminary step in Ca^{2+} accumulation by the vesicles is binding with the membrane, and this does not seem to require energy, as it can take place to some extent in the absence of ATP. Two subsequent steps have been characterised, namely, transport-without-storage and transport-with-storage. The latter process has been examined by the use of sodium oxalate as an intravesicular precipitating agent and is perhaps of limited physiological significance. In cardiac SR it was found[194] that the relative contribution of transport-with-storage, transport-without-storage, and true binding to total calcium uptake was $50:6:1$, with the latter estimated as about 0.8 μmol g^{-1} of reticular protein. The binding capacity reported by Katz and Rebke[196] for vesicles in the absence of oxalate was about six times greater than by these workers.

When oxalate is used, the product of concentrations of calcium and oxalate within the vesicles rose in some cases to about 5000 times that outside the vesicular membrane and it was found that Ca^{2+} uptake correlated with Ca-induced splitting of ATP[197]. An oxalate pump did not seem to operate here, but this anion reduced the activity of Ca^{2+} within the vesicles, thereby lessening energy requirement for its possible active transport. The appearance of the precipitate of calcium oxalate within the vesicles has been cited as evidence that Ca^{2+} is actually transported across the vesicular membrane and not merely bound to the surface of these particles. Onishi and Ebashi[198], however, pointed out that on the basis of available surface area such binding was feasible at a binding rate of 3 mol Ca per cm^2 s^{-1}. They also draw attention to an anomaly in relation to the stoichiometry of Ca^{2+} uptake to ATP breakdown. While at an ATP concentration of 5 mM, the Ca:ATP ratio was about $1:2$; this increased to as much as 10 when ATP was decreased below 0.2 μM[199]. This finding was subsequently contradicted by Weber and her associates[200], who found the Ca:ATP ratio unchanged over a wide range of ATP even below 0.1 μM.

ENERGETICS OF CALCIUM TRANSPORT

This question might be considered first in relation to the muscle fibre as a whole and then in relation to Ca^{2+} uptake by the SR.

If calcium ions were at equilibrium across the fibre membrane and the fibre was freely permeable to those ions then we could apply the Nernst equation to determine their passive distribution between sarcoplasm and external fluid, as follows:

$$- E_m = RT/2F \ln [Ca]_o/[Ca]_i \qquad (6.10)$$

In resting skeletal muscle we may write

$$\log Ca_o/Ca_i = -90/29 \text{ or } -3 \text{ approx. at } 20\,°C \qquad (6.11)$$

So, passive distribution required that Ca^{2+} be about 1000 times more concentrated inside the fibres than outside. If we accept the estimated ionic concentrations of 10^{-3} M for $[Ca]_o$ and 10^{-7} M for $[Ca]_i$, then the ratio $[Ca]_o/[Ca]_i$ will be 10000, which would be in equilibrium with a membrane potential of $+ 116$ mV. The electrochemical gradient tending to move Ca^{2+} ions into the fibre would be proportional to the difference between E_m and E_C, that is, $116 + 90$ mV. The work done in maintaining the low intracellular calcium concentration at rest would be

$$\Delta G = \frac{nF(116 + 90)\,(\text{efflux})}{1000} \text{ cal} \qquad (6.12)$$

In the steady-state, where influx and efflux are equal, we may calculate the energy consumption by the Ca pump using the estimated influx rate of 100·8 nmol g^{-1} h^{-1} [160], which yields an energy value of 1·09 cal kg^{-1} h^{-1}, or about 0·6% of the total energy output by the muscle. In smooth muscle, however[161], the Ca influx rate was 0·15 mmol g^{-1} h^{-1}, or about 1000 times greater than in skeletal muscle, and in this case it was estimated that the energy required for Ca transport was 540 cal kg^{-1} h^{-1}, or almost twice the resting metabolic rate. Although it is likely that steady-state conditions were not realised in these experiments, the discrepancy is still very large.

In the case of the isolated SR vesicles, it has been found that they can establish a $[Ca]_i : [Ca]_o$ ratio of about 10^5 across their membranes, which does suggest that active transport is involved in their accumulation of this cation. This view is substantiated by the fact that the uptake of each $2Ca^{2+}$ by the vesicles splits a molecule of ATP and this relation holds down to an external $[Ca^{2+}]$ of about 10^{-9} M, from which level no further accumulation can take place[197, 200].

COMPARATIVE STUDIES IN CALCIUM TRANSPORT BY SARCOPLASMIC
RETICULUM

It is believed that the SR in skeletal muscle is a more extensive and
elaborate structure than that of either cardiac muscle[201] or smooth
muscle[202]. Certainly, the yield of particulate material from the SR
has been less in the latter muscles. For example, Inesi and his colleagues[203] obtained 1·3 mg dry vesicles per gramme of bovine
cardiac muscle, compared with 2·0 mg dried vesicles per gramme of
rabbit skeletal muscle. In addition to the lower yield from cardiac
tissue, the vesicles from this source also showed a lower capacity
for calcium accumulation in the presence of ATP. Bovine cardiac
microsomes took up 29 μmol g^{-1}, compared with 59 μmol g^{-1} for
microsomes from rabbit skeletal muscle[203]. Weber and her associates[157] found that their best preparation obtained from cardiac
tissue accumulated Ca^{2+} at the rate of 26 nmol mg^{-1} reticular
protein per minute, which was only about one quarter the rate reported for skeletal muscle reticular protein[200]. Van der Kloot[195]
used a vesicular preparation from lobster muscle treated with
dithio-threitol. This accumulated Ca^{2+} at a rate of 39×10^{-8}
mol s^{-1} mg^{-1} protein, about a hundred times faster than Weber's
cardiac preparation. This rate of Ca^{2+} uptake would allow the
sarcoplasmic Ca^{2+} to be reduced from 10^{-6} to 10^{-7} within 14 ms,
well within the relaxing time of the muscle. The treated vesicles took
up Ca^{2+} about 19 times faster than the control preparation, which
is in agreement with the finding that untreated skeletal muscle
vesicles had five times the affinity for Ca^{2+} shown by cardiac
microsomes. Chicken heart muscles[157] from which an unusually
high yield of 2·4 mg reticular protein was isolated from 1 g wet
weight of tissue had a capacity for Ca^{2+} accumulation just sufficient
to account for the relaxation properties of the muscle. These measurements were made in the absence of oxalate.

In the case of smooth muscles, Carsten[204] found that, while the
Ca^{2+} binding capacity of SR obtained from this tissue was much
lower than that found in cardiac muscle, it might still be of physiological importance in relaxation, having regard for the extreme slow-
ness of relaxation in this muscle. When ^{45}Ca was injected *in vivo*
into dogs, up to half the vascular radioactivity was located in a
fraction of tissue homogenate having the same sedimentation
characteristics as microsomal vesicles[205]. Nevertheless, in spite of
such evidence for Ca^{2+} binding in smooth muscle microsomes, it
is still likely that extracellular calcium may play a more significant
role than internally stored calcium in activating the contractile
proteins in this tissue[206]. If the slow-exchanging Ca^{2+} are located

in the SR, the higher exchangeability of smooth muscle Ca^{2+} (Table 6.5) would suggest this.

CALCIUM BINDING IN 'SKINNED' MUSCLE FIBRES

While it is profitable to consider the properties of the microsomal preparations described in relation to the calcium pump and relaxation process, studies carried out on 'skinned' muscle fibres[207] are also worthy of mention in so far as they represent a more integrated system. Constantin and Podolsky[208], using fibres of frog semi-tendinosis from which the sarcolemma had been removed under oil, looked for the site of calcium sequestration under conditions where this cation was precipitated as oxalate. One of the techniques they employed was to perfuse their preparation first with calcium salt, which produced contraction followed by relaxation as the Ca^{2+} was removed from the myofibrils by relaxing factor. The preparation was then perfused with 10 mM sodium oxalate, either before or after fixing, and the location of the optically dense precipitate of calcium oxalate was observed by electronmicroscopy. These deposits were found to be generally confined to the terminal sacs of the SR, which was taken as evidence of their functioning as a Ca^{2+} sink during relaxation. It was taken for granted that this sink was intracellular. However, in view of the evidence discussed above that the lumen of the SR is extracellular[44], the Ca-pump in muscle may be situated in a membrane separating the sarcoplasm and external bathing fluid.

The reaccumulation and binding of Ca^{2+} by the muscle membrane may be significant from the point of view of permeability, as well as for relaxation of contractile protein. While membrane depolarisation and electrical stimulation of vesicles of SR^{209} produced a release of free Ca^{2+} into the sarcoplasm and incubation medium, this was accompanied by a period of more rapid exchange of labelled for unlabelled Ca^{2+} consistent with less firm binding. The reversible increase in exchangeability of membrane Ca^{2+} must reflect changes in structural rigidity and organisation of the membrane involving orientation and distribution of anionic polyelectrolytes, including phospholipids. There is some evidence that lecithin is important for the accumulation of Ca^{2+} by the microsomal fraction of muscle. The dried material obtained from bovine cardiac and rabbit skeletal muscle microsomes was found to contain about 70% protein and 20% phospholipid[203] (see Figure 6.16). About 50% of the latter appeared to be lecithin. Ebashi and Yamanouchi[199] found that removal of this substance from microsomes by the action of

256

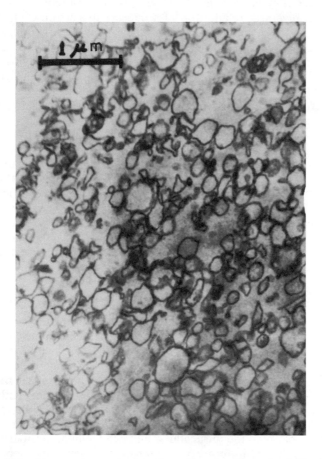

Figure 6.16. Electronmicrograph of cardiac microsomes (From Inesi, Ebashi, and Watanabe[203]; *courtesy Am. J. Physiol.)*

K

lecithinase C destroyed both their relaxing and Mg-ATP-ase activities. Martonosi[210] also demonstrated this loss of activity following enzymatic removal of 90% of microsomal lecithin, but this was reversed by addition of a miceller dispersion of synthetic lecithin to the preparation. Lecithin itself shows a high affinity for Ca^{2+}, but unlike the binding seen in microsomal particles, that in pure lecithin was competitively reduced by addition of Mg^{2+}, K, and ATP. It would perhaps be surprising if the specificity of synthetic lecithin was not modified by its combination with protein as part of the structure of the microsomal membrane, so conclusions drawn from studies with pure lecithin may be misleading.

Martonosi[211] subsequently used ATP^{32} labelled predominantly in the terminal position, and allowed this to react with isolated microsomes, during incubation in various media. In the steady state, the ^{32}P bound to the membranes in mixtures containing 5 mM Mg^{2+} and $[Ca^{2+}]_o$ ranging from 0 to 5 mM was inversely proportional to the ATPase activity. Maximum ATPase activity and minimal P^{32} binding was found in 0.01–0.1 mM $CaCl_2$. The concentration of bound ^{32}P increased by a factor of 10^2–10^3 in the phospholipase C-treated microsomes, suggesting that lecithin was required for the dephosphorylation step.

CONCLUSION

It has been necessary in writing this chapter to concentrate on ions that must be actively transported across the cell membrane, particularly as so much experimental work has been done over the last decade in connection with the sodium and calcium pumps in muscle. As was pointed out above, a number of ions including chloride, potassium, and bicarbonate seem to be close to equilibrium in their distribution across the cell membrane, so that their concentration gradients between the intracellular and extracellular compartments may be explained in terms of a Donnan system[2, 40] and depend to a large extent on the exclusion of sodium and calcium ions from the fibres resulting from active transport and low permeability. The question of whether hydrogen ions are at equilibrium across the membrane is far from settled[212, 213]. By means of pH-sensitive glass capillary electrodes, Caldwell[214] and Kostyuk and Sorokina[215] found the internal pH of muscle fibres to be as high as 7.15 and 7.12, respectively, where external pH was about 7.4. If hydrogen ions were distributed in a Donnan equilibrium, the internal pH under these conditions should be as low as about pH 6 [2]. This, combined with the fact that intracellular pH was not changed by membrane depolarisation produced by increase of $[K]_o$, suggested that the hydrogen ions are actively excreted from the fibres. On the other hand, more recently Carter and his colleagues[216], using an elaborate double-barelled electrode found a mean intracellular pH of 5·99 $\pm$ 0·14 for rat muscles where the calculated value based on the Donnan relation was 5·97 $\pm$ 0·04. The intracellular pH also varied with membrane depolarisation and hyperpolarisation produced electrically, in a manner that would be expected if the hydrogen ions were distributed across the membrane in a Donnan equilibrium. Criticism of these and other methods for measurement of intracellular have been reviewed[213], and there the matter rests.

Finally, it may be said that the transfer of phosphate and of organic metabolites across the cell membrane would not seem to be determined so much by considerations of membrane permeability as by their utilisation in the metabolism of the cell. While the distribution of some acidic and basic amino acids across the cell membrane may be dependent on membrane potential and intracellular potassium concentration[217, 218], the movement of some unchanged substances including sugars across the cell membrane seem to be related to active transport of ions in a manner that is not yet understood. It is possible that in the future much more attention will be paid to the possible relationship between ion transport on the one hand and transfer of metabolites on the other.

ENTRY OF SUGARS AND AMINO ACIDS INTO MUSCLE CELLS
(ADDENDUM BY E. J. HARRIS)

Glucose and some related sugars, as well as non-metabolised methylated derivatives, enter muscle cells by a saturable process corresponding to that described in Chapter 2. The net change in content depends on the difference between a saturable influx and a saturable efflux. Glucose is so rapidly phosphorylated that no free sugar is normally found in the cells. To study the kinetics and distribution, it has been convenient to employ the non-metabolised compound 3-O-methyl glucose. It has been shown that this compound (and related penetrant sugars) attains an inside level as if 500–650 µl water per gramme of cells was acting as solvent for the sugar at its outside concentration. The total tissue water was typically 760 µl g^{-1}. If the muscle was first loaded with a penetrant sugar, the latter could be displaced by glucose. The unnatural stereoisomer of glucose did not penetrate heart muscle. The hormone insulin has a property of accelerating sugar entry. This it does by combinations of increased V_{max}, and in some cases, increased K_m, depending on the tissue. After insulin has increased the permeability, the cell admits glucose sufficiently rapidly to permit existence of detectable free glucose inside[219, 220].

Some amino acids are concentrated in muscle tissue and, like the sugar uptake, the process is strongly enhanced by insulin. To varying degrees, the amino acids in muscle are constantly being incorporated into protein, that meanwhile is breaking down. On this account, studies with labelled amino acids meet difficulties of interpretation because of internal dilution.

Table 6.6 ACCUMULATION OF SOME AMINO ACIDS IN RAT DIAPHRAGM AND THE ENHANCEMENT BY INSULIN OF SOME ACCUMULATIONS*

Amino acid concn. applied		Accumulation ratio measured	
Insulin-sensitive		No insulin	With insulin
Alanine	16	1·85	2·31
	1000	1·84	2·52
Histidine	1000	2·97	3·36
Serine	1·1	3·94	4·55
Threonine	40	2·85	3·00
Insulin-insensitive			
Alanine	15	1·07	1·08
Tyrosine	1000	2·02	2·03
Arginine	1000	2·01	2·02

* Incubations for 1 h, insulin at 0·1 unit per ml.

However, the accumulation persists when protein synthesis has been inhibited by specific compounds (e.g. cycloheximide) and in certain examples the assayable amount of the amino acid has been shown to agree with that deduced from the radioactivity method. Table 6.6 summarises some of the values given by Manchester[221]. As in the other examples mentioned in Chapter 2, the amino acid concentration depends on the activity of the Na extrusion process, and for that reason is reduced if the tissue is incubated without K salt in the saline mixture[144]. There is generally an optimal K concentration associated with maintenance of the Na excretion because excess K, with its associated high metabolic activity, leads to exhaustion of the tissue when in the isolated condition.

REFERENCES

1. CONWAY, E. J., 'Nature and significance of concentration relations of potassium and sodium ions in skeletal muscle', *Physiol. Rev.*, **37**, 84 (1957)
2. BOYLE, P. J., and CONWAY, E. J., 'Potassium accumulation in muscle and associated changes', *J. Physiol.*, **100**, 1 (1941)
3. CONWAY, E. J., and MOORE, P. T., 'Cation and anion permeability constants for the muscle fibre membrane', *Nature, Lond.*, **156**, 170 (1945)
4. GRAHAM, J., and GERARD, R. W., 'Membrane potentials and excitation of impaled single muscle fibres', *J. cell. comp. Physiol.*, **28**, 99 (1946)
5. ADRIAN, R. H., 'The effect of internal and external potassium concentration on the membrane potential of frog muscle', *J. Physiol.* **133**, 631 (1956)
6. HARRIS, E. J., and MARTINS-FERREIRA, I., 'Membrane potentials in the muscles of

the South African frog *Leptodactylus ocelatus*', *J. exp. Biology*, **32**, 539 (1955)

7a. HODGKIN, A. L. and HOROWICZ, P., 'The influence of potassium and chloride ions on the membrane potential of single muscle fibres', *J. Physiol.*, **148**, 127 (1959)

7b. HODGKIN, A. L., and HOROWICZ, P., 'Movements of Na and K in single muscle fibres', *J. Physiol.*, **145**, 405 (1969)

8. BERNSTEIN, J., 'Untersuchungen zur Thermodynamie der bioelectrischen Ströme. Erster Theil'. *Pflügers Arch. ges.` Physiol.*. **92**. 521 (1902)

9. KERNAN, R. P., 'Resting potentials in isolated frog sartorius fibres at low external potassium concentrations , *Nature, Lond.*, **185**, 471 (1960)

10. KERNAN, R. P., 'Resting potential of isolated rat muscles measured in plasma', *Nature, Lond.*, **200**, 474 (1963a)

11. VAUGHAN-WILLIAMS, E. M., 'The effect of changes in extracellular potassium concentration on the intracellular potentials of isolated rabbit atria', *J. Physiol.* **146**, 411 (1959).

12. CONWAY, E. J., 'Evidence for a redox pump in the active transport of cations', *Int. Rev. Cytol.*, **IV**, 377 (1955)

13. HOLMAN, M. E., 'Membrane potentials recorded with high-resistance microelectrodes and the effect of changes in ionic environment on the electrical and mechanical activity of the smooth muscle of the taeniae coli of the guinea pig', *J. Physiol.*, **141**, 464 (1958)

14. BÜLBRING, E., and KURIYAMA, H., 'Effects of changes in the external sodium and calcium concentrations on the spontaneous electrical activity in smooth muscle of guinea-pig taeniae coli', *J. Physiol.*, **166**, 29 (1963)

15. CASTEELS, R., 'Calculation of the membrane potential in smooth muscle cells of the guinea-pig taeniae coli by the Goldman equation', *J. Physiol.*, **205**, 193 (1969)

16. CASTEELS, R., 'The physiology of intenstinal smooth muscle', *Am. J. dig. Dis.*, **12**, 231 (1967)

17. HUTTER, O. F., and NOBLE, D., 'Rectifying properties of heart muscle', *Nature, Lond.*, **188**, 495 (1960)

18. FALK, G., and LANDA, J. F., 'Effects of potassium on frog skeletal muscle in a chloride deficient medium'. *Am. J. Physiol.*. **198**, 1225 (1960)

19a. HUTTER, O. F., and WARNER, A. E., 'The pH sensitivity of the chloride conductance of frog skeletal muscle', *J. Physiol.*. **189**, 403 (1967a)

19b. HUTTER, O. F., and WARNER, A. E.. 'The effect of the pH on the ^{36}Cl efflux from frog skeletal muscle', *J. Physiol.*, **189**, 427 (1967b)

20. CARMELIET. E., and BOSTEELS. S., 'Chloride efflux in cardiac Purkyne fibers and hydrogen ion concentration , *Arch. Int. Physiol. Biochim.*, **76**, 147 (1968)

21. MEYER, K. H., and SIEVERS. J. F., 'La perméabilité des membranes. II. Essaies avec des membranes selectives artificielles', *Helv. chim. Acta*, **19**, 665 (1936)

22. PASSOW, H., 'Zusammenwirken von Membranstruktur und Zellstoffwechsel bei der Regulierung der Ionenpermeability roter Blutkörperchen', in *Biochemie des aktiven Transports (Colloquium Ges. physiol. Chem.*, *Mosbach/Baden)*, Springer, Berlin (1964)

23. HAGIWARA, S., GRUENER, R., HAYASHI, H., SAKATA, H., and GRIMELL, A. D., 'Effect of external and internal pH changes on K and Cl conductances in the muscle fibre membrane of a giant barnacle', *J. gen. Physiol.*, **52**, 773 (1968).

24. HUTTER, O. F., and WARNER, A. E., 'Action of some foreign cations and anions on the chloride permeability of frog muscle , *J. Physiol.*, **189**, 445 (1967)

25. PADSHA, S. M., 'The influence of anions on the membrane resistance of skeletal muscle . *J. Physiol.*, **137**, 26P (1957)

26. HUTTER, O. F., and PADSHA, S. M., 'Effect of nitrate and other anions on the membrane resistance of frog skeletal muscle , *J. Physiol.*, **146**, 117 (1959)

261

27. EDWARDS, C., HARRIS, E. J., and NISHIE, S., 'The exchange of frog muscle Na⁺ and K⁺ in the presence of the anions Br⁻, NO₃⁻, I⁻ and CNS⁻', *J. Physiol.,* **135**, 560 (1957)
28. HARRIS, E. J., 'Anion interaction in frog muscle', *J. Physiol.,* **141**, 351 (1958)
29. HUTTER, O. F., and WARNER, A. E., 'The anion discrimination of the skeletal muscle membrane', *J. Physiol.,* **194**, 61P (1968)
30. SPURWAY, N. S., 'The site of anion interaction in frog muscle', *J. Physiol.,* **178**, 51P (1965)
31. SPURWAY, N. C., 'Effects of pH-variation upon the anion permeability of frog muscle', *J. Physiol.,* **181**, 51P (1965)
32. CARR, C. W., 'Studies on the binding of small ions in protein solutions with the use of membrane electrodes. III. The binding of chloride ions in solutions of various proteins', *Arch. Biochem. Biophys.,* **46**, 417 (1953)
33. EISENBERG, R. S., and GAGE, P. W., 'Ionic conductances of the surface and transverse tubular membranes of frog sartorius fibers', *J. gen. Physiol.,* **53**, 279 (1969)
34. FENN, W. O., and COBB, D. M., 'Electrolyte changes in cat muscle during stimulation', *Am. J. Physiol.,* **121**, 595 (1938)
35. HEPPEL, L. A., 'The diffusion of radioactive sodium into the muscles of potassium deprived rats', *Am. J. Physiol.,* **128**, 449 (1940)
36. MARGOSHES, M., 'An introduction to flame photometry', in *Physical Techniques in Biological Research* (ed. W. Nastuk), Academic, New York. vol. IV (1962)
37. WILDE, W. S., 'The chloride equilibrium in muscle', *Am. J. Physiol.,* **143**, 666 (1945)
38. BOYLE, P. J., CONWAY, E. J., KONE, F., and O'REILLY, H. L., 'Volume of interfibre spaces in frog muscle and the calculation of concentrations in the fibre water', *J. Physiol.,* **99**, 401 (1941)
39. COTLOVE, E., 'Mechanism and extent of distribution of inulin and sucrose in chloride space of tissues', *Am. J. Physiol.,* **176**, 396 (1954)
40. HARRIS, E. J., 'Distribution and movement of muscle chloride', *J. Physiol.,* **166**, 87 (1963)
41. DESMEDT, J. E., 'Electrical activity and intracellular sodium concentration in frog muscle', *J. Physiol.,* **121**, 191 (1953)
42. TASKER, P., SIMON, S. E., JOHNSTONE, B. M., SHANKLY, K. H., and SHAW, F. H., 'The dimensions of extracellular space in sartorius muscle', *J. gen. Physiol.,* **43**, 39 (1959)
43. LING, G. N., and KROMASH, M. H., 'The extracellular space of voluntary muscle tissues', *J. gen. Physiol.,* **50**, 677 (1967)
44. BIRKS, R. I., and DAVEY, D. F., 'Osmotic response demonstrating the extracellular character of the sarcoplasmic reticulum', *J. Physiol.,* **202**, 171 (1969)
45. DYDYNSKA, M., and WILKIE, D. R., 'The osmotic properties of striated muscle fibres in hypertonic solutions', *J. Physiol.,* **169**, 312 (1963)
46. PEACHEY, L. D., 'The sarcoplasmic reticulum and transverse tubules of the frog's sartorius', *J. Cell. Biol.,* **25**, 209 (1965)
47. YAMADA, K., 'A histological study of frog sartorius muscles stored in cold low potassium media', *Kumamoto Med. J.,* **20**, 56 (1967)
48. LEV, A. A., 'Determination of activity and activity coefficient of potassium and sodium ions in frog muscle fibres', *Nature. Lond.,* **201**, 1132 (1964)
49. EISENMAN, G., RUDIN, D., and CASBY, J., 'Glass electrode for measuring sodium ion', *Science,* **126**, 831 (1957)
50. STEINBACH, H. B., 'Sodium extrusion by the sartorius of *Rana Pipiens*', *J. gen. Physiol.,* **44**, 1131 (1961)
51. KEYNES, R. D., and SWAN, R. C., 'The effect of external sodium concentration on the sodium fluxes in frog skeletal muscle', *J. Physiol.,* **147**, 591 (1959)
52. CAREY. M. J., and CONWAY, E. J., 'Comparison of various media for immersing

frog sartorii at room temperature and evidence for the regional distribution of fibre sodium'. *J. Physiol.,* **125**. 232 (1954)

53. CAREY, M. J., CONWAY, E. J., and KERNAN, R. P., 'Secretion of sodium ions by the frogs sartorius', *J. Physiol.,* **148**, 51 (1959)

54. HARRIS, E. J., 'The transfer of Na and K between muscle and the surrounding medium. 2. The Na flux', *Trans. Faraday Soc.,* **46**, 872 (1950)

55. DRAPER, M. H., FRIEBEL, H., and KARZEL, K, 'The changes in ionic composition and resting and action potentials in frog sartorius muscle maintained *in vitro*', *J. Physiol.,* **168**, 1 (1963)

56. NASTUK, W. L., and HODGKIN, A. L., 'The electrical activity of single muscle fibres', *J. cell. comp. Physiol.,* **35**, 39 (1950)

57. CREESE, R., and NORTHOVER, J., 'Maintenance of isolated diaphragm with normal sodium content', *J. Physiol.,* **155**, 343 (1961)

58. CREESE, R., 'Sodium fluxes in diaphragm muscle and the effects of insulin and serum proteins', *J. Physiol.,* **197**, 255 (1968)

59. FENN, W. O., and COBB, D. M., 'Electrical changes in muscle during activity', *Am. J. Physiol.,* **115**, 345 (1936)

60. MILLER, H. C., and DARROW, D. C., 'Relation of muscle electrolyte to alterations in serum potassium and the toxic effects of injected KCl', *Am. J. Physiol.,* **130**, 747 (1940)

61. CONWAY, E. J., and HINGERTY, D., 'Relations between potassium and sodium levels in mammalian muscle and blood plasma', *Biochem. J.* **42**, 372–376 (1948)

62. COOKE, R. E., SEGAR, W. E., CREEK, D. B., COVILLE, F. E., and DARROW, D. C., 'The extra renal correction of alkalosis associated with potassium deficiency', *J. clin. Invest.,* **31**, 798 (1952)

63. ORLOFF, J., KENNEDY, T. J., and BERLINER, R. W., 'The effect of potassium in nephrectomised rats with hypokalemic alkalosis', *J. clin. Invest.,* **32**, 538 (1953)

64. HINGERTY, D., 'Changes in muscle composition during potassium depletion and restoration'. *Ir. J. med. Sci.,* **5**, 375–380 (1963)

65. CONWAY, E. J., KERNAN, R. P., and ZADUNAISKY, J. A., 'The sodium pump in skeletal muscle in relation to energy barriers'. *J. Physiol.,* **155**, 263–279 (1961)

66. DEE, E., and KERNAN, R. P., 'Energetics of sodium transport in *Rana pipiens*', *J. Physiol.,* **165**, 550 (1963)

67. FRAZIER, H. S., and KEYNES, R. D., 'The effect of metabolic inhibitors on the sodium influxes in sodium loaded frog sartorius'. *J. Physiol.,* **148**, 362 (1959)

68. CONWAY, D., HARRINGTON, M. G., and MULLANEY, M., 'The nature of sodium exchanges in isolated frog sartorii, *J. Physiol.,* **165**, 246 (1963)

69. CONWAY, E. J., 'Critical energy barriers in the excretion of sodium', *Nature, Lond.,* **187**, 396 (1960)

70. KERNAN, R. P., and CONWAY, E. J., 'Muscle fibre membrane potentials in relation to extreme potassium concentration', *Abstr. 3rd Int. Congr. Biochem.,* 83 (1955)

71. MULLINS, L. J., and AWAD, M. Z., 'The control of the membrane potential of muscle fibres by the sodium pump', *J. gen. Physiol.,* **48**, 761 (1965)

72. DANIEL, E. E., and ROBINSON, K., 'The secretion of sodium and uptake of potassium by isolated uterine segments made sodium rich', *J. Physiol.,* **154**, 421 (1960a)

73. DANIEL, E. E., and ROBINSON, K., 'The relation of sodium secretion to metabolism in isolated sodium rich uterine segments', *J. Physiol.,* **154**, 445 (1960)

74. PAGE, E., and STROM, S. R., 'Cat heart muscle *in vitro*. VIII. Active transport of sodium in papillary muscles', *J. gen. Physiol.,* **48**, 957 (1965)

75. USSING, H. H., 'Transport of ions across cellular membranes', *Physiol. Rev.,* **29**, 127 (1949)

76. HARRIS, E. J., 'The dependence of efflux of sodium from frog muscle on internal sodium and external potassium', *J. Physiol.,* **177**, 355 (1965)

77. MULLINS, L. J., and FRUMENTO, A. S., 'The concentration dependence of sodium efflux from muscle', *J. gen. Physiol.*, **46**, 629 (1963)
78. KEYNES, R. D., 'Some further observations on the sodium efflux in frog muscle', *J. Physiol.*, **178**, 305 (1965)
79. KEYNES, R. D., and STEINHARDT, R. A., 'The components of the sodium efflux in frog muscle', *J. Physiol.*, **198**, 581 (1968)
80. KERNAN, R. P., 'The electrical coupling of sodium and potassium movements across muscle fibre membrane', *Proc. R. Ir. Acad.*, **6413**, 401 (1966)
81. BEAUGÉ, L. A., and SJODIN, R. A., 'The dual effect of lithium ions on sodium efflux in skeletal muscle', *J. gen. Physiol.*, **52**, 408 (1968)
82. GARRAHAN, P. J., and GLYNN, I. M., 'The behaviour of the sodium pump in red cells in the absence of external potassium', *J. Physiol.*, **192**, 159 (1967)
83. ZADUNAISKY, J. A., 'The location of sodium in the transverse tubules of skeletal muscle', *J. Cell Biol.*, **31**, C11 (1966)
84. HOROWICZ, P., TAYLOR, J. W., and WAGGONER, D. M., 'Fractionation of sodium efflux in frog sartorius muscles by strophanthidin and removal of external sodium', *J. gen. Physiol.*, **55**, 401 (1970)
85. KEYNES, R. D., 'The ionic fluxes in frog muscle', *Proc. R. Soc.* (B) **142**, 359 (1954)
86. KERNAN, R. P., 'Membrane potential changes during sodium transport in frog sartorius muscle', *Nature, Lond.*, **193**, 986 (1962)
87. CROSS, S. B., KEYNES, R. D., and RYBOVA, R., 'The coupling of sodium efflux and potassium influx in frog muscle', *J. Physiol.*, **181**, 865 (1965)
88. HASHIMOTO, Y., 'The effects of metabolic inhibitors, pH and temperature on sodium extrusion and potassium uptake in sodium-loaded amphibian sartorius muscles', *Kumamoto Med. J.*, **18**, 11 (1965)
89. FRUMENTO, A. S., 'Sodium pump—Its electrical effect in skeletal muscle', *Science*, **147**, 1442 (1965)
90. HARRIS, E. J., and OCHS, S., 'Effects of sodium extrusion and local anaesthetics on muscle membrane resistance and potential', *J. Physiol.*, **187**, 5 (1966)
91. ADRIAN, R. H., and SLAYMAN, C. L., 'Membrane potential and conductance during the transport of sodium, potassium and rubidium in frog muscle', *J. Physiol.*, **184**, 970 (1966)
92. RITCHIE, J. M., and STRAUB, R., 'The hyperpolarisation which follows activity in mammalian non-myelinated nerve fibre', *J. Physiol.*, **130**, 80 (1957)
93. FENG, T. B., and LIU, Y. M., 'The concentration effect relationship in the depolarisation of amphibian nerve by potassium and other agents', *J. cell. comp. Physiol.*, **34**, 33 (1949)
94. SANDOW, A., and H. MANDEL, 'Effects of potassium and rubidium on the resting potential of muscle', *J. cell. comp. Physiol.*, **38**, 271 (1951)
95. GOLDMANN, D. E., 'Potential, impedance and rectification in membranes', *J. gen. Physiol.*, **27**, 37 (1943)
96. YAMADA, K., and YONEMURA, K., 'Electrical constants of sartorius muscle fibres loaded in sodium and deficient in potassium', *J. physiol. Soc. Japan.*, **29**, 268 (1967)
97. GEDULDIG, D., 'A ouabain sensitive membrane conductance', *J. Physiol.*, **194**, 521 (1968)
98. KERNAN, R. P., 'Membrane potential and chemical transmitter in active transport of ions by rat skeletal muscle', *J. gen. Physiol.*, **51**, 204 (1968)
99. LING, G. N., and WOODBURY, J. W., 'Effect of temperature on the membrane potential and membrane resistance of human red cells', *J. cell. comp. Physiol.*, **42**, 407 (1949)
100. SJODIN, R. A., and BEAUGÉ, L. A., 'Strophanthidin sensitive components of potassium and sodium movement in skeletal muscle as influenced by the internal sodium', *J. gen. Physiol.*, **52**, 389 (1968)

101. GLYNN, I. M., 'Activation of adenosine triphosphatase activity in a cell membrane by external potassium and internal sodium', *J. Physiol.*, **160**, 18P (1962)

102. WHITTAM, R., 'The assymetrical stimulation of a membrane adenosine triphosphatase in relation to active cation transport', *Biochem. J.*, **84**, 110 (1962)

103. CALDWELL, P. C., 'Factors governing movement and distribution of inorganic ions in nerve and muscle', *Physiol. Rev.*, **48**, 1 (1968)

104. CHANCE, B., and MELA, L., 'Energy linked changes of hydrogen ion concentration in submitochondrial particles', *J. biol. Chem.*, **242**, 830 (1967)

105. KERNAN, R. P., 'Spectroscopic studies of frog muscle during sodium uptake and excretion', *J. Physiol.*, **169**, 862 (1963b)

106. POTTER, V. R., and REIF, A. E., 'Inhibition of an electron transfer component by antimycin A', *J. biol. Chem.*, **194**, 287 (1952)

107. NASON, A., VASINGTON, F., and DONALDSON, K. O., 'Tocopherol as a component of cytochrome *c* reductase', *Abstr. IV int. Congr. Biochem.*, Vienna, 61 (1958)

108. SLATER, E. C., 'The constitution of the respiratory chain in animal tissues', *Adv. Enzymol.*, **20**, 147 (1958)

109. BITTAR, E. E., 'Effect of inhibitors and uncouplers on the sodium pump of *Maia* muscle fibre', *J. Physiol.*, **187**, 81 (1966)

110. CREESE, R., EL-SHAFIE, L., and VRBOVA, G., 'Sodium movements in denervated muscle and the effects of antimycin A', *J. Physiol.*, **197**, 279 (1968)

111. SLATER, E. C., 'A respiratory catalyst required for the reduction of cytochrome *c* by cytochrome *b*', *Biochem. J.*, **45**, 15 (1949)

112. LOOMIS, W. F., and LIPMANN, F., 'Reversible inhibition of the coupling between phosphorylation and oxidation', *J. biol. Chem.*, **173**, 802 (1948)

113. ABOOD, L. G., KOKETSU, K., and NODA, K., 'Effect of dinitrophenol on phosphorylation and bioelectric phenomena of excitable tissue', *Am. J. Physiol.*, **200**, 431 (1961)

114. HALJAMÄE, H., 'Potassium transport in single mammalian muscle cells during in vitro incubation', *Acta physiol. scand.*, **78**, 201 (1970)

115. MANERY, J. F., GOURLEY, D. R. H., and FISHER, K. C., 'The potassium uptake and rate of oxygen consumption of isolated frog skeletal muscle in the presence of insulin and lactate', *Can. J. Biochem. Physiol.*, **34**, 893 (1956)

116. SMILLIE, L. B., and MANERY, J. F., 'Effect of external potassium concentrations in insulin and lactate on frog muscle potassium and respiratory rate', *Am. J. Physiol.*, **205**, 67–77 (1960)

117. REGAN, T. J., FRANK, M. J., LEHAN, P. H., and HELLMANS, H. K., 'Relationship of insulin and strophantidin to myocardial metabolism and function', *Am. J. Physiol.*, **205**, 790 (1963)

118. ZIERLER, K. L., 'Increase in resting membrane potential of skeletal muscle produced by insulin', *Science*, **126**, 1067 (1957)

119. ZIERLER, K. L., 'Effect of insulin on membrane potential and potassium content of rat muscle', *Am. J. Physiol.*, **197**, 515 (1959)

120. ANDRES, R., BALTZAN, M. A., CADER, G., and ZIERLER, K, 'Effect of insulin on carbohydrate metabolism and on potassium in the forearm of man', *J. Clin. Invest.*, **41**, 108 (1962)

121. KERNAN, R. P., 'The role of lactate in the active excretion of sodium by frog muscle', *J. Physiol.*, **162**, 129 (1962)

122. VALLEE, B. L., 'Metal and enzyme: Correlation of composition function and structure', in *The Enzymes* (eds P. D. BOYER, H. LARDY, K. MYRBACK), Academic New York, vol. 3, pt B (1960)

123. HERRERA, F. C., WHITTEMBURY, G., and PLANCHART, A., 'Effect of insulin on short circuit current across isolated frog skin in the presence of calcium and magnesium', *Biochim. biophys Acta*, **66**, 170 (1963)

K*

124. HERRERA, F. C., 'Effect of insulin on short circuit current and sodium transport across toad urinary bladder', *Am. J. Physiol.*, **209**, 819 (1965)
125. CRABBE, J., 'Stimulation of active sodium transport by insulin', *Proc. Int. Symp. on Polypeptide and Protein Hormones*, Milan, 377 (1967)
126. USSING, H. H., and ZERAHN, K., 'Active transport of sodium as the source of electric current in the short-circuited isolated frog skin'. *Acta physiol. scand.*, **23**, 110 (1951)
127. DOCKRY, M., KERNAN, R. P., and TANGNEY, A., 'Active transport of sodium and potassium in mammalian skeletal muscle and its modification by nerve and by cholinergic and adrenergic agents', *J. Physiol.*, **186**, 187 (1966)
128. HOROWICZ, P., and GERBER, C. J., 'Effects of sodium azide on sodium fluxes in frog striated muscle', *J. gen. Physiol.*, **48**, 515 (1965)
129. KOKETSU, K., 'Bioelectric potential and cellular metabolism. Effect of 2,4-dinitrophenol on the membrane potential of muscle fibres', *Perspect. Biol. Med.*, **9**, 54 (1965)
130. HOPFER, V., LEHNINGER, A. L., and THEPSON, T. E., 'Protonic conductance across phospholipid bilayer membranes induced by uncoupling agents for oxidative phosphorylation', *Proc. natn. Acad. Sci., U.S.A.*, **59**, 484 (1969)
131. MITCHELL, P., 'Chemiosmotic coupling and photosynthetic phosphorylation', *Biol. Rev.*, **41**, 445 (1966)
132. DYDYNSKA, M., and HARRIS, E. J., 'Consumption of high energy phosphates during active sodium and potassium interchange in frog muscle', *J. Physiol.*, **182**, 92 (1966)
133. KERNAN, R. P., and DOCKRY, M., 'A role of nerve in the active transport of ions by rat muscle', *Abstr. 2nd Int. biophys. Congress*, Vienna, 299 (1966)
134. AXELSSON, J., and THESLEFF, S., 'A study of supersensitivity in denervated mammalian skeletal muscle', *J. Physiol.*, **147**, 178 (1959)
135. MILEDI, M., 'The acetylcholine sensitivity of frog muscle after complete and partial denervation', *J. Physiol.*, **151**, 1 (1960)
136. NICHOLLS, J. G., 'The electrical properties of denervated skeletal muscles', *J. Physiol.*, **131**, 1 (1956)
137. HUBBARD, S. J., 'The electrical constants and the component conductance of the frog skeletal muscle after denervation', *J. Physiol.*, **165**, 443 (1963)
138. HARRIS, E. J., and NICHOLLS, J. G., 'The effect of denervation on the rate of entry of potassium into frog muscle', *J. Physiol.*, **131**, 473 (1956)
139. KLAUS, W., LULLMAN, H., and MUSCHOLL, E., 'Der Kalium-Flux des normalen und denervierten Rattenzwerchfells', *Pflügers Arch. ges. Physiol.*, **271**, 761 (1960)
140. GUTH, L., ' "Trophic" influences of nerve on muscle', *Physiol. Rev.*, **48**, 645 (1968)
141. KOVACS, T., VISSY, A., and WENT, E., 'The effect of denervation on ion transport in tonic and tetanic skeletal muscles of the rat', *Acta physiol. Hung.*, **33**, 55–68 (1968)
142. KEYNES, R. D., and SWAN, R. C., 'The permeability of frog muscle to lithium ions', *J. Physiol.*, **147**, 626 (1959)
143. KIPNIS, D. M., and NOELL, M. W., 'Stimulation of amino acid transport by insulin in the isolated rat diaphragm', *Biochim. biophys. Acta*, **28**, 226 (1958)
144. HARRIS, E. J., and MANCHESTER, K. L., 'The effect of potassium as well as denervation on protein synthesis and the transport of amino acids in mouse', *Biochem. J.*, **101**, 135 (1966)
145. ROBINSON, J. W. L., 'Les relations des ions de sodium et l'absorption intestinale d'acides amines', *Biochem. biophys. Acta*, **126**, 61 (1966)
146. INUI, Y., and CHRISTENSEN, H. N., 'Discrimination of single transport systems: The Na^+ sensitive transport of neutral amino acids in the Ehrlich cell', *J. gen Physiol.*, **50**, 203 (1966)

147. DEL CASTILLO, J., and KATZ, B., 'Interaction at end-plate receptors between different choline derivatives', *Proc. R. Soc.* (B), **146** (1957)

148. KATZ, B., *Nerve, Muscle and Synapse*, McGraw-Hill, London (1966)

149. HOFFMAN, B. F., and SUCKLING, E. E., 'Cardiac cellular potentials Effect of vagal stimulation and acetylcholine', *Am. J. Physiol.*, **173**, 312 (1953)

150. HUTTER, O. F., and TRAUTWEIN, S., 'Vagal and sympathetic effects on the pacemaker fibres in the sinus venosus of the turtle', *J. gen. Physiol.*, **39**, 715 (1956)

151. HARRIS, E. J., and HUTTER, O. F., 'The action of acetylcholine on the movement of potassium ions in the sinus venosus of the heart', *J. Physiol.*, **133**, 58P (1956)

152. RAYNER, B., and WEATHERALL, M., 'Acetylcholine and potassium movements in rabbit auricles', *J. Physiol.*, **146**, 392 (1959)

153. KEYNES, R. D., 'Energy requirement for sodium ion excretion from a frog muscle', *Proc. R. Soc.* (B), **142**, 383 (1954)

154. CHANCE, B., 'Reversal of electron transfer in the respiratory chain', *J. gen. Physiol.*, **45**, 595P (1962)

155. HARRIS, E. J., 'The stoichiometry of sodium ion movement from frog muscle', *J. Physiol.*, **193**, 455 (1967)

156. PORTER, K. R., and PALADE, G. E., 'Studies on the endoplasmic reticulum. III. Its form and distribution in striated muscle cells', *J. biophys. biochem. Cytol.*, **3**, 269 (1957)

157. WEBER, A., HERZ, R., and REISS, I., 'The regulation of myofibrillar activity by calcium', *Proc. Roy. Soc.* (B) **160**, 489 (1964)

158. BRIGGS, F. N., and FUCHS, F., 'The biosynthesis of a muscle relaxing substance', *Biochem. biophys. Acta*, **42**, 519 (1960)

159. PARKER, C. J., and GERGELY, J., 'Soluble relaxing factor from muscle', *J. biol. Chem.* **235**, 3449 (1960)

160. BIANCHI, P., *Cell Calcium*, Butterworths, London (1968)

161. BOZLER, E., 'Distribution and exchange of calcium in connective tissue and smooth muscle', *Am. J. Physiol.*, **205**, 686 (1963)

162. LANGER, G. A., 'Ion flux in cardiac excitation and contraction and their relation to myocardial contractivity', *Physiol. Rev.*, **48**, 708 (1968)

163. HEILBRUN, L. V., and WIERCINSKI, F. J., 'The action of various cations on muscle protoplasm', *J. cell. comp. Physiol.*, **29**, 15 (1947)

164. CALDWELL, P. C., and WALSTER, G. E., 'Studies on the micro-injection of various substances into crab muscle fibres', *J. Physiol.*, **169**, 353 (1963)

165. NIEDERGERKE, R., 'Local muscular shortening by intracellularly applied calcium', *J. Physiol.*, **128**, 12P (1955)

166. PODOLSKY, R. J., 'Local activation of striated muscle fibrils', *Proc. XXII Int. Congr. Physiol. Sci.*, II abstr., 902 (1962)

167. GILBERT, D. L., and FENN, W. O., 'Calcium equilibrium in muscle', *J. gen. Physiol.*, **40**, 393 (1957)

168. HARRIS, E. J., 'The output of ^{45}Ca from frog muscle', *Biochim. biophys. Acta*, **23**, 80 (1957)

169. SHANES, A. M., 'Electrochemical aspects of physiological and pharmacological action in excitable cells. Part 1: The resting cell and its alteration by extrinsic factors', *Pharmac. Rev.*, **10**, 59 (1958)

170. COSMOS, E., and HARRIS, E. J., 'In vitro studies of the gain and exchange of calcium in frog skeletal muscle', *J. gen. Physiol.*, **44**, 1121 (1961)

171. COSMOS, E., 'Factors influencing movement of calcium in vertebrate striated muscle', *Am. J. Physiol.*, **195**, 705 (1958)

172. HENROTTE, J. G., COSMOS, E., and FENN, W. O., 'Calcium exchange in isolated turtle ventricle', *Am. J. Physiol.*, **199**, 779 (1960)

173. VAN BREEMEN, C., and DANIEL, E. E., 'Influence of high potassium depolarisation

and acetylcholine on calcium exchange in the rat uterus`, *J. gen. Physiol.*, **49**, 1299 (1966)

174. NIEDERGERKE, R., and HARRIS, E. J., 'Accumulation of calcium (or strontium) under conditions of increasing contractibility', *Nature, Lond.*, **179**, 1068 (1957)

175. VAN BREEMEN, C., DANIEL, E. E., and VAN BREEMEN, D., 'Calcium distribution and exchange in rat uterus', *J. gen. Physiol.*, **49**, 1265 (1966)

176. ORENT-KEILES, E., and MCCOLLUM, E. V., 'Potassium and animal nutrition', *J. biol. Chem.*, **140**, 337 (1941)

177. WHANG, R., MOROSI, H. J., RODGERS, D., and REYES, R., 'The influence of sustained magnesium deficiency on muscle potassium repletion', *J. Lab. clin. Med.*, **70**, 895 (1967)

178. RYAN, M. P., and HINGERTY, D., 'Effect of magnesium deficiency on restoration of potassium and sodium levels in potassium depleted muscle', *Ir. J. med. Sci.* (7th series), **2**, 137 (1969)

179. BIANCHI, C. P., and SHANES, A. M., 'Calcium influx in skeletal muscle at rest, during activity and during potassium contracture', *J. gen. Physiol.*, **42**, 803 (1959)

180. FRANK, G. B., 'Inward movement of calcium as a link between electrical and mechanical events in contraction', *Nature, Lond.*, **182**, 1800 (1958)

181. EDMAN, K. A. P., and SCHILD, H. O., 'The need for calcium in the contractile response induced by acetylcholine and potassium in the rat uterus', *J. Physiol.*, **161**, 424 (1962)

182. POTTER, J. M., and SPARROW, M. P., 'The relationship between the calcium content of depolarised mammalian smooth muscle and its contractility in response to acetylcholine', *Aust. J. exp. Biol. med. Sci.*, **46**, 435 (1968)

183. AXELSSON, J., and THESLEFF, S., 'Activation of the contractile mechanism in striated muscle', *Acta physiol. scand.*, **44**, 55 (1958)

184. BIANCHI, P., 'The effect of caffeine on radiocalcium movement in frog sartorius'. *J. gen. Physiol.*, **44**, 845 (1961)

185. BLUM, J. J., CREESE, R., JENDER, D. J., and SCHOLES, N. W., 'The mechanism of ryanodine on skeletal muscle', *J. Pharmac. exp. Ther.* **121**, 477 (1957)

186. ISAACSON, A., and SANDOW, A., 'Quinine and caffeine effects on ^{45}Ca movement in frog sartorius muscle', *J. gen. Physiol.*, **50**, 2109 (1967)

187. NIEDERGERKE, R., 'Movements of calcium in frog ventricles at rest and during contractures', *J. Physiol.*, **167**, 515 (1963)

188. NIEDERGERKE, R., 'Movements of calcium in beating ventricles of the frog heart'. *J. Physiol.*, **167**, 551 (1963)

189. REUTER, H., and SEITZ, N., 'The dependence of calcium efflux from cardiac muscle on temperature and external ion composition', *J. Physiol.*, **195**, 451 (1968)

190. NIEDERGERKE, R., PAGE, S., and TALBOT, S. M., 'Calcium fluxes in frog heart ventricles', *Pflügers Arch. ges Physiol.*, **306**, 357 (1969)

191. BEELER, G. W., and REUTER, H., 'Membrane calcium current in ventricular myocardial fibres', *J. Physiol.*, **207**, 191 (1970)

192. BRADING, A., BÜLBRING, E., and TOMITO, T., 'The effect of sodium and calcium on the action potential of the smooth muscle of the guinea-pig taenia coli'. *J. Physiol.*, **200**, 637 (1969)

193. HAGIWARA, S., and TAKAHASHI, K., 'Surface density of calcium ions and calcium spikes in the barnacle muscle fibre membrane', *J. gen. Physiol.*, **50**, 583 (1967)

194. PALMER, R. F., and POSEY, V. A., 'Ion effects on calcium accumulation by cardiac sarcoplasmic reticulum', *J. gen. Physiol.*, **50**, 2085 (1967)

195. VAN DER KLOOT, W., 'Calcium uptake by isolated sarcoplasmic reticulum treated with dithiothreitol'. *Science.* **164**, 1294 (1969)

196. KATZ, A. M., and REBKE, D. I., 'Quantitative aspects of dog cardiac microsomal calcium binding and calcium uptake', *Circulation Res.*, **21**, 153 (1967)
197. HASSELBACH, W., 'Adenosinetriphosphate driven active transport in the membranes of the sarcoplasmic reticulum', *Proc. Roy. Soc.* (B), **160**, 501 (1963)
198. ONISHI, T., and EBASHI, S., 'The velocity of calcium binding of isolated sarcoplasmic reticulum', *J. Biochem.*, **55**, 599 (1964)
199. EBASHI, S., and YAMANOUCHI, I., 'Calcium accumulation and adenosinetriphosphatase of the relaxing factor', *J. Biochem., Tokyo*, **55**, 504 (1964)
200. WEBER, A., HERZ, R., and REISS, I., 'Kinetics of calcium transport by isolated fragmented sarcoplasmic reticulum', *Biochem. Z.*, **345**, 329 (1966)
201. FAWCETT, D. W., 'The sarcoplasmic reticulum of skeletal and cardiac muscle', *Circulation*, **24**, 336 (1961)
202. PEACHEY, L. D., and PORTER, K. R., 'Intracellular impulse conduction in muscle cells', *Science*, **129**, 721 (1959)
203. INESI, G., EBASHI, S., and WATANABE, S., 'Preparation of vesicular relaxing factor from bovine heart tissue'. *Am. J. Physiol.*, **207**, 1339 (1964)
204. CARSTEN, M. E., 'Role of calcium binding by the sarcoplasmic reticulum in the contraction and relaxation of uterine smooth muscle', *J. gen. Physiol.*, **53**, 414 (1969)
205. OVERBECK, H. W., and CONRAD, L. L., 'Intracellular distribution of ^{45}Ca in arteries of normal and hypertensive dogs', *Proc. Soc. exp. Biol. Med.*, **127**, 565 (1968)
206. SOMLYO, A. P., and SOMLYO, A. V., 'Pharmacology of excitation-contraction coupling in vascular smooth muscle and in avian slow muscle', *Fedn Proc. Fedn Am. Socs exp. Biol.*, **28**, 1634 (1969)
207. NATORI, R., 'The property and contraction process of isolated myofibrils', *Jikeikai Med. J.*, **1**, 119 (1954)
208. CONSTANTIN, L. L., and PODOLSKY. R. J., 'Calcium localisation and the activation of striated muscle fibres', *Fedn Proc. Fedn Am. Socs exp. Biol.*, **24**, 1141 (1965)
209. LEE, K. S., LADINSKY, H., CHOI, S. J., and KASUYA, J., 'Studies on the *in vitro* interaction of electrical stimulation and calcium movement in sarcoplasmic reticulum', *J. gen. Physiol.*, **49**, 689 (1966)
210. MARTONOSI, A., 'The activating effect of phospholipids on the adenosinetriphosphatase activity and calcium transport of fragmented reticulum', *Biochem. biophys. Res. Commun.* **13**. 273 (1963)
211. MARTONOSI, A., 'The role of phospholipids in the ATP-ase activity of skeletal muscle microsomes', *Biochem. biophys. Res. Commun.*, **29**, 753 (1967)
212. BITTAR, E. E., *Cell pH*, Butterworths, London (1964)
213. WADDELL, W. J., and BATES, R. G., 'Intracellular pH', *Physiol. Rev.*, **49**, 285 (1969)
214. CALDWELL, P. C., 'Studies on the internal pH of large muscle and nerve fibres', *J. Physiol.*, **142**, 22 (1958)
215. KOSTYUK, P. G., and SOROKINA, Z. A., 'On the mechanism of hydrogen ion distribution between cell protoplasm and the medium', in *Membrane Transport and Metabolism* (eds. A. KLEINZELLER and A. KOTYK), Academic, New York, 193 (1960)
216. CARTER, N. W., RECTOR, F. C., CAMPION, D. S., and SELDIN, D. W., 'Measurement of intracellular pH of skeletal muscle with pH-sensitive glass microelectrodes', *J. clin. Invest.*, **46**, 920 (1967)
217. ECKEL, R. E., POPE, C. E., and NORRIS, J. E. C., 'Lysine as a muscle cation in potassium deficiency', *Arch. Biochem. Biophys.*, **52**, 293 (1954)
218. IACOBELLIS, M. E., MUNTWYLER, E., and DODGEN, C. L., 'Free amino acid patterns of certain tissues from potassium and/or protein deficient rats', *Am. J. Physiol.*, **185**. 275 (1956)
219. NARAHARA, H. T., and OZAND, P., 'Studies on tissue permeability. IX. Effect of

insulin on the penetration of 3 methyl-glucose ^{3}H into frog muscle', *J. biol. Chem.*, **238**, 40 (1963)

220. MORGAN, H. E., REGEN, D. M., and PARK, C. R., 'Identification of a mobile corner-mediated transport system in muscle', *J. biol. Chem.*, **239**, 369 (1964)

221. MANCHESTER, K. L., 'The control by insulin of amino acid accumulation in muscle', *Biochem. J.* ,**117**, 457 (1970)

222. CONNELLY, C. M., 'Recovery processes and metabolism of nerve', *Rev. mod. Phys.*, **31**, 475 (1959)

7

THE ASCITES TUMOUR CELL

H. G. HEMPLING
Medical University of South Carolina, Charleston, U.S.A.

INTRODUCTION: MEMBRANE FUNCTION AND CELL GROWTH

The ascites tumour cell is a cell whose function is given over almost completely to growth and division, which can lead to a dramatic increase in the total mass of cellular material in the host animal. For example, a host mouse weighing 25 g can yield after 10 days, 20 g of ascitic fluid containing up to 4 g of tumour cells. The host then can produce 16% of its body weight in new living tissue from a starting inoculum of 0·2 ml injected intraperitoneally.

It should be appropriate, then, to ask what effect, if any, such anabolic demands have on the membrane permeability of the tumour cell. Since the cells of each new generation are like the cells of the preceding generation, each cell must incorporate into its substance the complete complement of water, electrolytes, and protein within the average generation time of 24 hours.

Table 7.1 contains a summary of some of the nutritive requirements for the cell and the extent to which these demands are met. The data are expressed as the fluxes required to maintain normal cell composition and the actual measured fluxes under reasonably physiological conditions. The margin of safety is the ratio of actual available flux to required flux. When calculating the demands for amino acids, the assumption has been made that all protein nitrogen is derived from free amino acids that have entered from the medium.

The general conclusion is that only amino acid transport and phosphate transport approach anything like a critical margin of safety, where small changes in transport could significantly influence the economy of the cell. It is no wonder, then, that Christensen[1] in his early studies pointed to the ascites tumour cell as one of the most vigorous accumulators of amino acids of any mammalian cell

Table 7.1 MEMBRANE FUNCTION AND TUMOUR ECONOMY

Component	Content per 10^7 cells	Net influx required	Fluxes measured	Margin of safety
Water	15×10^{-3} ml	7.5×10^{-4} ml per 10^7 cells per h	12·7 ml per 10^7 cells per h	17000
Potassium	2·5 μmol	0·12 μmol per 10^7 cells per h	2 to 3 μmol per 10^7 cells per h	20 to 30
Sodium	0·3 μmol	0·015 μmol per 10^7 cells per h	Fast: 6 μmol per 10^7 cells per h Slow: 0·35 μmol per 10^7 cells per h	20 to 400
Calcium	0·024 μmol	1×10^{-3} μmol per 10^7 cells per h	Fast: 0·13 μmol Slow: 3×10^{-3} μmol per 10^7 cells per h	3 to 100
Phosphorus	2 μmol	0·08 μmol per 10^7 cells per h	0·2 μmol per 10^7 cells per h	2·5
Nitrogen	46 μmol	2·3 μmol per 10^7 cells per h	5 μmol per 10^7 cells per h*	2

* Assumption that source of nitrogen derives from amino acid transport.

then studied. As for phosphate transport, I think we have here an example of a component of the cell that was always thought to be readily available to sites of energy turnover. However, Levinson's studies[2] would indicate that a key rate-limiting step exists at the membrane and that sites of phosphate metabolism important to the cell's economy may be present at the outermost margin of the cell. This possibility has been ignored all too often by investigators of systems of cell extracts.

However, a simple analysis such as that in Table 7.1 does not mean that we should neglect the importance of membrane permeability to other components, particularly electrolytes. As will be emphasised, the ascites tumour cell demonstrates coupling between amino acid transport and Na transport, although the details are still quite vague, and it is quite possible that portions of Na permeability may be critical because of this coupling and the margin of safety for this aspect of Na permeability may be far smaller than indicated.

Apart from the need to supply bulk amounts of constituents during cell growth, has the tumour cell any specific need that would make membrane permeability critical in the regulation of the economy of the cell? Lubin[3] has described how suspensions of Sarcoma 180 cells lost their ability to synthesise protein and DNA if their intracellular K was replaced with Na from the medium. Therefore, maintenance of the proper intracellular K concentration or the K–Na ratio may be of significance in protein synthesis.

In addition, Jung and Rothstein[4] could show that when mouse leukaemic lymphoblasts were grown synchronously, their cation content fluctuated during a single cell-cycle. For example, net loss of K occurred about $1\frac{1}{2}$ h after the period of rapid cell division, concomitant with a doubling of both unidirectional fluxes. Over the next five to six hours, the deficit in K was eliminated, while the unidirectional fluxes remained high until after the next cell division. The pattern of electrolyte regulation during cell division did not follow a simple function such as cell number, cell surface, or cell volume, but was proposed to be related to specific internal events in the cell.

ELECTROLYTE PATTERN IN THE MOUSE ASCITES TUMOUR CELL

Not all the observations on ion fluxes across the membrane of the ascites tumour cell made during the 1950s and early 1960s fitted the dogmas set by the holy trinity of muscle, nerve, and red cell. But, as membrane physiology moved out of the descriptive phase and investigators using different cell systems converged on mechanism, they found themselves faced with problems of compartmentalisation, coupling of transport systems, control of transport by metabolism, and its fascinating converse, control of metabolism by transport. Perhaps because of its precocity, the tumour cell met these problems much sooner than other cells and tissues, so that its electrolyte metabolism seemed to be the exception rather than the rule. Therefore, I propose to present certain features of ion transport in this type of cell that should not be neglected in any attempt to develop an overall scheme of transport for this cell or any others.

SODIUM

Let us look at the steady-state electrolyte pattern for these cells. Since Christensen and his associates[5] and more recently Eddy[6-9] have emphasised that concentration gradients of Na may provide the driving force for amino acid transport, it has been somewhat disconcerting that Aull and Hempling[10] described two compartments of Na in their populations of tumour cells. The internal content of Na was 25 mequiv per l cell water, and of this 65% was in a fast-exchanging compartment with a turnover of 20 per hour (about 12 pmol cm^{-2} s^{-1}) and 35% was in a slow compartment with a turnover of 3 to 4 per hour. Computer fits of the data based either on a parallel or a series model were equally good.

This fact raises problems of interpretation. If we choose a series model, then the two compartments would be cell compartments, such as outer cortical layer and inner cytoplasmic layer, or cytoplasm and nucleus, or cytoplasm and mitochondria. If the two compartments are in parallel, then they may be separate transport compartments in a single cell or two different cell populations. At the present time, this problem has not been handled adequately in any theories of coupling phenomena with amino acids, and yet it is particularly critical to a gradient hypothesis because any value to be assigned to the internal Na concentration has to apply to the same compartment as that for the amino acid, which occupies only one cell compartment.

CHLORIDE

Hempling[11], Hempling and Kromphardt[12], Kromphardt[13], and Aull[14] reported that the chloride content of ascites tumour cell was between 70 and 80 mequiv per kg cell water and about 60 to 70% was exchangeable with a turnover of 3 per hour. In other dimensions, the permeability of Cl was 1.3×10^{-7} cm s^{-1} at 25 °C. Aull in turn measured the p.d. across the membrane of the tumour cell and found an average value of 11.2 mV, inside ($-$). The p.d. was not eliminated with high K; NO_3 could replace chloride, but in SO_4 the potential could be reversed. What are the implications if only the exchangeable chloride contributed to the membrane p.d.? From the data of Aull, a Nernst p.d. would predict 33.5 mV, inside ($-$). It was thought that Na might act as a shunting cation to reduce the Nernst potential to the measured value. Aull calculated that sufficient Na flux was available from the fast compartment to provide the shunt. Kromphardt used a different strain of tumour cell and worked with cells at 37 °C. Her chloride distribution ratios predicted a p.d. of 11.6 mV inside ($-$) but no measurements of p.d. were made. However, she made the interesting observation that 2,4-dinitrophenol at 10^{-3} M inhibited both unidirectional fluxes of the chloride ion.

In some very recent work, Lassen and his associates have simplified the problem of chloride binding. They attribute the reported nonexchangeable component to an artifact in the measurement of total chloride content from aqueous extracts. It is their claim that the Cotlove titrimetric method, or any method that uses $AgNO_3$, will measure an excess of Cl because residual protein in the titration mixture will adsorb Ag and give a falsely high reading. Therefore, the measurements made by Hempling, Kromphardt, and Aull of chloride content in their extracts were too high exactly by the amount of the nonexchangeable component, so that the isotope distribution reflects 100% exchangeability. Since both groups of investigators agreed on the content of exchangeable Cl, there is no residual disagreement on the value of the chloride fluxes or the chloride concentration that must be used in the calculation of Nernst potential differences. Levinson[15] recently measured the chloride content in Ehrlich Lettré diploid cells extracted with perchloric acid, and estimated that only 9% of the total cell chloride remained after exposure to a chloride-free medium. The ratio of internal to external chloride was 0.365 at 37 °C, implying a p.d. of 26 mV inside ($-$). Lassen[16] has made some new measurements of the p.d. across the ascites tumour cell and has found a value of 30 to 40 mV inside ($-$), which was maintained for approximately 40 ms and then decayed to an average value of 11 mV ($-$). It is his view then that the true p.d. is due to the distribution of Cl and no shunting hypothesis is necessary to explain the data.

POTASSIUM

The average potassium content in the ascites tumour cell varies between 125 and 150 mequiv per kg cell water. From measurements of the initial influx during the first two minutes after exposure to K^{42}, one can detect a small, rapidly exchanging compartment less than 10% of the total exchangeable potassium[11]. The remainder of the potassium pool exhibited conventional two-compartment kinetics between the external medium and a well-mixed internal cell compartment. When this steady-state flux was studied over a range of external K concentrations, it was found that the fluxes *decreased* with increase in external K, passed through a minimum between 15 and 25 mequiv l^{-1} and then increased with increase in external potassium[11]. On the other hand, the *initial* influx increased with increase in external K according to saturation kinetics, with a K_m of 4 mM at 25 °C. Heinz and his associates[17] measured a K_m of 2 mM at 37 °C in their measurements of initial influx.

One hypothesis to explain the phenomenon is based upon a model of fixed charges. Weiss's work on the electrophoretic mobility of the ascites cell[18] supports the view that negative fixed charges predominate at the surface of the cell and form a Helmholtz layer at least 100 nm in thickness, and Eylar and Wallach[19] have correlated it with the presence of sialic acid in the outer coat. This layer of negative fixed charges in the outer cortex of the cell may provide the small compartment of rapidly-exchanging K we measure. When the concentration of K in the medium is increased, more negative charges may be occupied by the cation, which may then act as an impediment to like-charged K ions on their way to a carrier mechanism. Such a model would explain why cells *pre-equilibrated* with increasing concentrations of K *have lesser steady-state fluxes*. When the concentration has been increased enough to force K past the barrier by mass action and to reach the carrier system, fluxes would increase again and provide an explanation for the minimum observed at concentrations between 15 and 25 mequiv l^{-1}.

CALCIUM

Levinson and Blumenson[20] have made a kinetic analysis of calcium fluxes in Ehrlich–Lettré diploid ($n = 41$) tumour cells. Exchangeable calcium represented 11·9% of the total, or about 2·8 nmol per 10^7 cells. The exchangeable calcium in turn could be analysed into approximately two equal compartments in series. Fluxes between environment and the first compartment averaged 133 nmol per

10^7 cells per hour, while the mean flux between the middle and end compartments of the series was only 2·9 nmol per 10^7 cells per hour. Expressed on the basis of cell water, total cell Ca^{2+} was 2·2 mM, and the exchangeable compartments were each 0·14 mM respectively.

The distribution of phosphate across the ascites tumour cell membrane depends upon the species of anion. Levinson has shown that the divalent species, HPO_4^{2-} fits a Donnan distribution, as does Cl^-, but that the monovalent species, $H_2PO_4^-$, is out of electrochemical equilibrium and may be the species transported through a mediated pathway.

TRANSITION EXPERIMENTS WITH THE ASCITES TUMOUR CELL

A transition experiment is one in which metabolism is depressed by cooling the cells, so allowing them to lose K and gain Na, and then transferring them to room temperature or to 37 °C to initiate the reaccumulation of K and extrusion of Na. Historically, in the 1940s and 1950s, this was the classical experiment used by Conway, Kernan, Steinbach, and Denstedt in muscle[21] and Danowski and Ponder in erythrocytes to initiate net fluxes and to study the role of metabolism on these active movements. Perhaps because they grew weary of waiting 5 to 10 days until the erythrocytes lost enough K and gained enough Na to do a meaningful transition experiment, the red cell people have gone over to unidirectional fluxes and reverse haemolysis experiments to elucidate mechanisms of transport[22]. There was a revival of interest in transition experiments in muscle when Kernan, and then Keynes, and Mullins and his associates, used it to provide evidence for an electrogenic pump[23-27].

In contrast to erythrocytes and muscle, it is possible to alter the electrolyte content of the ascites tumour cell quite readily at low temperature. It is equally easy during a transition experiment to determine both isotope and chemical contents of the cell for K and Na and to calculate the apparent fluxes of these two ions. My choice of 'apparent' recognises the fact that when net fluxes occur in a multicompartment system, one does not know how much each compartment has contributed to the net flux without direct access to it. For K, this is not a serious problem, since between 90 and 100% of the flux behaves as if it were an exchange between two well-mixed compartments. The problem is more acute with Na, where we have already seen the evidence for a three compartment system.

Two important observations came out of the early work on transition experiments[28]. The first was the observation that the net fluxes obtained were related to the energy demands required for pumping. The longer the stay at low temperatures (0 to 4 °C) the more K left the cell and the more Na entered. Therefore, the smaller the gradient against which pumping occurred the greater were the net fluxes under these conditions. It was as if a packet of energy had been reserved for active transport and one could move more ions if the gradients were not large.

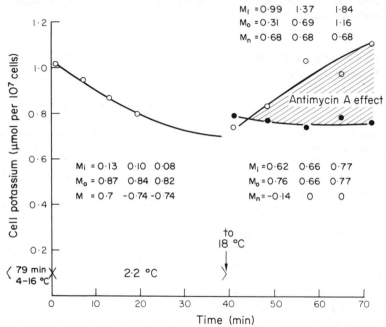

Figure 7.1. Illustration of a transition experiment. M_i, *unidirectional influx,* M_o, *unidirectional efflux, M,* M_n, *net flux. Each triad of fluxes (*M_i*,* M_o*,* M_n*) is shown above or below the point on the curve when fluxes were calculated.* ○, *control.* ●, *antimycin A*

The second observation had to do with the coupling between Na and K. Even in those early observations, there was no evidence of obligatory coupling between the two ions. In transition experiments, coupling of net fluxes between Na and K has always been variable, and, as we will note below, we can arrange the conditions such as to extrude Na against an electrochemical gradient simultaneous with the net loss of K down its electrochemical gradient. In a sense, the ascites tumour cell was ahead of its time in its demonstration of variable coupling between ions, for this was the period where dogma from muscle and red cell required tight and stoichiometric coupling. Current models of transport now seek to emphasise the flexibility in coupling between fluxes.

With the ascites tumour cell, we can calculate unidirectional fluxes at the same time that we measure net fluxes. Therefore, we can determine what fraction of the unidirectional flux is net flux, and in so doing follow the coupling between exchange flux and the active component which results in net gain, for example of potassium. Figure 7.1 illustrates several of the points I have been making. On the ordinate is cell content of potassium, on the abscissa is time, after transition from 4 °C to 18 °C being indicated by the arrow. Note that in a period of two hours the cells have dropped to 25% of the normal K content. This net loss is the result of a depression of the flux. When transition occurs, there is a dramatic increase in unidirectional fluxes of K and coupled to them is the initiation of a net gain of K. Antimycin A, which is an effective block to the electron transport system and therefore to oxidative phosphorylation, depresses the level of ATP in the cell. In response to the inhibition of respiration and the absence of ATP, coupling of the exchange system to the active reaccumulation of K is eliminated, no net gain of K occurs, but the exchange system itself is still present.

AMINO ACID TRANSPORT

Any discussion of amino acid transport across the membrane of the ascites tumour cell has to deal with the following participants:

1. the amino acid and its particular carrier system;
2. sodium concentrations in the cell and in the medium;
3. sodium fluxes and their associated pathways,
4. potassium fluxes and their associated pathways; and
5. the direct and indirect energy sources for steady-state amino acid transport.

It is beyond the scope of this chapter to deal with any one of these items in great detail and only the highlights will be noted.

The concentration dependence of the initial flux or uptake of neutral amino acids in the ascites tumour cells is fairly well described by the equation

$$J = \frac{J_{max}[C]_e}{K_m + [C]_e} + k\{[C]_e - [C]_i\}$$

in its simplest form, although Jacquez[29, 30] has proposed a more general description. The first term in the equation defines classical saturation kinetics, while the second term may be thought of as diffusion through passive channels, although one cannot discount a saturation system of high K_m. Christensen, Oxender, and their many associates have presented evidence for the existence of several distinct mediating systems for the transport of neutral amino acids. Through the use of competition and inhibition experiments, they have selected natural amino acids or synthesised special ones that served to delineate one or the other mediated pathways. Christensen has summarised the characteristics of these pathways[31] and they are reproduced with his permission.

THE A SYSTEM

Substrates: all neutral amino acids; characterising amino acids are α-amino-isobutyric acid (AIB) and its methylated analogue (MeAIB); it serves as a primary pathway for the transport of glycine, proline, sarcosine, serine, methionine, and norleucine; poorly transported by this pathway are valine and other branched-chain, apolar amino acids.

Kinetic behaviour: V_{max} rather constant for all but a few substrates; dependence on Na is first-order, with one Na presumed to migrate per amino acid molecule; the pH sensitivity is scarcely detectable at pH 5; there are weak exchange properties in *cis–trans* homologue and heteroexchange.

THE L SYSTEM

Substrates: most neutral amino acids, with reactivity in proportion to hydrocarbon mass of side chain. Excluded are AIB, N-methylated amino acids. Characteristic substrate: BCH (2-aminobicylo-(2.2.1)-hexane-2-carboxylic acid).

Kinetic behaviour: V_{max} is characteristically variable; no dependence on Na; pH sensitivity is minimal; exchanging properties very strong so that net operation shown only with special care.

THE ASC SYSTEM

Substrates: characterised by 3- and 4-carbon aliphatic and hydroxy-aliphatic amino acids, proline, cysteine. No inhibition detected by MeAIB and therefore used to discriminate from A system.

Kinetic behaviour: V_{max} characteristically variable, with several amino acids inhibiting without migrating detectably by this pathway; first-order dependence on Na; exceptional stereospecificity; exchange properties are weak.

Jacquez[32] has questioned whether one need assume separate systems to transport different species of amino acid or whether one could think of the distinction of the *A* and *L* systems, for example, as a distinction between transport and exchange. Jacquez cites as the most striking evidence for the existence of two distinct systems the demonstration by Inui and Christensen[33] that the Na-dependent portion of the initial influx of methionine was approximately equal to the portion of the initial flux inhibitable by AIB. Yet Johnstone and Scholefield[34] had shown that methionine *homoexchange* is independent of Na, but that approximately half of the *transport flux* is Na-dependent. Further, Belkhode and Scholefield[35] have shown that heteroexchange between 1-methionine and 1-amino-cyclopentane-carboxylic acid is independent of Na concentration. Note that Christensen himself conceives of the *A* system as one for net transport and the *L* system as one for exchange. At this stage of study, it is perhaps too soon to distinguish between two carriers or two forms of one carrier.

In a series of papers, Eddy[6-9] proposed that glycine uptake in his strain of mouse ascites tumour cell (strain LS, lymphosarcoma) depended on the gradient of Na across the cell membrane and was mediated by a carrier with two sites, one for glycine and one for Na. Basing his kinetic analysis primarily on measurements of the one-minute influx of glycine, as a function of extra and intracellular Na and K, he proposed that the carrier had to be in the form ENaGly to allow glycine uptake. Transfer coefficients for EGly or ENa were considered to be negligibly small. Further, he proposed that K could not substitute for Na, but instead inhibited the uptake of glycine dependent on Na. It is important to note that the cells contained only 2% of their normal ATP content under the conditions of these experiments, since the author starved the cells and then added cyanide and/or 2-deoxyglucose as metabolic inhibitors prior to measurements of glycine uptake. Under such conditions, he showed an inverse, linear relation between the log of the sodium distribution ratio and the log of the glycine distribution ratio, such that the direction of the sodium gradient determined the direction of net glycine transport and was independent of energy supply.

Although much has been made of the coupling between Na and amino acids, in actual fact almost all careful measurements have dealt with the response of amino acids to changes in Na concentration, and I have found no concise or complete kinetic analysis of Na transport in the ascites tumour cell under these conditions. Much of the difficulty lies in the multicompartmental distribution of Na and its very rapid isotopic equilibration. The key question requiring an answer is: How do we apportion Na fluxes into amino-acid-dependent and -independent components, when the cells are in the steady state and have developed an asymmetric distribution of amino acids?

Eddy's approach, which is to use metabolic inhibitors to unmask the amino acid transport dependent on the Na gradient, does not assist in the analysis of the response of Na transport to amino acid transport. Aull and I[36] have shown that metabolic inhibitors will expand the Na compartments in our strain A of Ehrlich mouse ascites tumour cell, accelerate passive exchange fluxes, and inhibit several of the active fluxes. To demonstrate coupling of Na influx to amino acid influx on such a high background of unidirectional fluxes will be no easy task. Eddy's correlation between the log of the distribution ratio of Na and that of glycine is a valuable observation for the gradient hypothesis, but it offers no clue to the site of coupling or the point of control of amino acid transport. We must still learn whether we can control amino acid transport at the entry-point, coupling through Na influx, or at the exit point, coupling through Na efflux.

An analysis of the response of K transport to amino acid transport has been more informative, for most of K behaves like a two-compartment system and the kinetics are easier to interpret. Hare and Hempling[37] ruled out the potassium gradient as a source of free energy for amino acid transport, but went on to show that glycine could alter the pattern of potassium transport[38]. When the ascites tumour cell in the steady state was exposed to K[42] in the presence of glycine, unidirectional fluxes of the ion increased by a small but significant amount. Pyridoxal effectively blocked the stimulatory action of glycine. However, in a transition experiment, when a net gain of potassium occurred, glycine and pyridoxal, together or alone, reduced influx, efflux, and net flux of the ion, with pyridoxal being more effective.

At one time, Christensen had looked on pyridoxal as a potential carrier[39, 40] when he observed that 1 mM pyridoxal increased the distribution ratio of glycine between cell and environment by 75%, but he abandoned the view when both he[41] and Bittner and Heinz[42] observed that pyridoxal inhibited the efflux of amino acids and not the influx. Nevertheless, the significance of pyridoxal in the analysis of coupling phenomena lies in its similar mode of action on potassium and on amino acids.

The ability to concentrate amino acids is not confined to naturally occurring substances but the D-isomers move more slowly (about $\frac{1}{30}$ of the rate) towards the final concentration than the natural L-isomers. Values of the concentration ratios attained, taken mostly from the work of Christensen and his associates, are given in Table 7.2.

Table 7.2 CONCENTRATION RATIOS ATTAINED BY VARIOUS AMINO ACIDS BETWEEN THE CELL WATER AND THE SUSPENSION MEDIUM, USING ASCITES CELLS

Amino acid	Final distribution ratio (concn. in cell/concn. outside cell)
glycine	25
L-alanine	14
L-methionine	13
L-leucine	3·5
L-valine	4·5
D-alanine	11
D-valine	4
L-tert-leucine	5
D-tert-leucine	3·5
α-aminoisobutyric acid	30

COUPLING BETWEEN K TRANSPORT AND AMINO ACIDS: SOME SPECULATIONS

To bring together these several observations on potassium transport, glycine, and the role of pyridoxal, we may refer to Figure 7.2. I would consider that the tumour cell has two transport pathways for potassium:

1. An active influx balanced by a passive unidirectional efflux.
2. An exchange system. Figure 7.1 had shown an example of how low temperature reduced the unidirectional influx dramatically but affected the unidirectional efflux by only 20%.

One may conclude that exchange diffusion does not contribute much to steady-state fluxes at 25 °C, because otherwise one would expect that both influx and efflux would decrease to the same extent when the temperature was lowered. Further, Weinstein and Hempling[43] and Hempling[44] had demonstrated how important ATP was to the steady-state unidirectional influx of K. These results would imply, therefore, that influx was predominantly active and ATP-dependent at 25 °C.

Evidence for an exchange system for potassium transport was illustrated in Figure 7.1 during the transition portion of the experiment. In all cases, the unidirectional fluxes were considerably greater than the net flux. Further, it is a general observation in transition experiments that during the initial phases of the transition, unidirectional fluxes exceed overall steady-state fluxes and then taper off toward steady-state values, as net fluxes approach zero. Taken together, a considerable fraction of the transfer of potassium during a transition occurs initially without accumulation and is independent of ATP, but with time more of the flux is associated with net accumulation.

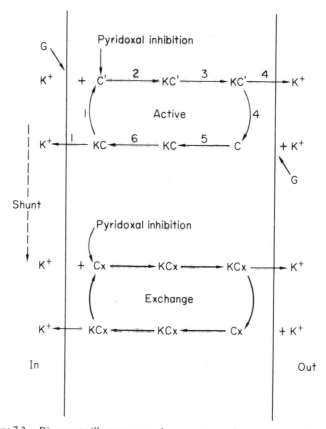

Figure 7.2. Diagram to illustrate an exchange system and a pump system for potassium and possible sites of coupling to amino acid transport. See text for details

I would propose that at low temperatures components of the exchange system (Cx) are being produced at the expense of the components that mediate net gain of potassium (C). When the temperature is raised, both the exchange system and the pump system show themselves. Initially, most of the carrier system is in the form of exchange carrier (Cx), but with time pump carrier is reformed (C) and a greater portion of the flux is associated with net accumulation, until a major fraction of the carrier is pump-carrier-coupled to a leak pathway.

To explain the behaviour of glycine on potassium fluxes, I suggest that glycine may participate in the carrier system, mediating the accumulation of potassium (C), since in a transition experiment, the amino acid inhibited influx and net flux of potassium. On the other hand, in the steady state, glycine can stimulate potassium fluxes[37]. This observation would imply that potassium ions that have been displaced by glycine from the active system would be available to the exchange system, which, although only a small proportion of the total carrier, turns over more rapidly and is now made evident by the potassium ions displaced by the glycine. The ability of pyridoxal to reduce efflux and influx of potassium during a transition experiment or in the steady state is consistent with its action to reduce the efflux of amino acids reported by Bittner and Heinz[42] and by Christensen and his associates[41].

In summary, we have two transport systems, one for the net accumulation of potassium, and one for its rapid exchange, which share several features with amino acids. Could the pump system (carrier C) be Christensen and Oxender's A system, coupled to Na transport, while carrier Cx is Christensen's L system operating as an exchange system?

COUPLING OF METABOLISM AND POTASSIUM TRANSPORT

Two popular hypotheses have served to explain the linkage between metabolism and the transport of electrolytes across cell membranes. In one, an ATPase, sensitive to sodium and potassium, converts the free energy of ATP into the directed movements of sodium and potassium. In the other, the free energy of ATP may be diverted into directional movements of electrons along a redox chain coupled to the movement of ions.

In ascites tumour cells, the ATP content must still be at normal levels to maintain ion fluxes and content[44, 43]. However, with the metabolic inhibitor oligomycin, fluxes could be reduced and the ion content of the cell altered even though ATP levels were normal. The results would indicate that some high-energy intermediate step, sensitive to oligomycin, exists between the energy of ATP and ion transport. The intermediate may be a step in oxidative phosphorylation, occurring at a mitochondrial site or part of a complex in the cell membrane.

Levinson and Hempling[45] have demonstrated that external potassium can trigger the breakdown of this high-energy intermediate and produce a stimulation of respiration comparable to that with 2,4-dinitrophenol. For example, cell suspensions transitioned from 2 °C to 23 °C into a K-free medium respire at rates only one sixth the rate of cells normally maintained at 23 °C with 6 mmol K in the medium. Addition of K to the external medium stimulates respiration. The relation between respiration rate and external K followed saturation kinetics, with a $K_m = 0.43$ mM and a $V_{max} = 1.0$ μmol per 10^7 cells per hour. Ouabain inhibited the respiration stimulated by K at a concentration effective against the active transport of Na and K. Finally, they showed that external K and 2,4-DNP affected a common source of respiratory intermediate, but acted at different sites, primarily because 2,4-DNP could overcome the respiratory inhibition produced by oligomycin, but external potassium could not.

It is worth speculating that the site of action of external K is in the membrane and is associated with the ATPase action of directed transport, while the site of action of 2,4-DNP is at the high-energy intermediate which provides the ATP for transport. At present, it is not clear whether the intermediate is a part of the cell membrane or mitochondrial in origin.

PERMEABILITY OF THE TUMOUR CELL TO WATER AND TO NON-ELECTROLYTES

The ascites tumour cell, because of its spherical shape, is ideal for a study of the rate at which the cell changes volume when water leaves the cell in response to an osmotic gradient produced by the addition of small volumes of 5 M NaCl to the medium. A value for the permeability of the membrane to water, L_p, may be calculated from the equation of Lucke', Hartline, and McCutcheon[46]. I found it useful[47] to use an analog computer to generate curves with different values for L_p and to fit the best curve to the experimental data, obtained with the Parpart densimeter.

Although morphologically indistinguishable, different strains of Ehrlich ascites tumours differ in their membrane permeability to water. One strain, which I have labelled the A-strain, has a value for L_p of 3 μm^3 water crossing 1 μm^2 of surface area in 1 min under an osmotic gradient of one atmosphere, at 20 °C. This particular strain was in continuous transplant for 10 years at the time of the measurements. The parent strain, called the Lewis Ehrlich strain, was maintained in the deep freeze with only intermittent reinstitution in mice for two or three transplants during this same period. Its value for L_p was 8 μm^3 μm^{-2} min^{-1} atm^{-1}, and is among the highest values reported for mammalian cells, being almost double that of the mammalian erythrocyte. In addition, strains could differ in their energies of activation for water transfer. Early transplants showed values of 9·6 kcal mol^{-1}, but after 10 years of prolonged transplantation, the value has decreased to 5·7 kcal mol^{-1}.

Further, the ascites tumour cell has proved to be a valuable tool with which to demonstrate the applicability of irreversible thermodynamics to membrane permeability. The permeability of the membranes of several strains of Ehrlich mouse ascites tumour cells to a series of homologous glycols was analysed with the equations of Kedem and Katchalsky[48] by means of an analog computer and values for σ, L_p, and ωRT were calculated. These parameters could form a functional profile of the tumour cell. The glycols cross the membrane with energies of activation ranging from 16 kcal mol^{-1} for ethylene glycol to 20 kcal mol^{-1} for triethylene glycol. Entropies of activation ranged from 17 to 27 entropy units. These data imply that even diffusion across the membrane will produce considerable bond breaking and structural change in the membrane. Stein[49] has discussed these types of data in other cells.

This type of kinetic analysis has provided additional dividends. It is one of the few examples with a mammalian cell where the kinetics of non-electrolyte transport have been described completely by the equations of irreversible thermodynamics. It has provided the means to distinguish how much of the transport occurred through porous paths in the membrane and how much involved solution within the matrix of the membrane.

REFERENCES

1. CHRISTENSEN, H. N., and RIGGS, T. R., *J. biol. Chem.* **194**, 57 (1952)
2. LEVINSON, C., *Biochim. biophys. Acta* **120**, 292 (1966)
3. LUBIN, M., *Nature, Lond.,* **213**, 451 (1967)
4. JUNG, C., and ROTHSTEIN, A., *J. gen. Physiol.,* **50**, 917 (1967)
5. WHEELER, K. P., INUI, Y., HOLLENBERG, P. F., EAVENSON, E., and CHRISTENSEN, H. N., *Biochim. biophys. Acta* **109**, 620 (1965)
6. EDDY, A. A., *Biochem. J.,* **108**, 195 (1968)
7. EDDY, A. A., *Biochem. J.,* **108**, 489 (1968)
8. EDDY, A. A., MULCAHY, M. F., and THOMSON, P. J., *Biochem. J.* **103**, 863 (1967)

9. EDDY, A. A., and HOGG, M. C., *Biochem. J.,* **114**, 807 (1969)
10. AULL, F., and HEMPLING, H. G., *Am. J. Physiol.,* **204**, 789 (1963)
11. HEMPLING, H. G., *J. cell. comp. Physiol.,* **60**, 181 (1962)
12. HEMPLING, H. G., and KROMPHARDT, H., *Fedn Proc. Fedn Am. Socs exp. Biol.,* **24**, 709 (1965)
13. KROMPHARDT, H., *Eur. J. Biochem.,* **3**, 377 (1968)
14. AULL, F., *J. cell. comp. Physiol.,* **69**, 21 (1967)
15. LEVINSON, C., *Biochim. biophys. Acta,* **203**, 317 (1970)
16. LASSEN, U. V., Personal communication
17. GROBECKER, H., KROMPHARDT, H., MARIANI, H., and HEINZ, E., *Biochem. Z.,* **337**, 462 (1963)
18. WEISS, L., 'The Cell Periphery, Metastasis, and Other Contact Phenomena', North Holland Publishing Co., Amsterdam (1967)
19. WALLACH, D. F. H., and EYLAR, E. H., *Biochim. biophys. Acta* **52**, 594 (1961)
20. LEVINSON, C., and BLUMENSON, L. E. (in press) (1971)
21. WHITTAM, R., *Transport and Diffusion in Red Blood Cells,* Arnold, London (1964)
22. GLYNN, I. M., *Brit. med. J.,* **24**, 165 (1968)
23. ADRIAN, R. H., and SLAYMAN, C. L., *J. Physiol,* **184**, 970 (1966)
24. CROSS, S. R., KEYNES, R. D., and RYBOVA, R., *J. Physiol.,* **181**, 865 (1965)
25. FRUMENTO, A. S., *Science,* **147**, 1442 (1965)
26. KERNAN, R. P., *Nature, Lond.,* **193**, 986 (1962)
27. MULLINS, L. J., and AWAD, M. Z., *J. gen. Physiol.,* **48**, 761 (1965)
28. HEMPLING, H. G., *J. gen. Physiol.,* **41**, 565 (1958)
29. JACQUEZ, J. A., *Proc. natn. Acad. Sci. U.S.A.,* **47**, 153 (1961)
30. JACQUEZ, J. A., *Biochim. biophys. Acta,* **79**, 318 (1964)
31. CHRISTENSEN, H. N., *Adv. Enzymol.,* **32**, 1 (1969)
32. JACQUEZ, J. A., SHERMAN, J. H.. and TERRIS, J.. *Biochim. biophys. Acta,* **203**, 150 (1970)
33. INUI, Y., CHRISTENSEN, H. N., *J. gen. Physiol.,* **50**, 203 (1966)
34. JOHNSTONE, R. M., and SCHOLEFIELD, P. G., *Biochim. biophys. Acta,* **94**, 130 (1965)
35. BELKHODE, M. Z., and SCHOLEFIELD, P. G., *Biochim. biophys. Acta,* **173**, 290 (1969)
36. AULL, F., and HEMPLING, H. G., Unpublished observations
37. HEMPLING, H. G., and HARE, D., 1961, *J. biol. Chem.,* **236**, 2498 (1961)
38. HEMPLING, H. G., and HARE, D., *J. cell. comp. Physiol.,* **65**, 419 (1964)
39. CHRISTENSEN, H. N., *Adv. Protein Chem.,* **15**, 239 (1960)
40. CHRISTENSEN, H. N., RIGGS, T. R., and COYNE, B. A., *J. biol. Chem.,* **209**, 413 (1954)
41. CHRISTENSEN, H. N., in *Amino Acid Pools* (ed. J. T. HOLDEN), Elsevier, Amsterdam, 527 (1962)
42. HEINZ, E.. in *Biochemie des aktiven Transports,* Springer. Berlin, 167 (1961)
43. WEINSTEIN, S. W., and HEMPLING, H. G., *Biochim. biophys. Acta* **79**, 329 (1964)
44. HEMPLING, H. G., *Biochim. biophys. Acta,* **112**, 503 (1966)
45. LEVINSON, C., and HEMPLING, H. G., *Biochim. biophys. Acta,* **135**, 306 (1967)
46. LUCKE', B., HARTLINE, H. K., and MCCUTCHEON, M., *J. gen. Physiol.,* **14**, 405 (1931)
47. HEMPLING, H. G., *J. cell. comp. Physiol.,* **70**, 237 (1967)
48. KEDEM, O., and KATCHALSKY, A., *Biochim. biophys. Acta,* **27**, 229 (1958)
49. STEIN, W. D., *The Movement of Molecules Across Cell Membranes,* Academic, New York (1967)

8

MITOCHONDRIA

E. J. HARRIS

Department of Biophysics, University College, London

INTRODUCTION

Mitochondria are the intracellular organelles concerned with the oxidation of the acids of the tricarboxylic acid cycle, fatty acids, hydroxybutyric acid and glycerol phosphate. To carry out these oxidations they are endowed with·a set of specific enzymes called dehydrogenases, located either in their inner membrane or internally. The oxidation reaction is usually coupled to the production of ATP* from ADP via a set of enzymes and cofactors called the respiratory chain (Figure 8.1). About $64.5\,kJ$ are required to phosphorylate 1 mol ADP with the concentrations of ADP, ATP, phosphate and Mg^{2+} in the reaction system usually studied. This is approaching twice the standard free energy change of the reaction since the ADP concentration attained becomes low compared with that of ATP[1, 2]. The Mg^{2+} ions favour the process because chelation with ATP lessens the energy requirement for ATP formation. Discussion of the energetics of this and related reactions has extended to the question of the heats of solvation of the components[3].

An essential property of most mitochondria is that the rate of oxidation is controlled by the availability of ADP and phosphate, so that the oxidisable substrates are not wastefully consumed in relation to the cell's requirement for ATP. The mitochondria maintain the ATP supply for demands such as muscle contraction,

---

* Abbreviations used in this chapter: ATP adenosine triphosphate; DNP dinitrophenol; EDTA ethylene diamine tetracetic acid; EGTA ethylene glycol bis(2-aminoethyl)tetracetic acid; FCCP trifluoromethoxycarbonylcyanide phenylhydrazone; NAD nicotinamide adenosine dinucleotide; NADP nicotinamide adenosine dinucleotide phosphate; TMPD tetramethylphenylene diamine; TTFB trifluormethyltetrachlorbenzimidazole; ADP adenosine diphosphate.

L

ion secretion in kidney, and syntheses in liver. In general, the oxidative process runs in parallel with eithei glycolysis or glyconeogenesis in the cytoplasm. Sugar is converted to pyruvate (and lactate) by glycolysis; the pyruvate enters the mitochondria and is both oxidised directly and built into larger molecules which themselves can either oxidise or emerge into the cytoplasm. An understanding of how alternative mitochondrial fuels, fatty acids, and amino acids derived from protein breakdown can be converted to the intermediates from which glucose can be synthesised requires knowledge of the distributions of the intermediates between cytosol and mitochondrial interior and the permeabilities to the intermediates, in addition to more generalised enzymology. The reader is referred to a clear and well documented review for a discussion of intracellular compartmentation between cytosol and mitochondria[4]. In the following various special aspects of mitochondrial permeability are considered; many of these are without obvious relevance to the control of metabolism in the whole cell, because the properties of the membrane have their own interest. It is indeed only in the last few years that biochemists have come to realise the importance of controls exerted by mitochondrial membrane properties and it is still a matter for questioning whether data obtained with isolated mitochondria apply directly to conditions pertaining when the particles are in the cell.

CONTROL OF RESPIRATION BY ADP AND PHOSPHATE

In presence of a substrate mitochondrial, respiration can be controlled either by phosphate in presence of excess ADP, or by ADP in presence of excess phosphate. Klingenberg[5] proposed that the function: ADP $\times$ phosphate/ATP, was the control, in which intramitochondrial concentrations were to be used. For half maximal stimulation of respiration the function has the value about 60 μM. The ratio between ATP and ADP concentrations in the mitochondria depends on the metabolic rate and on whether ATP is being exported in exchange for entering ADP. The exchange occurs so as to preserve a constant adenine nucleotide content irrespective of the state of phosphorylation. This has been shown by use of labelled nucleotides[6].

The mechanism by which the availability of ADP and phosphate control respiration is unknown. One proposal is that an allosteric modification of the respiratory chain occurs[2] but it is necessary to accommodate the possibility of control by either substance. The older hypothesis involved mediation of a chemical intermediate,

290

often denoted by $\sim$. There is some evidence that cytochrome b in a dimeric occurs as an intermediate[7, 8]. Any theory has to allow for the fact that respiration can be stimulated by other energy consuming processes, such as ion movement, and that such processes will continue in presence of substances which inhibit phosphorylation, like oligomycin. The latter fact requires that control is before the step that adds phosphate onto ADP.

An attractive alternative hypothesis was adduced by Mitchell (reference 9 in revised form). This proposed that the process of conversion of chemical energy into synthesis of polyphosphate, or its use for ion movement, depended upon the joint action of transmembrane gradients of pH and electrical potential. The implication that the membrane must have a low permeability to protons led to experimental demonstrations that artificial membranes became more conductive when agents were added that also remove control of mitochondrial respiration. These agents are referred to as 'uncoupling substances' or 'uncouplers'. The correlation between uncoupling action and effect on bilayer membranes has been underlined by the work of Skulachev et al.[10]. Much controversy has occurred over Mitchell's theory. It may be remarked here that pH adjustments can readily be mediated by transfers of the differently protonated forms of phosphate and phosphate is always present in mitochondrial suspensions if only by reason of leakage from the particles. It is difficult to understand how energy from added ATP can be transferred back up the respiratory chain (see next section) to cause reduction by reversed electron flow. For example, NAD^+ can be reduced by the combined action of succinate oxidation and ATP, and ubiquinone can be reduced from cytochrome c with ATP. There is evidence that a gradient of pH consistent with a transmembrane potential does subsist across the membrane but these appear to be different manifestations of the asymmetry due to internal charged substances.

THE RESPIRATORY CHAIN

The reaction between the dehydrogenase enzymes and their substrates generate reducing equivalents that pass along a series of enzymes collectively called the respiratory chain (Figure 8.1). Ideally, the passage of two electrons down the chain to oxygen generates three ATP molecules from three ADP molecules. Three separate sites of phosphorylation have been located in the chain. The oxidation of oxoglutarate furnishes an additional triphosphate because one of its primary oxidation products is succinyl coenzyme

291

A. The latter reacts with the endogenous guanosine diphosphate of the mitochondrial matrix to form GTP. A nucleotide diphosphokinase can use the GTP to phosphorylate ADP so the overall result is that four ATP molecules can be produced per oxoglutarate oxidised. The enzyme responsible for oxoglutarate oxidation, like that for pyruvate oxidation, involves lipoic acid as a cofactor[11].

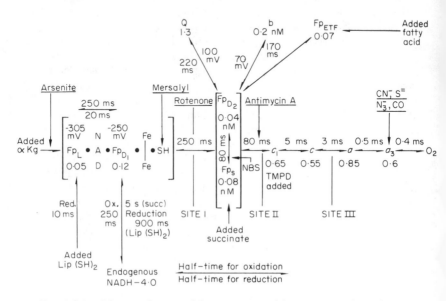

Figure 8.1. Schematic diagram of the components of the respiratory chain of rat liver mitochondria, showing half-times for oxidation taken from oxygen pulse studies; half-times for reduction of the flavoproteins taken from dihydrolipoamide studies; potential measurements taken from acetoacetate/β-hydroxybutyrate titrations; rapidly reacting amounts of the components in nmol per mg protein, taken from oxygen pulse studies. Other units are indicated on the figure. Specific inhibitions are underlined, and an arrow points to where inhibition occurs. Symbols: αkg, α-ketoglutarate; $Lip(SH)_2$, reduced form of lipoic acid; F_{p_L}, flavine transferring electrons from lipoamide to NAD; $F_{p_{D_1}}$, low-potential (high-energy) flavine acceptor of NAD dehydrogenase; $F_{p_{D_2}}$, high-potential (lower-energy) ditto; Fe–Fe. SH Non-heme iron involved in transfer from $F_{p_{D_1}}$ to $F_{p_{D_2}}$; Q, ubiquinone; F_{p_s}, succindehydrogenase; b, c_1, c, a, a_3, cytochromes; $F_{p_{ETF}}$, electron transfer flavine oxidising keto acids formed by fatty acid oxidation. (*From Chance* et al.[49]; *courtesy Academic Press*)

Oxidation of succinate is coupled to a flavine below the first phosphorylation site so the maximal ATP yield by its oxidation is only two molecules per atom of oxygen respired.

292

Some substances having the right range of redox potential will couple directly to the membrane bound citochrome c. For each reduced cytochrome c only a single molecule of ATP is obtained. The cytochrome c is peculiar in being readily extracted from the membrane and it can be restored. The reactivity towards substances in the medium[12] points to its being located on the outside of the inner membrane.

The final site of phosphorylation is bypassed if the cytochrome c is oxidised with ferricyanide when the ATP yield per electron pair becomes one less than when the electrons pass to oxygen.

The cytochromes are so large that it would be possible for the cytochrome b on the inside of the membrane to enter into oxidation reduction reactions with the c on the outer side, which in turn could react with its oxidases (cytochromes a and a_3) on the inside. This alternating pattern is one of the features of Mitchell's proposals[13], but its existence does not prove the theory as a whole. Margoliash et al.[14] have adduced interesting correlations between the capacity of cytochrome c to bind various ions in its oxidised and reduced states and mitochondrial ion transport. Some yeasts yield mitochondria lacking the first site of phosphorylation[15]; these, in common with mitochondria prepared from plant cells, have the ability to oxidise exogenous NADH without phosphorylation. Yeast, plant and insect mitochondria are insensitive to Ca^{2+}, in marked contrast to those derived from mammalian cells. Whether the insensitivity to Ca^{2+} is related to lack of the first site has still to be established. Specialised mitochondria derived from brown fat cells appear to have lost their phosphorylating ability and instead oxidise fat to produce heat[16, 17].

INTERNAL LOCATION OF ENZYMES AND THEIR SUBSTRATES

The location of dehydrogenases within mitochondria is important because the differing accessibilities determine whether or not permeation of the membrane can become a limiting factor to rapid turnover. Mitochondria have the ability to accumulate anions, so raising the internal concentrations to many times those outside. The concentrations applied to internal enzymes may differ, therefore, from the outside concentrations. Many interactions exist between substrates at the permeation step. Some of these are important in the control of cellular metabolism. In the laboratory it is often possible to find an analogue of a substrate that competes with it for movement and usually as well for internal accumulation.

To show that the effects take place at the membrane controls are made with sonicated particles, with which membrane effects should be absent.

CONTENTS AND RETENTION OF CATIONS

Most substrates are anions; the mitochondrial interior has a buffer capacity ionised as an anion to neutralise much of the total internal cation but there remains a balance to made up with mobile penetrating anions. These can exchange amongst themselves and play a major role in metabolic control. The cations comprise about $100\,nmol\,K^+$ $20\text{--}35\,nmol\,Mg^+$, $6\text{--}15\,nmol\,Ca^{2+}$, and a few nmol Na^+ per mg (for rat liver particles). The Mg^{2+} content is remarkably constant and insensitive to exposure to chelating agents such as EGTA but can be lessened by about 10 nmol per mg after exposure to EDTA. The Ca content depends critically on the level of chelating agent used during the isolation from the cell. This is because disruption of the cell frees considerable Ca, which tends to accumulate in the mitochondria; the chelating agents compete with the mitochondria for this external source of Ca. It is not possible to say what is the Ca content of mitochondria *in situ*. The Na seems to be not fully exchangeable and presumably is occluded in lipid material[18].

Retention of the full complements of K and Ca by the particles depends on a residual metabolism providing energy. The oxidation of endogenous material including fatty acids derived from the membranes occurs. If such oxidation is inhibited, a slow loss of K occurs because the membrane is comparatively impermeable to the ion. The loss of K is somewhat accelerated by the ionophores, which give the lipid part of the membrane a high K permeability but to obtain a high rate of K loss it is necessary to provide a permeability either to another cation, or to protons. The latter effect is obtained by use either of a mixture of valinomycin and uncoupling agent or by one of the acidic ionophores such as nigericin (Chapter 1). When energy supply is withdrawn part of the mitochondrial Ca^{2+} is rapidly lost (Figure 8.2). Restoring energy supply (oxygen supply in the example) leads to uptake of the Ca^{2+} that had been lost. This can lead to membrane disruption.

Sonication, or treatments that disrupt the membranes, leads to shedding of cations though the resulting vesicles carry trapped cations and will display an energy-dependent uptake of Ca^{2+} [19, 1]. Such vesicles are permeable to the more lipophilic anions such as iodide and thiocyanate[20].

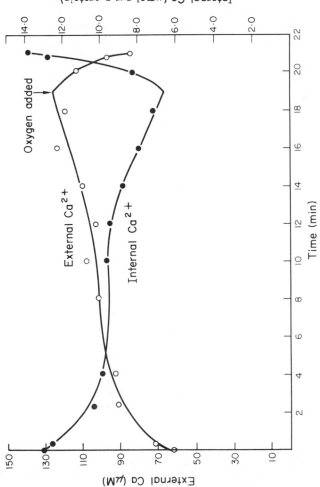

*Figure 8.2. The release of Ca^{2+} to the medium that occurs over a few minutes when energy supply is cut off by withdrawal of oxygen from the suspension. When oxygen is restored the movements are reversed and Ca is reaccumulated. (From Thomas, Manger and Harris, Europ. J. Biochem., **11**, 413 (1969); courtesy Europ. J. Biochem)*

VOLUME CHANGES ASSOCIATED WITH ION MOVEMENTS

When cations are gained in the limited amount (about an additional 50 nequiv per mg protein) that can be neutralised by further ionisation of the internal buffer there is no important change of volume of the particles; limited uptake of Ca^{2+} can even cause shrinkage, perhaps because it precipitates internal phosphate. Uptakes of cations along with anions, such as Ca^{2+} with acetate, or K^+ and

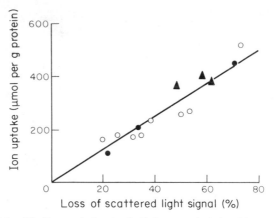

Figure 8.3. The linear relation that holds between the induced ion uptake and the loss of signal from light scattered at about 340° to the incident beam from a suspension with 5 mg protein per ml. The ion uptakes were induced with: ●, valinomycin; ▲, dinactin; and ○, gramicidin

acetate, or phosphate under the influence of an ionophore give rise to an osmotic shift of water. The swelling is related to extent of the ion uptake (Figure 8.3). The response is not necessarily linear because the difference in refractivity between the particles and the medium is a factor. The osmotic nature of the swelling can be shown by use of different osmolarities in the medium but there may be two separate spaces to consider, the one between the two membranes that carries colloidal material, and the matrix space inside the inner membrane. Expansion of the matrix in response to uptake of K^+ has been shown[21] (Figure 8.4c and d). The importance of the colloid can be inferred when comparisons are made between specimens fixed in the absence and in the presence of external colloid (Figure 8.4a and b).

The optical effect of swelling is complex. It is not necessarily linearly related to volume because the difference in refractive index between particle and medium is a factor. There may be two independently variable volumes to consider; the space between the membranes, which undoubtedly contains colloids, and the matrix space. The matrix expands when K and anions are gained, but there is a requirement for external colloid to balance the internal colloid of the intermembrane space.

SEPARATION OF MEMBRANES; MEMBRANE ENZYMES

When the matrix is made to expand by exposure of the particles to hypotonic phosphate, the outer membrane is ruptured. The resulting suspension of broken outer membranes and matrix material can be separated because the higher lipid content of the outer membrane gives it a lower density. A density gradient centrifugation yields a pure outer membrane fraction[22, 23]. The outer membrane resembles in several ways the reticular membrane, but the enzymes attached to the two differ. Notably, the reticular membrane carries a phosphatase and a NADPH–cytochrome c reductase, in contrast to the outer mitochondrial membrane that carries monamine oxidase, a rotenone-insensitive NADH–cytochrome c reductase, as well as NADH–cytochrome b_5 reductase.

The inner membrane carries succinic dehydrogenase, hydroxybutyrate dehydrogenase, the respiratory chain, and ATPase. The latter is associated with material that in the electronmicrographs appear as knobs, about 7 nm diameter, attached to the inner surface of the membrane by stalks[24]. The knobs can be detached by sonication of particulate material and then have an oligomycin insensitive ATPase activity. They can be combined with the lipid-extracted membrane protein to give an amorphous product devoid of ATPase activity. Addition of phospholipid restores ATPase activity and confers oligomycin sensitivity[25–27].

Mitochondrial succinoxidase is inhibited by thenoyltrifluoro-acetone on account of chelation of the iron of cytochrome b. This is effective also when added to particulate preparations obtained by sonication. However, if these are treated with lipid solvents (e.g. pentane), so removing the ubiquinone and lipids, then the sensitivity to the inhibitor is lost[28]. It is restored by adding back ubiquinone and phospholipid. Similarly, coupling of NADH oxidation to reduction of ubiquinone requires the presence of phospholipid[29].

297

L*

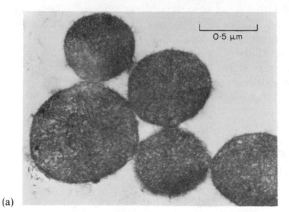

(a)

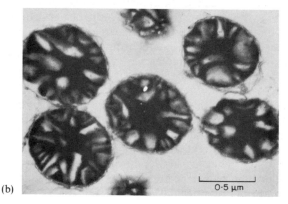

(b)

Figure 8.4. Electronmicrographs of sections of rat liver mitochondria fixed (a) in presence of 2% albumin and (b) without the albumin. The difference is ascribed to the effect of the external colloid acting to modify the membrane permeability but whether this depends on the fixation reaction or preexists is not known

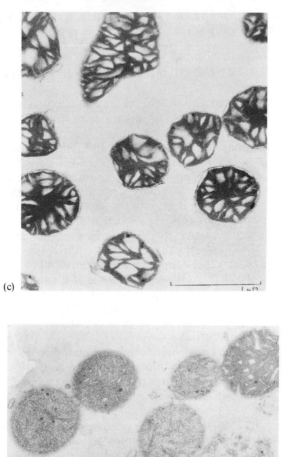

(c)

(d)

Figure 8.4 (cont.). Electronmicrographs of sections fixed without serum albumin showing the change in appearance accompanying valinomycin-induced K uptake; (c) is a specimen taken from 5 mM KCl in 250 mM sucrose +. 20 mM tris chloride with substrate, (d) is from the same medium after valinomycin addition. The matrix (stained dark) remains more uniformly spread within the confine of the outer membrane in this condition (Thanks are due to Miss C. Tate for these illustrations)

This indicates that the enzymes and their inhibitors are brought into closer proximity when phospholipid is present.

THE IMPORTANCE OF MEMBRANE INTEGRITY

In its organised condition, the mitochondrion can use energy obtained either by oxidation of substrates or from exogenous ATP to perform a number of 'energy-linked functions'. These include synthetic reactions, ion accumulation, and reduction of NADP by NADH. When ATP is supplied to provide energy, it becomes split to ADP and phosphate, so the mitochondria behave as an ATPase whose integrated activity depends on the load imposed. The rate at which energy is fed into the particles depends on the rate at which positive charges (Ca^{2+} or, when an ionophore is present, K^+ or other cation) reach the locus for ion movement within the membrane. The positive charges are required to maintain an electrostatic balance between the more negative ATP, which enters, and the less negative ADP, which leaves. If a carrier for positive charges, such as an uncoupler, is provided, the effect is to allow protons to accompany the ATP, and a high ATPase activity results[30]. Such dissipation of energy is not controlled, because the H^+ can cycle in and out.

This concept puts the mechanism of uncoupler action and the dissipation of energy for cation movement as basically similar processes. The essential feature is the provision of mobile charges to allow passage of ATP^{-3} across the membrane in exchange for ADP^{-2} without appearance of an electrostatic counter-potential; it is necessary, however, to accept the postulate that membrane bound nucleotide functions in the uncoupler action since no nucleotide has to be added to the medium. It is relevent to recall here the fact that the ion gradient can be used to generate ATP[31].

SWELLING NOT ATTRIBUTABLE TO ION MOVEMENTS

Certain readily autoxidisable substances, such as ascorbate and glutathione, lead to mitochondrial swelling provided (1) that oxygen is present and (2) that chelating agents are absent. These provisos formerly led to misconceptions about the process, since it was inferred that oxygen energised the process yet the chelating property of ATP inhibited it. The simple explanation is that autoxidation in presence of traces of Fe (inevitably present in mitochondrial suspensions) leads to peroxides[32,33]. With these the membrane lipids react to form surface active substances that destroy the membrane selectivity. Then the effect of the internal colloid is to induce swelling (as in the Donnan system). Peroxides added directly to the mitochondria in minute amounts are potent swelling agents.

ATPase ACTIVITY

Intact mitochondria do not hydrolyse ATP unless some linked process occurs. For example, external ATP is consumed to move cations into the matrix, either Ca, or K under the influence of an ionophore. It may also be consumed to produce a more electro-positive compound (e.g., NADH or NADPH) at the expense of a less electropositive one (a flavine or cytochrome or NADH). The behaviour of ionophores can be conveniently investigated using the ATPase associated with presence of cations (for example, reference 34). Use of ATP for NADPH formation has been described by Ernster[34a].

Uncoupling agents lead to a dissipation of the potential energy stored as ATP. In conditions of low ionic strength, the uncouplers are less effective as inducers of ATPase activity. Perhaps because of a counter-potential set up between the particles and the medium[35-37].

Impairment of the membrane leads to appearance of ATPase activity. For this reason the addition of Ca^{2+} in excessive quantity provokes a continued ATP-splitting. Sonicates also have a high ATPase activity, but certain preparations still conduct phosphory-lation[38, 39]. If oligomycin, an inhibitor of phosphorylation, is added to such preparations at a carefully chosen concentration (about 0·5 µg per mg protein), the efficiency of phosphorylation is im-proved. This seems to be because the parallel ATPase activity has been lessened.

WATER CARRIED BY MITOCHONDRIA, INTERNAL SPACES

A pellet of mitochondria separated either by centrifugation or by filtration inevitably carries suspension medium with it, so the total water, which can be found by drying, includes a part external to the particles. When visualised in the electronmicroscope, two mem-branes can be seen. The outer is permeable to a wider spectrum of substances than the inner.

The volume of water carried outside the particles can be measured by finding how much ^{14}C-labelled carboxydextran added to the medium is carried with the pellet. This large molecule does not penetrate the outer membrane. If, instead, a somewhat smaller molecule is chosen, such as sucrose, inulin, EDTA, NAD, or even chloride, it appears to enter the water in space between the two membranes as well. This means that the space can be found as a difference between the sucrose and the dextran-permeable volumes. Finally, the total water can be deduced from the content of tritiated water carried out of a suspension. This total includes the water in the matrix, plus that between the membranes and that outside. By taking differences between tritiated water and sucrose accessible volumes, the content of matrix water is found (Table 8.1).

Table 8.1 ADHERENT WATER AND INTERNAL WATER IN MITOCHONDRIA
Values are in ml g^{-1} for biuret protein for tonicity 250–300 mosmolar

Material	Reference	Adherent water (dextran space)	Internal water accessible to sucrose (between outer membrane and matrix)	Internal water inaccessible to sucrose (bounded by inner membrane)
Rat liver	40	1·0–1·6	0·3–0·8	0·65–0·77
Rat liver	41	2·1 (centrifugation through silicone)	0·5	0·5–0·8
Rat liver	42	1·2–1·4	0·6–1·0	1·65
Rat kidney	42	?	3·0	0·66
Rat heart	41	2·7 (centrifugation through silicone)	0·9–1·4	0·9

Within a limited range the matrix compartment responds to the tonicity of the medium (Figure 8.5). The graph of the volume versus the reciprocal of the tonicity does not pass through the origin because of the presence of internal osmotically unresponsive material[43–45].

When a salt is accumulated, the mitochondrial matrix swells. The change has been shown by electronmicroscopy[21, 46] and by measurement of the difference between the total water and the sucrose-accessible water[44]. Figure 8.6 illustrates the changes and their association with K$^+$ uptake and loss. Besides those uptakes associated with use of ionophores to increase permeability to alkali cations, the uptake of Ca^{2+} along with acetate also leads to swelling,

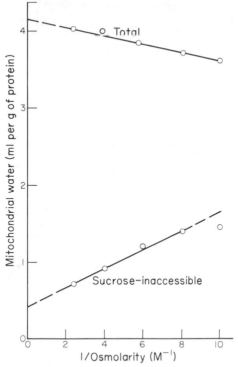

Figure 8.5. Response of the total water content and the sucrose-impermeable water of a mitochondrial pellet to changes in osmolarity. Note that more than half the total water is carried outside the particles as shown by other measurements using ^{14}C-labelled dextran. Rat liver mitochondria (6 mg protein per ml) were incubated at 23 °C in a medium containing 50 mM sucrose, 2·5 mM tris chloride, pH 7·0, 5 mM phosphate buffer, pH 7·0, 1 μg of rotenone per ml, 2 mM EDTA, 10 mM $MgCl_2$, and trace amounts of ^{14}C-sucrose and tritiated water. The osmolarity of the medium was varied by further addition of KCl (From Harris and van Dam[44]; courtesy Biochem. J.)

no ionophore being necessary. Under some initial conditions swelling of the matrix can lead to expulsion of sucrose-containing solution from the space between the matrix and the outer membrane.

It has been argued that the conformation of the mitochondrial crystal system responds to changes in phosphorylation activity[47, 48]. Certainly, the appearance of the material after fixation can differ according to the state of respiration, that is, depending on presence of ADP or ATP in the medium. However, differences are also found within in one pellet after fixation, and the response of the configurational change has a much longer timescale than do the changes in redox condition of the cytochromes[49]. This puts in doubt the validity of equating the observed differences to pre-existing differences in the membrane structure. Firmer evidence for some change is that there is a slight shrinkage during phosphorylation; this can be shown by recording the optical density during ADP cycles and was noted by Lehninger[50].

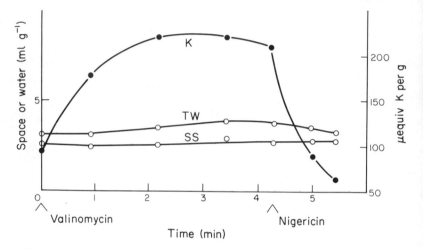

Figure 8.6. The net gain of mitochondrial K^+ caused by addition of valinomycin and the release following addition of nigericin. The total water (TW) carried down by the protein pellet through a silicone layer was measured using tritiated water and the water acting as solvent for sucrose (SS) measured using ^{14}C-sucrose. Both quantities include the water carried external to the particles. Note that the difference $TW - SS$ increases and decreases with the K^+ content. It is the water inaccessible to sucrose that is considered to be held in the matrix and to be the solvent for the K^+ salt which is gained

The arrangement of the cristae in the mitochondrion depends on the cell type and the species from which it was derived. Insect mitochondria have very regular cristae with perforating holes. (For illustrations, see reference 51).

CATION MOVEMENTS

The retention by mitochondria of their normal complement of K and Ca involves a small energy dissipation included in the so-called 'resting' metabolism. Even when stored at $0°$ C, there is a small energy consumption. The endogenous ATP is a further store of energy and has to be consumed before release starts; this can be shown by adding oligomycin to prevent use of the ATP: then the cations are more promptly released when respiration is stopped.

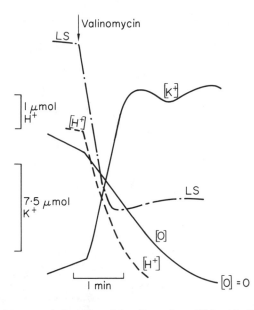

Figure 8.7. Multiparameter recording of the effects of an addition of valinomycin to a suspension of rat liver mitochondria in a medium having 5 mM KCl, 20 mM tris acetate, 3 mM each tris glutamate and malate 250 mM sucrose. Protein 20 mg in 10 ml. The light scattering (LS) falls because there is gross swelling. There is an uptake of K from the medium shown by the rising potential reading from the K-selective glass electrode, there is an output of protons into the medium shown by the falling potential from the pH glass electrode, and the oxygen consumption rate, deduced from the oxygen electrode, increases

Release of cations is accompanied by release of phosphate and organic anions and some uptake of protons. The ion movements are much accelerated if an uncoupling agent is added to make the membrane effectively permeable to protons and stimulate respiration to consume the substrates. There is then a loss of potassium ions along with inorganic phosphate derived from breakdown of ATP.

305

Mitochondria that have lost K after energy depletion can regain it if a source of energy is provided[52, 53] Since the concentration of K in the medium need only be 5 mM and the apparent concentration

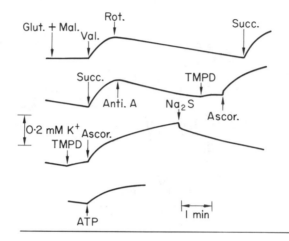

Figure 8.8. To illustrate the dependence of K uptake (shown by an upward deflection) in presence of valinomycin (Val.) unspecifically upon a supply of energy. At the commencement, the energy is provided by oxidation of the NAD-coupled glutamate and malate; this is inhibited by rotenone (Rot.) and K emerges. Energy is restored by adding succinate whose oxidation is insensitive to rotenone; K again enters. Succinate oxidation is stopped by antimycin A; the K emerges. Energy is supplied by a combination of the cytochrome C-coupled TMPD in small quantity and ascorbate to keep the TMPD reduced; K enters. Oxidation is again inhibited, this time with sulphide which reacts with the cytochrome-C oxidase; K emerges. Finally, exogenous ATP is added and K again enters

in the matrix water rises to about 100 mM the movement is against a considerable concentration gradient. Untreated mitochondria take up K^+ readily when their membranes are made permeable to this ion with valinomycin or a similar neutral ionophore. Associated with the uptake is an appearance of H^+ in the medium, a swelling and an enhanced respiration, the changes are shown in the multiparameter recording (Figure 8.7); for this, selective glass electrodes were used to monitor the K^+ and H^+ concentrations in the medium and a photoelectric cell was used to measure light reflected from an

306

incident beam; swelling is shown by a reduced light scatter. The energy for K^+ uptake in presence of an ionophore can be derived from NAD-coupled substrates, or from succinate, or from TMPD (plus ascorbate to keep it reduced) or from ATP. The successive use of these sources is shown in Figure 8.8 which is a record of K concentration in the medium as different energy sources are supplied and then made ineffective by a selective inhibitor.

Studies of the heat change associated with the K^+ flux[54] indicate that the internal K^+ is ionised; this agrees with the observation that water moves along with it when a penetrating anion is present. Some binding of the original content of K^+ may, however, exist. The influx is energy-consuming and is balanced in the steady state by an outward leak. The turnover of K^+ has been measured using ^{42}K by Harris, Catlin and Pressman[55], who showed the influx was sustained even when the content of K^+ had reached its steady value. The ionophore-induced influx depends on the K^+ concentration and on the amount and nature of the ionophore used. Such large quantities of K^+ can be moved from the medium into the particles that a large shift in concentration occurs. To obtain data not disturbed by this factor, use can be made of a 'K$^+$-stat', a device having a K^+-sensitive electrode to operate a titrimeter which adds more K^+ salt to the medium[53].

Though entry of K^+ normally is found to require energy, it might be expected that it could alternatively be driven in from a sufficiently high external concentration. In the presence of valinomycin mitochondria in a medium 120 mM in KCl take up both KCl and water without energy requirement[56]. It is possible that the KCl and water are now accumulating in the sucrose-accessible space, rather than in the matrix. When energy is supplied to this system, there is an expulsion of KCl and water, so the consequence is just opposite to that observed in media having low concentrations of K^+ salt.

Once the particles are suspended in a K^+-free medium, the internal K^+ represents a store of potential energy. By its discharge under the influence of valinomycin, it is possible to synthesise ATP from ADP and phosphate[57, 58].

STOICHIOMETRY OF K UPTAKE

The spontaneous gain of K that occurs when a mitochondrial suspension is added to a K-containing medium with substrate has not been related to a specific consumption of energy, but this may be for technical reasons. Certainly, the ionophore-induced uptake can be related to extra oxygen consumption[53], or, if driven from ATP, to ATP hydrolysis. About four K^+ ions will enter per ATP split or per electron pair passing through a phosphorylation site.

The acidification of the medium accompanying K uptake can be accounted for by the shift of anions, including the leaked endogenous phosphate, which accompanies K^+ movement. The tendency for doubly charged phosphate to be preferentially accumulated (on account of the greater electrostatic attraction) leaves excess protons in the medium (see section on internal buffer, Donnan behaviour).

MOVEMENTS OF ANIONS WITH K^+ AND Ca^{2+}

Induced uptakes of K tend to bring about co-movement of anions into the mitochondria. This has importance in showing the capacity to oxidise substrates faster when they are more concentrated within the particles[59]. Table 8.2 shows the ratios between the equivalents of anion moved and K moved. Certain acids such as hydroxybutyrate do not seem to enter the matrix, and their oxidation is not stimulated in the same way as is that of the penetrant acids. Evidently, the entry of substrates with K^+ can be mimicked by increasing the external substrate concentration, which also leads to a high internal concentration. This means that the effect of the ionophore-induced K uptake on maximal respiration rate is best seen at low concentrations of substrate.

In the contrary sense to what has just been described, it is found that treatments that discharge cations also lessen or abolish anion uptakes. Exposure to uncoupler before the anions to be tested stops uptake[60]. Uptake can be restored if a cation gain is brought about. For example, if a source of utilisable energy (ATP or TMPD in presence of ascorbate) is added with a small quantity of Ca^{2+} or Ca^{2+} plus Mn^{2+} salt, the divalent ion carries substrate in (Table 8.3; and see references 61, 62).

INTERNAL BUFFER, DONNAN BEHAVIOUR

When electrodes are inserted into the large mitochondria of *Drosophila* salivary gland cells, a positive potential is measured; the value depends on the pH, as would be expected if the ionisation of large molecules was in question[63]. There is, indeed, a considerable content of titratable buffer in mitochondria that provides anionic groups at pH above 7. To shift the pH by 1 unit up, takes about 50 nequiv of OH^- per mg protein and the figure increases above

Table 8.2 ADDITIONAL ANION GAINED WITH K^+ WHEN VALINOMYCIN IS ADDED TO RAT LIVER MITOCHONDRIA

The anion gained is expressed as the ratio of equivalents, assuming full dissociation to those of K^+ which are gained and are values attained at the plateau of the process.

Anion	Concentration (mM)	K^+ gained after valinomycin (μequiv g^{-1})	Equivalents anion gained/ equivalents K^+ gained	Conditions
Acetate	20	303	0·99	
Citrate	1	63	0·92 ⎫	Rotenone, TMPD, and ascorbate present
DL-Isocitrate	1	85	0·92 ⎭	
Malonate	1	37	1·20	Glutamate (3 mM) present
L-Glutamate	1	105	0·22	Includes oxidation products
α-Oxoglutarate	1	65	0·32	Arsenite and β-hydroxybutyrate present
Succinate	1	95	0·32	Antimycin, TMPD, and ascorbate present
L-Malate	1	66	0·36	Rotenone, TMPD, and ascorbate present
L-Malate*	1	60	0·61	Rotenone and succinate present
[2,^{14}C] Pyruvate	1	104	0·22	Includes acetate formed
L-Leucine	1	67	0·00	Rotenone and succinate present
Proline	1	102	0·01	Rotenone and succinate present
[^{14}C] ATP	2	66	0·01	Includes ADP formed
DL-Hydroxybutyrate	1	63	0·01	Includes acetoacetate formed

* In this experiment the labelled malate was added 1·5 min after the valinomycin-induced uptake of K^+, together with (unlabelled) succinate.

Table 8.3 DEPENDENCE ON Ca^{2+} OR $Ca^{2+} + Mn^{2+}$ OF THE REACCUMULATION OF SOME SUBSTRATES BY ENERGY-DEPLETED MITOCHONDRIA IN PRESENCE OF ADDED ENERGY SOURCES

The medium contained 120 mM KCl, 20 mM tris chloride buffer, pH 7·4, and in experiments nos. 1, 3, 5 and 6, 1 mM tris ascorbate. Additions were made sequentially. The values for contents were corrected for the amount of substrate carried down in the sucrose-accessible water of the pellet. They may include some adsorbed material.

Expt. no.	Anion being measured and concentration added	Addition	Mitochondrial content of the anion (µmol per g M protein)
1	Malate (0·6 mM)	Rotenone (2 µg ml^{-1})	1·4
		TMPD (20 µM)	1·7
		Ca^{2+} (7 µmol per g protein)	4·5
2	Malate (0·4 mM)	Rotenone (2 µg ml^{-1})	2·5
		Succinate (1 mM)	3·4
		Ca^{2+} (16 µmol per g protein)	5·1
3	Succinate (0·7 mM)	Antimycin (2 µg ml^{-1})	3·3
		TMPD (20 µM)	3·8
		Ca^{2+} (7 µmol per g protein)	5·1
4	β-Hydroxybutyrate (0·6 mM)	Rotenone (2 µg ml^{-1})	5·8
		Succinate (1 mM)	5·8
		Ca^{2+} and Mn^{2+} (8 µmol of each per g protein)	7·4
		Plus 10 µmol of each per g protein	10·8
5	Glutamate (0·8 mM)	Rotenone (2 µg ml^{-1})	2·3
		TMPD (20 µM)	2·6
		Ca^{2+} and Mn^{2+} (3·5 µmol of each per g protein)	3·5
		Plus 5·3 µmol of each per g protein	4·0
6	α-Oxoglutarate (1·3 mM)	Rotenone (2 µg ml^{-1})	3·4
		TMPD (20 µM)	3·4
		Ca^{2+} and Mn^{2+} (3·5 µmol of each per g protein)	6·4
		Plus 5·3 µmol of each per g protein	8·2

pH 8 (see Figure 8.9[63a] and reference 64, Figure 6). The values appear lower in absence of an uncoupler or detergent because of ion shifts across the membrane.

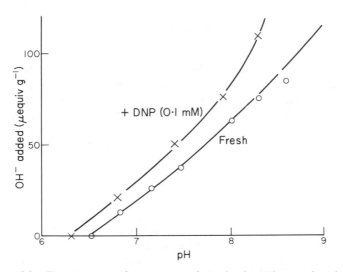

Figure 8.9.   Titration curve of a suspension of mitochondria. The particles (about 20 mg dry wt.) were suspended in 0·25 M sucrose (final volume 5 ml) at 20°C; titration was by means of a Radiometer T T T 1 automatic titrimeter; the burette contained 0·02 M NaOH (From Harris, Judah, and Ahmed[63a]; courtesy Academic Press)

The contribution of the internal buffer to provision of internal anions is made the more important by the Donnan-like situation that holds; protons tend to leave the mitochondrion, making the interior positive and alkaline. The electrostatic balance between cations and anions requires that if the cation content remains constant then the anionic charge from the internal buffer, plus that from the penetrating internal anions should be constant. As pH is changed towards acidity a diminished ionisation of the buffer has to be compensated by entry of more small anions, a fact borne out experimentally though not always explained in this simple manner.

Following the previous argument, it is clear that when cations are gained more anionic charge is required; in part this can come from an increased ionisation of internal buffer as the interior moves alkaline, and in part by entry of such mobile anions as are present in the external medium. The pH difference across the membrane of a Donnan system is equal to the logarithm of the ratio of the

311

concentrations of a given internal singly charged mobile anion of a weak acid to the external concentration of that anion and to the root of the corresponding ratio for a doubly charged anion, and so on (see p. 80).

$$\Delta pH = \log \frac{[A^-]_{int}}{[A^-]_{ext}} = \frac{1}{2} \log \frac{[B^{2-}]_{int}}{[B^{2-}]_{ext}}$$

As more anion moves in, the protein gradient increases (so long as swelling does not lead to maintenance of a constant concentration). The internal alkalinisation due to Ca^{2+} uptake (which does not cause swelling in absence of penetrant anion such as acetate), or due to valinomycin-induced K^+ uptake (that leads to swelling and hence less shift of concentration) has been described[65, 66].

OTHER UNIVALENT CATIONS

Most attention has been paid to K^+ permeability because this ion is a major component of the mitochondrial composition. Normally, liver mitochondria do not admit other cations to their matrix space, though these do enter the sucrose-permeable space between the membranes. After depletion of Mg^{2+} by exposure to chelating agents (EDTA or citrate) at pH exceeding 7·5 the cations Li^+ and Na^+ appear to enter along with phosphate, acetate or lactate[67]. Addition of Mg stops this effect. Similar ion shifts have been described for heart mitochondria[68]. Since entry of the salts is accompanied by water, there is a swelling which is usually used as an indicator of the process. Under some conditions gain of salt and water will be followed by a spontaneous discharge and further uptake for several cycles[69].

Specific ionophores (for details see Chapter 1) endow the membrane with permeability to Na^+ and even to the larger (hydrated) ion Li^+. To obtain a net uptake a high rate of energy dissipation is required; this is because of the imperfect selectivity of the ionophores, so that there is a complex uptake of K^+ and H^+ along with the Na^+, and a discharge of internal K^+.

UPTAKE OF Ca

Much experimental work has been done on the reaction between Ca^{2+} and mitochondria. There appear to be three processes by which Ca^{2+} becomes attached to, or taken into, the particles;

although different authorities do not agree on the amounts, they distinguish the following different modes of binding. First, there is a small capacity (1–2 nmol per mg protein) that half-saturates with only 10^{-8} M Ca^{2+}. It may represent binding to cardiolipin; energy is not required but uncoupling agents are inhibitory. This latter point suggests that a conformation change is induced by the uncoupler. The Ca^{2+} uptake is inhibited competitively by Mn^{2+} and by tri- and quadrivalent cations. It is abolished by heat treatment but is insensitive to 200 mM ethanol. Next, there is a low-affinity binding, probably to protein that does not require energy; this is insensitive to uncoupling agents. The capacity is about 50 nmol per mg, but the Ca^{2+} concentration required to half-saturate it is 200 µM. There is no proton liberation accompanying this Ca^{2+} adsorption, but K^+ and Mg^{2+} compete[70]. The low-affinity binding is distinguishable from the high-affinity binding by its insensitivity to heat as well as to uncouplers. Third, there is a high-affinity energy requiring process that is half-saturated at 1 µM Ca^{2+} and about 60 nmol capacity per mg protein. The rate of uptake is some 6 nmol per mg per second. The energy drain imposed by this Ca^{2+} uptake is imposed remarkably quickly; the cytochrome b is half-oxidised by Ca^{2+} addition in 30 ms.

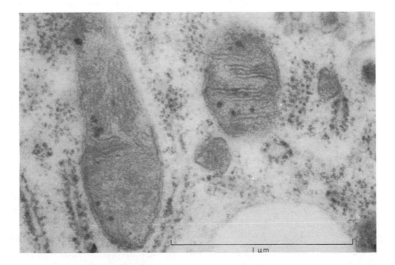

Figure 8.10.   A section of rat mammary gland with mitochondria showing dark internal granules due to Ca deposits (Courtesy Mrs A. Vincent)

313

The energised Ca^{2+} uptake is accompanied by liberation of one proton per Ca^{2+}, up to a maximum of about 50 nmol per mg. When either acetate or phosphate is present, the Ca^{2+} uptake can proceed to a much greater amount than without the penetrating anion, indeed in the presence of phosphate and adenine nucleotide the Ca gain can reach 3 µmol per mg protein. The Ca^{2+} phosphate becomes localised in granules scattered in the matrix (Figure 8.10). In this condition, normal phosphorylation will still take place. It is thought that such calcium phosphate granules may be extruded successively through the mitochondrial and cell membranes to form calciferous deposits as in bone formation and the calcification of connective tissue and blood vessels. It is important that for protection against damage by Ca^{2+} in presence of phosphate some exogenous source of adenine nucleotide be present unless, alternatively, oligomycin or Mn^{2+} are present. When Ca phosphate is precipitated intramitochondrially in presence of nucleotide, about one adenine nucleotide per 13 Ca^{2+} is taken and becomes associated, it is thought, with the Ca phosphate.

If mitochondria are exposed to Ca^{2+} in absence of energy, the low-affinity binding operates to take up Ca^{2+}, but there is no H^+ production. When energy is given in this situation, the Ca^{2+} is moved inside with the usual respiratory burst and H^+ efflux[71].

The energised uptake of Ca^{2+} will proceed in quantity, using acetate as counter-ion. In this case, the Ca^{2+} acetate that accumulates is associated with its osmotic equivalent of water, showing that the compound is ionised when inside[72]. Adding phosphate to the Ca^{2+} acetate-swollen particles causes a shrinkage as the Ca^{2+} is removed as insoluble phosphate.

Uptake of Ca^{2+} when energy is available takes precedence over phosphorylation; it also causes oxidation of the respiratory carriers more rapidly than does ADP. The concentration of ionised Ca that can exist in contact with energised mitochondria is extremely low, less than will give a reaction with murexide in a system that will detect Ca^{2+} at 0·5 µM. Ca^{2+} is taken from solutions of chelating agents such as EGTA into the mitochondria because the small concentration of ionised Ca^{2+} present with the chelate is continuously used to feed the ion into the particles.

The protective effect of Mn^{2+} mentioned above[73] may be related to the formation by Mn^{2+} of a different kind of ATP chelate than Ca. Combination with Mn^{2+} activates several intramitochondrial enzymes (isocitrate dehydrogenase, pyruvate carboxylase, and phosphoenolpyruvate carboxykinase). Pyruvate carboxylase produces oxalacetate from pyruvate; then either PEP or citrate, according to the availability of GTP or acetyl CoA, can be formed.

When citrate is produced, it chelates divalent cations so Ca can be changed to a form which leaves the particles. The presence of Mn allows this process to proceed at Ca concentrations that otherwise would have caused inhibition of the key enzyme, pyruvate carboxylase.

STOICHIOMETRY

Uptake of Ca^{2+} from either sucrose or other media having a low salt concentration and from any medium containing phosphate proceeds with a ratio of nearly two Ca taken in per O passing a phosphorylation site[74]. NAD-linked substrates energise entry of six Ca per O. Since two Ca ions appear to have the same energy requirement for entry as does phosphorylation of one ADP molecule, an intimate link between Ca and phosphorylation may turn up. From media having high salt concentrations and no phosphate, higher Ca^{2+}/O ratios are obtainable; they reach four to five Ca per O per phosphorylation site (see Figure 8.9 and reference 75). This Ca gain is associated with acidification of the medium, which can be attributed to uptake of the leaked endogenous phosphate. Another situation in which Ca is taken up, but this time without obvious extra energy consumption, is obtained if the ion is slowly added so as to keep its concentration below 5 µM. The respiratory rate then remains unaffected despite passage of the Ca into the particles, so presumably the resting energy production can be diverted to the Ca transport[74]. In media with high salt concentration the Ca does not remain permanently bound, but is subject to release followed by renewed uptake[76]. There is no corresponding oscillation of respiration, so it appears that the secondary uptakes are not coupled to energy requirement.

INTERNAL pH

A useful technique for demonstrating changes of acidity in some internal space of the mitochondria is afforded by staining the particles with bromthymol blue[65]. The adsorbed dye shows a move to alkalinity when energised Ca uptake occurs, but the change is transient, lasting a matter of a minute. Since the Ca^{2+} entry is accompanied by anion there is a higher inside/outside ratio for anions including HPO_4^{2-}, and this gives rise to alkalinisation inside, leaving the medium acid. The pH shift is impeded by the lanthanides $(La^{3+}, Y^{3+}, Pr^{3+})$ which may either act by attachment to the Ca^{2+}

binding sites or by their property of forming highly insoluble phosphates. The movement of Ca^{2+} into the particles is facilitated by certain local anaesthetics (butacaine, nupercaine) which also keep Ca^{2+} from becoming bound to membrane sites[49].

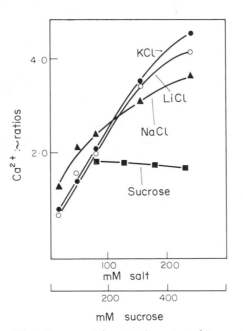

Figure 8.11. Effect of sucrose, K^+, and Li^+ on the Ca^{2+} : ∼ ratio. The test system (1·9 ml) contained 5·0 mg mitochondrial protein, 10 mM sodium succinate, 10 mM tris chloride (pH 7·4), and salts or sucrose as shown. The temperature was 24–25 °C. To start the jump, 390 nmol $^{45}CaCl_2$ were added. Data were calculated on the basis of extra oxygen uptake

Another method that has been applied to obtain a quantitative estimate of internal pH is to measure the distribution of the weak acid dimethyloxazolidinedione[77]. This substance, with pK 6·13 at 37 °C, is more ionised in the alkaline interior than in the medium. It is accumulated in the ionic form, while the nonionised form is presumed to be equally concentrated inside and out. From application of the ionisation equation, the internal pH comes out to be 0·5 unit higher than the outside, the difference rises to 1 unit after addition of Ca^{2+}. This change is consistent with the rise in anion accumulation accompanying the Ca^{2+} entry. A smaller alkalinisation accompanies ionophore-induced K^+ uptake[66]. In this case, the water movement that accompanies the K^+ salt lessens the concentration ratio attained and so there is less gradient of pH.

It is worth noting that the alkalinisation obtained by Ca uptake can move the interior to a pH (over 8·5) which is inhibitory to some of the enzymes concerned in energy production. This may account for inhibition of respiration by high uptakes of Ca.

Mg^{2+}

Liver mitochondria carry 20–30 nmol Mg per mg protein and retain this through exposure to chelating agents, though it is freed during exposure to 10 µM bilirubin or to ADP and DNP together. After depletion of energy reserves and internal K, they will take up Mg along with phosphate when provided with glutamate as energy source[78]. Heart mitochondria, even without prior treatment, take up Mg and phosphate and the ion uptake competes with phosphorylation for energy. The Mg uptake into heart mitochondria is facilitated by Zn^{2+} ions[79]. Since the Zn interferes with the first two phosphorylation sites, the Zn^{2+}-stimulated process has to be driven from TMPD ascorbate or ATP. Mercurials inhibit the energy transfer from TMPD to the membrane-bound cytochrome of the particles, so they are inhibitory with this energy source. However, if ATP is provided, the mercurials still further stimulate Mg^{2+} entry.

Sr^{2+}, Ba^{2+}, AND Mn^{2+}

The divalent cations, Sr, Ba, and Mn, are taken up by metabolising mitochondria with consumption of additional respiratory energy. The rates of accumulation are somewhat less than that of Ca (given as 6·2 nmol $mg^{-1} s^{-1}$), being respectively 4·8, 2·0, and 2·2 units. There is a marked interaction between Mn^{2+} and Ca^{2+}, such that the rate of accumulation for the mixture is intermediate between the values holding for the separate ions. Oxidation of the respiratory carriers occurs in response to exposure to the ions, as in the case of Ca; for cytochrome b the half-times are 30 ms for either Ca or Sr and 1 s for Mn^{2+} and for Ba^{2+} [10].

While uptake of Ca^{2+}, Sr^{2+}, and Ba^{2+} is impeded by the tri- and quadrivalent cations, a small amount of trivalent cation stimulates Mn^{2+} uptake and greater amounts are inhibitory. This last effect may be due to freeing of membrane-bound Ca by the multivalent cations, so that a combination of Ca and Mn is present to pass into the interior.

317

ANIONS

The permeability of the inner membrane of the mitochondrion to anions has a quite different spectrum to that of the cell membrane. It allows the passage of many of the metabolically important compounds while excluding small anions like chloride. Some of the substrates appear to have asymmetric permeation properties so that they can, for example, emerge when formed internally but not enter from the cytosol. Also, specific exchange requirements may exist and the degree of swelling certainly alters the permeability properties. Intact liver mitochondria will not oxidise external NADH but will do so when sufficiently swollen in hypotonic media. Oxoglutarate oxidation by rat liver mitochondria is particularly sensitive to increased tonicity and is also prevented if colloid is added to the medium. The latter point is specially interesting because it suggests that water distribution, as determined by the presence of colloids, is important in modulating the permeability.

Anions can only be gained if some compensating movement of other charged particles takes place. For example, an anion can enter with a proton, and the latter may then attach to an anionic group of the internal buffer rendering it neutral. This change would be likely to occur when a suspension is acidified, and it is experimentally observed that acidity favours anion entry. Alternatively, one anion can exchange for such other mobile ones as are present internally already[80, 81]. Still a third possibility is for the anion to accompany a cation (see pages 308 and 309 and Figure 8.12).

RANKING ANIONS IN ORDER OF PERMEABILITY

The permeability of the inner membrane can be studied by a variety of methods which involve different concentrations of test substance. Perhaps for this reason they do not always give consistent results. A method based on observation of whether the mitochondria swell in solutions of the ammonium salt of the acid recalls studies made in the same way of erythrocyte permeability. It has been applied by Chappell[82, 83]. The alternative is to see whether the substrate causes changes in redox state of the respiratory carriers; here it is used at low concentration.

From his studies, Chappell (reference 83 for a summary and see Table 8.4) concluded that monocarboxylic acids enter without a special requirement, while dicarboxylic acids appear to require phosphate; the latter is supposed to enter and then exchange against the dicarboxylate. Nevertheless, a reduction of endogenous pyridine

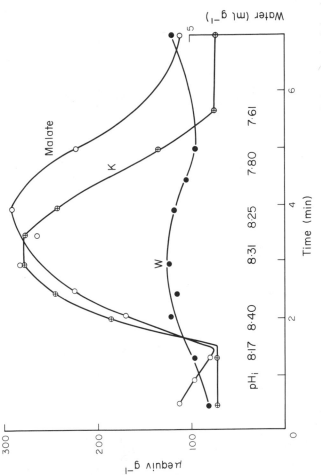

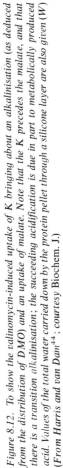

Figure 8.12. To show the valinomycin-induced uptake of K bringing about an alkalinisation (as deduced from the distribution of DMO) and an uptake of malate. Note that the K precedes the malate, and that there is a transition alkalinisation; the succeeding acidification is due in part to metabolically produced acid. Values of the total water carried down by the protein pellet through a silicone layer are also given (W) (From Harris and van Dam[44]; courtesy Biochem. J.)

Table 8.4 REQUIREMENTS FOR VARIOUS ANIONS TO ENTER LIVER MITOCHONDRIA

These were found by combinations of the swelling test made with the ammonium salt and by observations of changes in redox state of the endogenous pyridine nucleotides

Anion	Requirement, if any, for entry	Inhibitors of entry
Acetate	None	
Phosphate	None	Mercurials, formaldehyde, N-ethyl maleimide
Arsenate	None	Phosphate competes
Hydroxybutyrate	None	
Pyruvate	None	Lactate competes
Glutamate	None	Avenociolide, hydroxyglutamate, asportate
Malate	Phosphate	Mercurials inhibit exchange with phosphate, butyl malonate inhibits exchange with other dicarboxylates
Succinate	Phosphate	
Oxalacetate	Phosphate (slow entry)	
Oxoglutarate	Malonate or malate	Aspartate slows the reaction with endogeneous NAD
Citrate	Malate	
isoCitrate	Malate	2-ethyl citrate, benzenetricarboxylic acids
Aconitate	Malate	

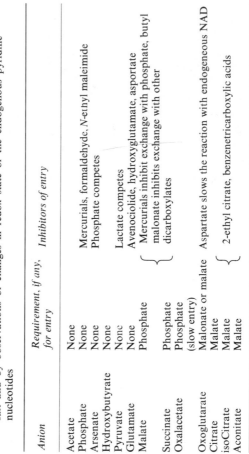

320

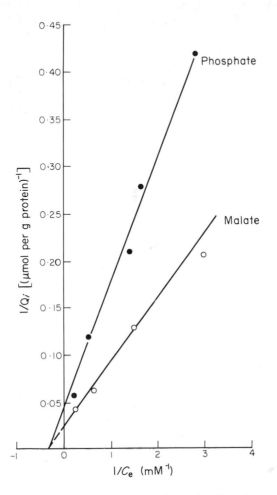

Figure 8.13. An example of the adsorption isotherms describing the malate accumulation as a function of the applied concentration using a double reciprocal plot. The competition offered by phosphate is shown by the lessened accumulation in the presence at 2 mM (From Harris and Manger[80]; courtesy Biochem. J.)

M

nucleotide by added dicarboxylate does not require more phosphate than is present inevitably from leakage. Under carefully chosen conditions, addition of a small concentration of phosphate will increase uptake of a dicarboxylate; this may be associated with a swelling of the matrix compartment, while more phosphate displaces the dicarboxylate competitively (Figure 8.13). Other competitions occur between dicarboxylates (Figure 8.14) and inert dicarboxylates lessen both accumulation and rate of oxidation of the

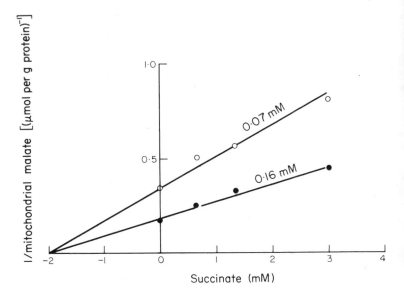

Figure 8.14. An example of intersubstrate competition for occupation of the mitochondrial interior. The two lines were obtained at 0·07 mM and 0·16 mM malate concentrations: succinate was the competitor (From Harris and Manger[80]; courtesy Biochem. J.)

active ones (Figure 8.15). Oxoglutarate seems to have special properties, depending in part on the origin of the mitochondria in question. Its entry into rat liver mitochondria is inhibited by colloids or by hypertonicity, but it readily penetrates heart mitochondria. When formed internally in liver mitochondria, it tends to remain inside until another dicarboxylate enters in exchange. The tricarboxylates behave as if they will only move in exchange for a dicarboxylate. While a little malate increases citrate uptake, more competitively ousts the citrate (Figure 8.16). Unless given a preliminary depletion, liver mitochondria carry sufficient endogenous anions to permit some tricarboxylate utilisation, but the rate of maximum respiration can usually be increased by adding malate.

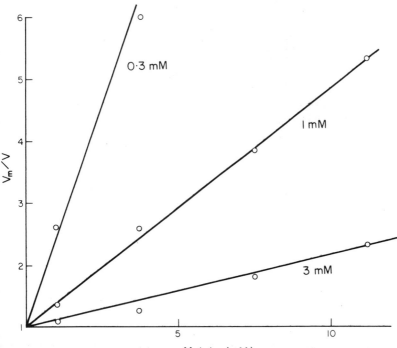

Figure 8.15. The rate of oxidation of succinate (V) plotted inversely against the concentration of malate added in presence of rotenone. Succinate was at 0·3, 1, or 3 mM. V_m was the rate of succinate oxidation in absence of malate. The form of the results shows that malate, which is not oxidisable in presence of rotenone, is acting as a competitor in the system. Taken with the result of experiments like those shown in Figure 8.14, it is inferred that the effect is due to a lessened concentration of succinate at the site of the enzyme (From Harris and Manger[80]; courtesy Biochem. J.)

Prior incubation with uncoupler converts endogenous phosphates to inorganic phosphate[84] which will exchange for dicarboxylate when the latter is offered in high concentration[85]. The uncoupler treatment discharges both Ca and phosphate. The particles will then only take up most substrates if given with Ca^{2+} and an energy source (page 309), though citrate only requires energy in presence of dicarboxylic acid[86] (Figure 8.17). The energy supply can come from the dicarboxylic acid itself. For reduction of the endogenous pyridine nucleotide, as distinct from accumulation, only the di-carboxylic acid is required[87] (Figure 8.18).

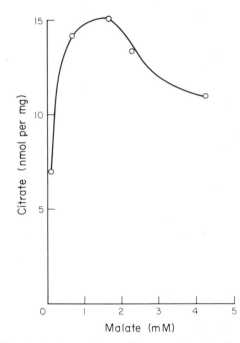

Figure 8.16. *Mitochondrial accumulation of citrate as a function of malate concentration. As malate is added its facilitatory effect is first seen, then mutual competition sets in. External citrate at 170 μM*

There is some doubt about the penetration of oxalacetate. It seems unlikely that it can move through the membrane of mito-chondria in the cell, because very different ratios between oxalace-tate and malate persist as between cytoplasm and mitochondrial matrix. There is no reason to expect malate not to be freely per-meable, so oxalacetate should not be. Under conditions of the ammonium salt swelling test a penetration has been recorded[88]. It appears that exchanges between di- and tricarboxylates, when on a large scale, are on a 1-for-1 molecular basis[89], so a pH shift is required to maintain neutrality.

Heart mitochondria have a different permeability pattern to those from liver. Citrate is poorly permeable, but oxoglutarate is

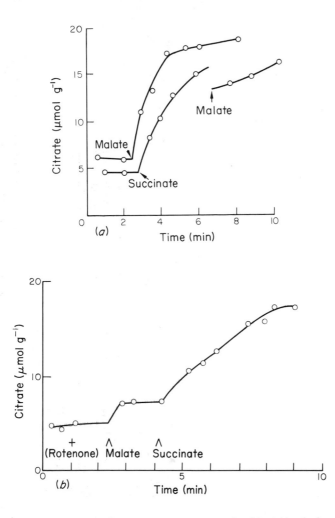

Figure 8.17. (a) To show that citrate entry into mitochondria depleted of energy by prior exposure to uncoupler is restored by dicarboxylic acids (succinate or malate) which yield energy. If rotenone is present to stop oxidation of malate (b) there is little consequent citrate uptake but succinate, which can still be oxidised to provide energy, still brings about the uptake (From Harris[86]; courtesy Biochem. J.)

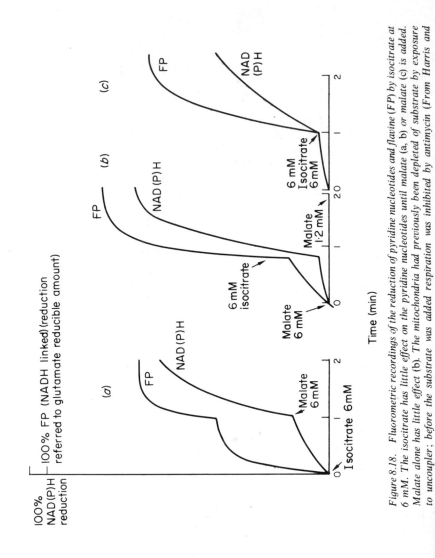

Figure 8.18. Fluorometric recordings of the reduction of pyridine nucleotides and flavine (FP) by isocitrate at 6 mM. The isocitrate has little effect on the pyridine nucleotides until malate (a, b) or malate (c) is added. Malate alone has little effect (b). The mitochondria had previously been depleted of substrate by exposure to uncoupler; before the substrate was added respiration was inhibited by antimycin (From Harris and

freely permeable so long as another dicarboxylate can move in exchange. For some reason, aspartate tends to be exported when it is formed but it does not enter when applied externally. For recent work involving heart mitochondrial permeability as a control of metabolism, see reference 90.

KINETICS OF ANION EXCHANGE

Observation of the reduction of NAD by substrate shows that the process has the necessary rate to account for the observed oxidation rates found in the oxygen electrode. It is difficult to apply analytical methods with good time resolution because there is a delay in separating the particles from the medium. A method that has been used involves the addition of a 'stopping' agent shortly after the anion has itself been added. Ideally, the stopping agent inhibits further movement without disturbing what has already penetrated. There is necessarily some question whether the stopping agent fulfils these conditions. but Quagliariello, Palmieri, and Klingenberg[91] have obtained data for succinate entry that are consistent with expectations based on knowledge of respiration rates. At 1 °C they found the rate constant for succinate was $1.25 \, min^{-1}$ and it rose to $18 \, min^{-1}$ at 20 °C. The flux of succinate was related to the applied concentration by a saturation law having a pH-dependent K_m. At pH 6.35 the K_m was 0.83 mM and at pH 7.1 it was 1.5 mM.

COMMON ANION RATIOS

The matrix, with its content of buffer, has a capacity for anions that depends on the ionisation of its buffer and on the content of cations. It appears that the internal buffer is acting as anion for about 70 nmol of the internal K per mg at pH 7.4 in rat liver mitochondria; internal material may well provide more anion in heart mitochondria. The conclusions come from consideration of the total anion content found in the particles, which amounts to about 30 nequiv per mg in liver and less for heart compared with 100 nequiv K per mg. For a non-energy consuming distribution of the anions the relations between the internal and external concentrations should depend on the anion's charges when they are present together (see p. 80).

This relation is valid for a number of anions distributed between interior and solution for rat liver mitochondria[92] (Table 8.5). Phosphate in both singly and doubly charged forms also fits[93] and so sets the internal pH in relation to that outside. The anions oxoglutarate and aspartate do fit in this scheme. To what extent anions distribute in heart mitochondria in an analogous way is still under investigation.

Table 8.5 USE OF TOTAL INORGANIC PHOSPHATE RATIO INSIDE/OUTSIDE TO FIND n AND COMPARISON WITH VALUES OF n FOUND FROM ORGANIC ANION RATIOS

Extl pH	Extl phosphate (mM)	Ratio of inorganic phosphate inside/outside	Ratio of citrate	Ratio of dicarboxylic acid	n from P_1	n from cit*	n from DCA*	n from pyr*
7·4	0·5	18·0	57	15·4 (mal*)	4·8	3·8	3·9	—
7·4	0·4	5·7	20	5·3 (suc*)	2·5	2·7	2·3	—
7·4	0·26	12·1	46	13·7 (mal*)	3·8	3·5	3·5	—
7·4	0·22	30·4	129	—	6·3	5·05	—	—
6·5	0.16	10·2	73	—	4·7	4·2	—	—
7·0	0·076	20·6	82·5	15·9 (mal*)	5·6	4·3	4·0	5·7

* cit: citrate; DCA: dicarboxylic acid; pyr: pyruvate; mal: malate; suc: succinate

COMPETITION OR INHIBITION BY ANALOGUES AND OTHER AGENTS

The movement of phosphate is inhibited if the mitochondria are exposed to mercurials such as mersalyl[94, 95]. The entry is restored by compounds which remove the mercurial, such as cysteine.

The movements of the organic anions is interfered with by analogues in general. To varying extents these may enter the mitochondria, so that besides competing for passage through the membrane the foreign compounds may compete for accommodation in the interior. Recent work shows that octyl malonate enters mitochondria despite its large alkyl group. Nevertheless, much published work depends on using the smaller butyl malonate, or phenyl succinate, as transport inhibitors without attention to the amounts which accumulate internally. Robinson and Williams[96] found better results using iodobenzyl malonate or pentyl malonate than the butyl malonate.

Movement of citrate or isocitrate is hindered by 2-ethyl citrate and by tricarbyllate, as well as by benzene 1,2,3 tricarboxylic acid[97].

Glutamate uptake is interfered with by the hydroxyanalogue, by hydroxyaspartate, and by aspartate itself, as well as by the naturally occurring avenacialide[98].

The reduction of internal pyridine nucleotide by added external oxoglutarate in presence of an inhibitor of respiration has been reported to be inhibited by aspartate[99]. With oxoglutarate, in presence of ammonia and arsenite to stop oxidation, the rate of pyridine nucleotide oxidation becomes an index of entry of the reactants, which form glutamate. The method has been applied to show that malate and malonate facilitate the process[100]. As mentioned elsewhere, the inward movement of oxoglutarate depends crucially on the state of swelling of the particles[101].

The entry of anions being discussed proceeds to the same degree, whether or not energy supply is cut off at the time of measurement, but does decline if energy is cut off before applying the test anion. A factor to be considered is that after energy is no longer available the endogenous phosphates hydrolyse and inorganic phosphate tends to leak from the particles[84, 102]; it is then no longer available inside to exchange against anions. This may not be the whole consequence, however, because the loss of phosphate and endogenous K (p. 294) leads to shrinkage and a lowered membrane permeability.

A MECHANISM FOR ION UPTAKE BY MITOCHONDRIA

Any attempt to explain the uptake of ions by mitochondria must allow for the fact that by giving a trace of an ionophore there can be caused a massive uptake of both positive and negative ions. The usual cation[50] is taken in K. The uptakes of Ca, and of K when it is induced by an ionophore, are clearly linked to energy consumption. The ionophore renders the lipid part of the membrane readily permeable to a cation or cations, as it does artificial bilayer membranes. It seems then correct to infer that the energy-consuming ion-moving process occurs on the inside of the lipid membrane and operates on such cations as reach it. A purely electrostatic effect on cations would not account for anion uptake; one needs that an energised opening of a system of fixed positive and negative sites should occur so as to provide regions to which both cations and anions can move. Experimental evidence is that cations tend to precede the anions and in many places in this chapter it has been pointed out that the anion uptakes can be

329

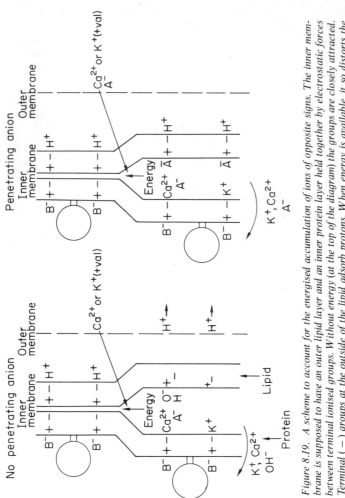

Figure 8.19. A scheme to account for the energised accumulation of ions of opposite signs. The inner membrane is supposed to have an outer lipid layer and an inner protein layer held together by electrostatic forces between terminal ionised groups. Without energy (at the top of the diagram) the groups are closely attracted. Terminal (−) groups at the outside of the lipid adsorb protons. When energy is available, it so distorts the membrane system that the layers tend to cleave to the extent that cations (Ca or K) can reach the (−) sites; OH^- groups are attracted to the (+) sites freed, but these are mainly satisfied by a migration of the outer charge into the lipid membrane. This entails a detachment of the adsorbed protons proportional to the cation gained and an alkalinisation of the interior. This is as observed either when Ca is gained or, if K is gained, in presence of an ionophore to increase the permeability of the lipid membrane. When, as on the right, a penetrating anion A^- is available, it replaces the requirement for internal OH and there is no longer an alkalinisation, nor are protons desorbed from the exterior

accounted for electrostatically, thus leaving cation movement as the primary electrogenic component of the process. A scheme in which layers of oppositely charged sheets cleave to provide sites is sketched in Figure 8.19. An important part of the process is the migration of charge within the membrane.

ADENINE NUCLEOTIDES

Adenosine diphosphate and triphosphate have their own specific carrier system. Primarily, this operates to exchange the internal phosphates in the proportion in which they are inside[103] since external ADP taken up preferentially to ATP. That the mechanism is an exchange has been shown by the use of labelled nucleotides, tritiated nucleotide inside and C-14 ADP outside. External ATP in the absence of ADP exchanges at about half the rate of the latter. At 10 °C, 38 μmol per g ADP exchanges per minute; the rate constant is $3–5·4 \, min^{-1}$, depending inversely on the total content of nucleotides. If external ATP is added with ADP it is a poor competitor; only 4–6 units move in as compared with 28 of ADP when both are applied at 0·15 mM. This behaviour accords with the requirement for mitochondria to export ATP when receiving ADP from the cytoplasm. At the same time, the lack of preference between the nucleotides emerging depresses the rate of phosphorylation by keeping up a continuous in-and-out cycling of ADP.

The ADP–ARP exchange ensures that the adenine moiety of the nucleotide is not transferred. The monophosphate (AMP) is nearly impermeable on the exchange carrier. External AMP has to react with myokinase and ATP to form ADP. Internally, AMP can only be phosphorylated by reacting with GTP and under the influence of nucleotide monophosphokinase to form ADP. Of the other nucleotides, only the synthetic deoxyadenosine analogues of ADP and ATP permeate[104]. It is important to note that the guanosine phosphates do not transfer; yet they are normal constituents of mitochondria; GDP is converted to GTP when succinyl CoA is hydrolysed to succinate and CoA.

ATRACTYLATE

The adenine nucleotide exchange process is specifically blocked by a steroid compound occurring as the K salt. It is known as atractylate or atractyloside, for it is also a glycoside. It occurs in a Sicilian thistle. The addition of 10 μM atractylate stops the exchange between external ADP and internal nucleotide. The compound has been used as an aid in determining the kinetics of the process[105], since it provides a means of immediately stopping the exchange. At low levels, atractylate actually favours the the phosphorylation of the internal nucleotides because it stops the export of ATP (in exchange for ADP). It does, however, stop the phosphorylation by substrates oxidised outside the matrix, such as hydroxybutyrate, because then the matrix ATP can only be generated by a cycling of ADP and ATP within the membrane. Atractylate binds to membrane material and competes with ATP in so doing. This indicates that the atractylate inhibition of the adenine nucleotide exchange stems from its occupation of the carrier[106]. Since atractylate inhibits APT transfer, it prevents incorporation of external ATP into mitochondrial RNA[107].

AUROVERTIN

Aurovertin binds to the mitochondrial membrane and imparts to it a characteristic fluorescence. The compound inhibits phosphorylation without delay (a distinction from clogomycin). It seems to act by altering the relative affinates of the ATPase to ATP and ADP, so that ADP becomes a much less effective competitor for the enzyme sites and consequently does not protect the ATP from hydrolysis. The K_i for ADP is raised from 8·6 to 180 μM by 2·5 nmol aurovertin per mg particle protein[108]. The V_{max} of the hydrolysis is left unchanged, but the K_m for ATP is halved. These facts serve to explain why the substance does not inhibit completely the use of added ATP for induced ion movement or for reverse electron flow[38]. Provided sufficient ATP is present to saturate the ATPase, some energy can be derived from the surplus entering.

OLIGOMYCIN

Oligomycin inhibits phosphorylation in intact mitochondria with a lag of seconds to minutes depending on the concentration applied. It also inhibits ATPase by reducing the V_{max}. However, phosphorylation by particles prepared by sonication of mitochondria can be enhanced by about 0·5 μg oligomycin per mg protein; this presumably shows that their ATPase activity is reduced more than the phosphorylating activity. The ATPase of the particle preparation is only oligomycin-sensitive when it includes the F_0[27]. Oligomycin does not inhibit use of respiratory energy to drive ion uptakes or reverse electron flow (e.g. reduction of NAD by succinate). This observation was the basis on which it was decided that the energy is produced at a site removed from the actual phosphorylation. Uncouplers are active in presence of oligomycin, though their activity is usually lessened in its presence.

GUANIDINES

Guanidine, and more especially its alkyl derivatives, which are more lipophilic, inhibit phosphorylation without fully inhibiting uncoupler-stimulated respiration. They appear to compete with Mg in the membrane[109, 110].

The hypoglycaemic agent phenyl ethyl biguanide inhibits phosphorylation. Its effect on ADP-stimulated respiration depends on the rate of respiration at the time it was added[109, 110] This suggests that with an active substrate more of the compound is taken into the matrix there to act on the phosphorylation system.

BONGKREKIC ACID

Henderson and Lardy[111] have described an inhibitor of phosphorylation called bongkrekic acid that has an effect similar to atractylate; it does not prevent phosphorylation of the internal adenine nucleotides but prevents export of ATP in exchange for ADP and stops the incorporation of ^{32}P into external ATP. Klingenberg[112] used labelled nucleotides to show that the compound causes a reduction of the K_m of the membrane adenine nucleotide carrier to such an extent that the ADP, when associated with the carrier, cannot be detached. The mode of action differs from the competitive binding found with atracylate[113].

TEMPERATURE EFFECTS ON PHOSPHORYLATION AND ON THE ADP–ATP EXCHANGE

The rate of the ADP–ATP exchange has a temperature sensitivity between 4 and 10 °C, corresponding to an activation energy of 24 kcal per mol. The rate at which endogenous nucleotide is phosphorylated only increased by about 1·5 in the same span. When ATP is exported in exchange for ADP, two factors determine the rate; these are the exchange permeability and the internal ATP concentration. The latter effectively sets the outward ATP gradient when external ATP is negligible. Different substrates sustain different respiration rates and ATP levels during phosphorylation. In an example with glutamate plus malate, the internal ATP during exposure to an ATP-trapping system was 1·7 nmole per mg at 10 °C and this provided a rate of 7 nmole per mg per min. Warming to 20 °C increased the internal ATP to 3 nmole per mg and the rate to 30 nmole per mg per minute. The changes correspond to the specific exchange permeability rising from $7/1·7 = 4·1$ to $30/3 = 10$, so part of the increased phosphorylation rate is from permeability, part from higher internal ATP.

Between 20 and 30 °C, the ADP for ATP exchange is increased by the same factor as between 10 and 20 °C, but the rate of ATP export is only doubled. This is in part because the internal ATP does not rise and probably also because limitation of the rate by diffusion sets in; the ADP and ATP have to reach the membrane and this involves diffusion in the solution.

Ratios between ATP and ADP inside and outside the mitochondria

Operation of the exchange between external ADP and internal ADP and ATP (in proportion to their prevalent internal ratios) brings about phosphorylation external nucleotides to a certain extent. With external inorganic phosphate at 1 mM the ratio ATP/ADP in the external medium may reach 30^{103} while inside, where phosphate was estimated to be at 4 mM, the ratio was between 3 and 5. The conclusion is that the equilibrium ratio for the conversion of ADP to ATP is subject to change by the membrane potential difference. At about pH 7·4, the reaction is $ADP + P_i + 0·7 H^+ \rightarrow$ ATP; the figure 0·7 is pH-dependent because it represents the difference between the ionisations of the various acids groups on the right and the left (for values see Nishimura et al.[114]). Klingenberg, Wulf, Heldt and Pfaff[30] have favoured the idea that internal negativity tends to expel the ATP rather than the ADP. However, the total charges on the phosphate ADP exceed those on the ATP. They prefer to invoke a P_i for hydroxyl exchange as a neutral process. It may be preferable to regard transmembrane ATP output as an energy-consuming process depending as for rate on the concentration of internal ATP. There is a linear relation between the output obtained in presence of excess external ADP and substrate and the internal concentration[115] (Figure 8.20).

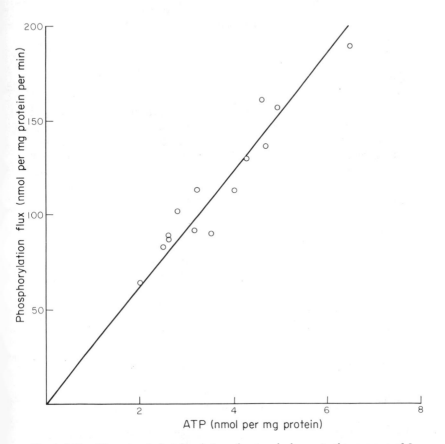

Figure 8.20. The rate of phosphorylation of external glucose in the presence of 5 units hexokinase and 120 μM ADP and the internal ATP content of the mitochondria. The points were obtained using different substrates. Protein concentration was 0·7–4 mg ml^{-1}; higher rates were from lower protein concentrations. T = 20 °C (From Harris[115]; courtesy J. Bioenergetics)

At 20 °C for rat liver mitochondria it is found that about 30 nmol ATP will be exported per min per mg protein for each 1 nmol ATP per mg in the particles. Raising the temperature to 30 °C increases the factor to about 70. This factor can be regarded as a measure of the outward permeability to ATP in the presence of excess external ADP and phosphate. The value is reduced by certain agents that have been found to hinder phosphorylation; these include atractylate, bongkrekic acid and fatty acids, the latter only when albumin is absent.

ATP–PHOSPHATE EXCHANGE

Exposure of mitochondria to phosphate in presence of Mg leads to loss of internal ATP and gain of phosphate[116, 117]. In absence of Mg, the ATP loss is very low, the same as occurs in a medium free from added phosphate. The ATP–phosphate exchange is relatively slow; 30 to 60% of the ATP can be lost in 10 min at 30 °C. Agents that use energy, so that internal ATP is converted to ADP, lessen the rate of nucleotide loss, because it is ATP which is preferentially lost. Examples of such agents are uncouplers, valinomycin, and oligomycin. Atractylate hinders the ATP loss, from which it is inferred that the same carrier is involved as is used for the ADP–ATP exchange. Sulphate replaces phosphate in exchanging with internal ATP.

Added external ADP at a concentration only one twentieth that of the sum of the internal ADP + ATP stops the ATP loss. It appears that the carrier bearing ATP derived from inside will preferentially exchange it for external ADP but given enough time will release ATP in exchange for phosphate or sulphate. It is not clear how the electrical balance is maintained when ATP is exchanging for phosphate.

TRANSFER OF FATTY ACIDS

Fatty acids in small quantities will penetrate freshly prepared liver mitochondria without needing a cofactor. The effect can best be seen on adding the acid to a suspension in presence of ADP and phosphate when a burst of respiration results. The fatty acids are like uncoupling agents or calcium ions; in quantity, they destroy the membrane and leave an active ATPase. This surface activity effect can be obviated if the acid is added after a sufficient level of serum albumin. The latter binds the acids[118] and liberates them at a sufficiently low concentration not to be destructive.

After storage or incubation of the mitochondria, the fatty acids with more than 10 C atoms no longer penetrate the membrane unless both external ATP and carnitine are added.

The ATP in presence of carnitine serves to activate the fatty acid, perhaps via a membrane-bound part of the total CoA pool, to form the acyl carnitine. The latter passes through the membrane readily, but the carnitine moicty does not enter the matrix; it returns to the outside, leaving inside the acyl CoA made now with endogenous CoA. There are separate internal and external carnitine–CoA transacylases. Hence, carnitine behaves as a membrane carrier for the acyl group; this property may be related to the fact that the acyl compound is neutral and so does not require an oppositely charged ion to accompany it.

Some details of the acyl transfer are discussed in reference 119.

CYCLIC OSCILLATIONS OF METABOLIC RATE AND ION CONTENT

The combination of energy requirement imposed by uptake of cations with the fact that the cations carry in substrate anions to provide more fuel can lead to an oscillatory condition in which respiration periodically builds up as the load increases and then collapses when the oxidative system is unable to provide energy at a rate sufficient to maintain the internal cation content with its associated anions. After the collapse, there is more or less dead time during which a small residual penetration of substrate can restart the process.

These oscillations can either be maintained without damping for long period or die away in a few minutes, according to the conditions used.

With phosphate present, a damped oscillation follows addition of valinomycin when the K concentration is 10–50 mM. In absence of phosphate, a combination of valinomycin with a trace of oleate leads to a sustained oscillation critically dependent on the K concentration (about 5 mM) and stopped by phosphate or other penetrating anion. This oscillation does not involve the mitochondrial ATP, which is at a low level. It can be protected from such lowering by oligomycin which does not interfere with the oscillation. Another mode of undamped oscillations is obtained in presence of succinate and low ionic strength; valinomycin is added to the anaerobic mixture and the oscillation is triggered with hydrogen peroxide which furnishes oxygen thanks to the endogenous catalase activity of liver mitochondria[120]. This oscillation involves the endogenous ATP, which rises and falls 90° out of phase with the mitochondrial K; the process is stopped by oligomycin.

Table 8.6 DISTRIBUTION RATIOS FOR ANIONS BETWEEN MITOCHONDRIA AND MEDIUM DETERMINED WHEN SEVERAL ARE PRESENT TOGETHER

	Anions measured and concentrations used (μM)									
	Citrate $n = 3$	Succinate $n = 2$	Malate $n = 2$	Malonate $n = 2$	Oxoglutarate $n = 2$	Acetate $n = 1$	Glutamate $n = 1$	Pyruvate $n = 1$	DL-Hydroxy-butyrate $n = 1$	Phosphate pH 7·4
No. 1.	40		200					500		
Ratio	263		27					7·8		
1/nth root	6·4		5·2					7·8		
No. 2	200	150	150					500		
Ratio	19·7	5·7	5·7					2·45		
1/nth root	2·7	2·4	2·4					2·45		
No. 3	60	120	180				760			
Ratio	276	6·9	5·0				2·7			
1/nth root	3·0	2·6	2·2				2·7			
No. 4	100	120			150	6000				
Ratio	26·5	10·3			3·5	5·5				
1/nth root	3·0	3·2			1·9	5·5				

No. 5	70	130	90	90		260	
Ratio	106	8·25	8·2	3·55		3·1	
1/nth root	4·7	2·8	2·8	1·9		3·1	
No. 6	40	130		200		375	
Ratio	134	27·8		7·6		11·4	
1/nth root	5·1	5·2		2·7		11·4	
No. 7	40	100		320	140	280	60
Ratio	414	35·8		15·1	5·8	9·1	8·9
1/nth root	7·4	6·0		3·9	5·8	9·1	
No. 8	14	30				1000	400
Ratio	20·3	5·35				3·35	5·7
1/nth root	2·7	2·3				3·35	

To test the Donnan-like behaviour, the 1/nth roots of the ratios are compared, where n is the presumed net charge. The ratios are of the mitochondrial content (μmol per g protein) corrected for the amount of solute carried down in the sucrose accessible space divided by the external concentration in mM. In experiments 1–7, rotenone was present at 0·15 μg per mg and in No. 8, antimycin was present at 0·25 μg per mg protein. Anions which do *not* appear to fit in the scheme are shown in roman.

PYRIDINE NUCLEOTIDES

The inner membrane of liver mitochondria is almost impermeable to pyridine nucleotides[121]. The matrix contains both nicotinamide adenosine diphosphate and the phosphorylated derivative. The two compounds serve as redox indicators and are generally more reduced in the matrix than are the corresponding compounds in the cytosol. The reason for the difference in the redox state lies in the differing concentrations of reactants of the various redox reactions as between cytosol and the mitochondrial interior. For example, the production of pyruvate by glycolysis occurring exclusively in the cytosol poises the equilibrium: $Pyruvate_c/Lactate_c = K_{LDH} \times NAD_c/NADH_c$, and the activity of the malic enzyme determines the degree of reduction of cytoplasmic NADP by:

$$[Pyruvate] \quad [CO_2]/[Malate_c] = K_{ME}[NADP]/NADPH)_c$$

The mitochondrial processes determining the ratios for the pyridine nucleotides are for NAD and NADH:

$$Acetoacetate_m/Hydroxybutyrate_m = K_{HbDH}NAD_m/NADH_m$$

and equally:

$$Oxalacetate_m/Malate_m = K_{MDH} \cdot NAD_m/NADH_m$$

while the NADP and NADPH are determined by:

$$\frac{[Oxoglutarate]_m[Ammonium \ ion]_m}{[Glutamate_m]}$$

$$= K_{GIDH}[NADP]_m/[NADPH]_m$$

To evaluate the ratios, it is necessary to have information about the distribution of the reactants between cytosol and mitochondria.

This problem is where the Donnan relations between a number of the anion ratios is specially useful. Analysis of whole cells give only total contents of the substrates, which have to be partitioned in some way between cytoflam and mitochondria. Analyses of mitochondria for NAD, etc., do not give useful information, because part of their pyridine nucleotide is membrane-bound and does not enter into the above equilibria.

Undamaged liver, kidney, and heart mitochondria do not oxidise external NADH directly, whereas those from plant cells and some yeasts do so. The indirect oxidation of NADH can be achieved by providing some enzymes and cofactors to carry reducing equivalents into the particles. For example, liver mitochondria supplemented with oxoglutarate, aspartate, malic dehydrogenase, and oxalacetate–glutamate transhydrogenase oxidise added NADH by importing malate and glutamate and exporting aspartate and oxoglutarate; these react outside and the oxalacetate is reduced by the external NADH.

$$\text{Aspartate} + \text{Oxoglutarate} \overset{\text{GOT}}{\rightleftharpoons} \text{Glutamate} + \text{Oxalacetate}$$

$$\text{Oxalacetate} + \text{NADH} \overset{\text{MDH}}{\rightleftharpoons} \text{Malate} + \text{NAD}$$

MICROSOMES

When cells are disrupted, as in the ordinary course of preparation of mitochondria, there are three primary fractions obtainable by centrifugation. The first precipitated at low speed is nuclei and cell walls, and second, obtained at about 40 000 g min, is predominantly mitochondria, while in the hazy supernatent there remain small vesicles that can be precipitated at 10^6 g min or more. These vesicles are a mixture; part are the lysosomes containing lytic enzymes that are so kept isolated in the cell itself to prevent self-digestion, and part are derived from infolding of fragments of the reticular membrane system. The latter vescicles are filled with fluid derived from the cell sap more or less admixed with the homogenisation medium. This means that they contain small concentrations of enzymes (besides the enzymes bound on the reticular membranes) and ions, notably K and Ca.

The reticular membrane has enzymes for protein synthesis, and it has also a peculiar respiratory chain concerned with oxidation of foreign substances and sterols by hydroxylation. NADPH is oxidised along with the conversion of RH to ROH. A reductase couples NADPH to a cytochrome usually referred to as $p.450$. It is one of the b group, but is peculiar in having an absorption at 450 nm when combined with carbon monoxide.

341

REFERENCES

In the following list, five symposium volumes are extensively quoted; to avoid repeating the details in full these are denoted by:

A Mitochondria—Structure and Function (eds L. ERNSTER and Z. DRAHOTA), Academic, London (1969)
B Mitochondrial Structure and Compartmentation (eds E. QUAGLIARIELLO et al.) Adriatica, Bari (1967)
C The Energy Level and Metabolic Control in Mitochondria (eds S. PAPA et al.), Adriatica, Bari (1969)
D Citric Acid Cycle (ed. J. LOWENSTEIN), M. Dekker, New York (1969)
E Regulation of Metabolic Processes in Mitochondria (eds J. M. TAGER et al.), Elsevier, London (1966)

1. COCKRELL, R. C., and RACKER, E., Biochem. biophys. Res. Commun., 35, 414 (1969)
2. SLATER, E. C., in A, 205; and in C, 255 (1969)
3. GEORGE, P., Biochim. biophys. Acta, 223, 1 (1970)
4. GREVILLE, G. O., p. 1 in D (1969)
5. HELDT, H. W., in B, 260 (1967)
6. KLINGENBERG, M., in C, 189 (1969)
7. SLATER, E. C., Biochim. biophys. Acta, B.11, 345 (1970)
8. WEGDAM, H. J., BERDEN, J. A., and SLATER, E. C., Biochim. biophys. Acta, B.11, 365 (1970)
9. MITCHELL, P., in A, 219 (1969)
10. SKULACHEV, V. P., JASAITIS, A. A., NAVICKAITE, V. V., YAGUZHINSKY, L. S., LIBERMAN, E. A., TOPALI, V. P., and ZOFINA, L. M., in A, 275; and in C, 283 (1969)
11. REED, L. J., in The Enzymes (eds P. D. BOYER, H. A. LARDY, and K. MYRBACK), Academic, London, vol. 3B, p. 195 (1960)
12. JACOBS, E. E., and SANADI, D. R., Biochim. biophys. Acta, 38, 12 (1960)
13. MITCHELL, P., and MOYLE, J., F.E.B.S. Letters, 6, 309 (1970)
14. MARGOLIASH, E., BARLOW, G. H., and BYERS, V., Nature, Lond., 228, 723 (1970)
15. OHNISHI, T., KRÖGER, A., HELDT, H. W., PFAFF, E., and KLINGENBERG, M., Eur. J. Biochem., 1, 301 (1967)
16. VAN DEN BURGH, S. G., MODDER, C. P., SOUVERIJN, J. H. N., and PIERROT, H. C. J. N., in A, 137 (1969)
17. HITTELMAN, K. J., CANNON, B., and LINDBERG, O., in A, 145 (1969)
18. GAMBLE, J. L., Biochim. biophys. Acta, 66, 158 (1963)
19. CHRISTENSEN, R. O., LOYTER, A., and RACKER, E., Biochim. biophys. Acta, 180, 207 (1969)
20. LIBERMAN, E. A., TOPALI, V. P., ZOFINA, L. M., JASAITIS, A. A., and SKULACHEV, V. P., Nature, Lond., 222, 1076 (1969)
21. BUTLER, W. A., and JUDAH, J. D., Biochem. J., 118, 887 (1970)
22. PARSONS, D. F., WILLIAMS, G. R., THOMPSON, W., WILSON, D., and CHANCE, B., in B, 23 (1967)
23. ERNSTER, L., and KUYLENSTIERNA, B., in A, 5 (1969)
24. FERNANDEZ-MÓRAN, H., Circulation, 26, 1039 (1962)
25. RACKER, E., and HORSTMAN, L. L., J. biol. Chem., 242, 2547 (1967)
26. RACKER, E., Scientific American, Feb., 32 (1968)
27. PULLMAN, M. E., and SCHATZ, G., A. Rev. Biochem., 36, 539 (1967)
28. RACKER, E., and BRUNI, A., in Membrane Models and the Formation of Biological Membranes (eds L. BOLIS and B. A. PETHICA), North Holland Publishing Co., Amsterdam, 138 (1968)
29. ERNSTER, L., LEE, C.-P., and NORLING, B., in C, 195 (1969)

30. KLINGENBERG, M., WULF, R., HELDT, H. W., and PFAFF, E., in *A*, 59 (1969)
31. ROSSI, E., and AZZONE, G. F., *Eur. J. Biochem.*, **12**, 319 (1970)
32. HUNTER, F. E., SCOTT, A., HOFFSTEN, P. E., GUERRA, F., WEINSTEIN, J., SCHNEIDER, A., SCHUTZ, B., and SMITH, E., *J. biol. Chem.*, **239**, 604, 632 (1964)
33. FLOHÉ, L., and ZIMMERMAN, R., *Biochim. biophys. Acta*, **223**, 210 (1970)
34. CRAVEN, S. N., LARDY, H. A., JOHNSON, D., and RUTTER, A., *Biochemistry*, **5**, 1729 (1966)
34a. ERNSTER, L., LEE, C.-P., and TORNDAL, V. B., in *C*, 439 (1969)
35. CEREIJO-SANTALO, R., *Can. J. Biochem.*, **46**, 55 (1967)
36. LARDY, H., and WELLMAN, H., *J. biol. Chem.*, **201**, 357 (1953)
37. SLATER, E. C., VELDSENA-CURRIE, R. D., KRAAYENHOF, R., and AMONS, R. in *C*, 309 (1969)
38. ERNSTER, L., LEE, C.-P., and JANDA, S., in *Biochemistry of Mitochondria*, Academic, London (1967)
39. LEE, C.-P., and ERNSTER, L., in *E*, 218 (1966)
40. AMOORE, J. E., and BARTLEY, W., *Biochem. J.*, **69**, 223 (1968)
41. HARRIS, unpublished
42. PFAFF, E., KLINGENBERG, M., RITT, E., and VOGELL, W., *Eur. J. Biochem.*, **5**, 222 (1968)
43. KLINGENBERG, M., and PFAFF, E., in *Regulation of Metabolic Processes in Mitochrondria*, Elsevier, Amsterdam, 180 (1966)
44. HARRIS, E. J., and VAN DAM, K., *Biochem. J.*, **106**, 759 (1968)
45. BENTZEL, C. J., and SOLOMON, A. K., *J. gen. Physiol.*, **50**, 1547 (1967)
46. PACKER, L., WRIGGLESWORTH, J. M., FORTES, P. A., and PRESSMAN, B. C., *J. cell. Biol.*, **39**, 382 (1968)
47. HACKENBROCK, C. R., *J. Cell Biol.*, **30**, 269 (1966)
48. GREEN, D. E., p. 126 in *B* (1967)
49. CHANCE, B., AZZI, A., LEE, I. Y., LEE, C.-P., and MELA, L., in *A*, 233 (1969)
50. LEHNINGER, A. L., *Physiol. Rev.*, **42**, 467 (1962)
51. SMITH, D. S., *J. Cell Biol.* **19**, 115 (1963)
52. CHRISTIE, G., AHMED, K., MCLEAN, A., and JUDAH, J. D., *Biochim. biophys. Acta*, **94**, 432 (1965)
53. HARRIS, E. J., COCKRELL, R. C., and PRESSMAN, B. C., *Biochem. J.*, **99**, 200 (1966)
54. POE, M., *Arch. Biochem. Biophys.*, **128**, 725 (1968)
55. HARRIS, E. J., CATLIN, G., and PRESSMAN, B. C., *Biochemistry*, **6**, 1360 (1967)
56. AZZI, A., and AZZONE, G. F., *Biochim. biophys. Acta*, **131**, 468 (1967)
57. COCKRELL, R. C., HARRIS, E. J., and PRESSMAN, B. C., *Biochemistry*, **5**, 2326 (1966)
58. ROSSI, E., and AZZONE, G. F., *Eur. J. Biochem.*, **12**, 328 (1970)
59. HARRIS, E. J., HÖFER, M. P., and PRESSMAN, B. C., *Biochemistry*, **6**, 1360 (1967)
60. HARRIS, E. J., in *C*, 135 (1969)
61. HASLAM, J. M., and GRIFFITHS, D. E., *Biochem. J.*, **115**, 645 (1968)
62. HARRIS, E. J., and BERENT, C., *Biochem. J.*, **115**, 645 (1969)
63. TUPPER, J. T., and TEDESCHI, H., *Science*, **166**, 1539 (1970)
63a. HARRIS, E. J., JUDAH, J. D., and AHMED, K., in *Current Topics in Bioenergetics* (ed. D. SANADI), Academic, London, vol. 1 (1966)
64. MITCHELL, P., and MOYLE, J., *Biochem. J.*, **104**, 588 (1967)
65. CHANCE, B., and MELA, L., *J. biol. Chem.*, **241**, 4588 (1966)
66. HARRIS, E. J., *F.E.B.S. Letters*, **5**, 50 (1969)
67. PACKER, L., UTSUMI, K., and MUSTAPHA, M., *Arch. Biochem. Biophys.*, **117**, 381 (1966)
68. SETTLEMIRE, C. T., HUNTER, G. R., and BRIERLEY, G. P., *Biochim. biophys. Acta*, **162**, 487 (1968)
69. PACKER, L., and WRIGGLESWORTH, J. M., in *A*, 125 (1969)

70. VAINIO, H., MELA, L., and CHANCE, B., *Eur. J. Biochem.*, **12**, 387 (1970)
71. CARAFOLI, E., ROSSI, C. S., and LEHNINGER, A. L., *J. biol. Chem.*, **240**, 2254 (1965)
72. RASMUSSEN, H., CHANCE, B., and OGATA, E., *Proc. Natn. Acad. Sci. U.S.A.*, **53**, 1069 (1965)
73. LINDBERG, O., and ERNSTER, L., *Nature, Lond.*, **173**, 1038 (1954)
74. ROSSI, C. S., CARAFOLI, E., DRAHOTA, Z., and LEHNINGER, A. L., in *E*, 317 (1966)
75. CARAFOLI, E., GAMBLE, J. L., ROSSI, C. S., and LEHNINGER, A. L., *J. biol. Chem.*, **242**, 1199 (1967)
76. CARAFOLI, E., GAMBLE, J. L., and LEHNINGER, A. L., *J. biol. Chem.*, **241**, 2644 (1966)
77. ADDANKI, S., CAHILL, F., and SOTOS, J. F., *J. biol. Chem.*, **243**, 2337 (1968)
78. JUDAH, J. D., AHMED, K., and CHRISTIE, G., *Biochim. biophys. Acta*, **94**, 452 (1965)
79. BRIERLEY, G. P., JACOBUS, W. E., and HUNTER, G. R., *J. biol. Chem.*, **242**, 2192 (1967)
80. HARRIS, E. J., and MANGER, J. R., *Biochem. J.*, **109**, 239 (1968)
81. HARRIS, E. J., and MANGER, J. R., *Biochem. J.*, **113**, 617 (1969)
82. CHAPPELL, J. B., in *Metabolic Roles of Citrate* (ed. T. GOODWIN), Academic, London (1968)
83. CHAPPELL, J. B., *Br. med. Bull.*, **24**, 150 (1969)
84. TEPLEY, L. J., *Arch. Biochem.*, **24**, 383 (1949)
85. CHAPPELL, J. B., and JOHNSON, R. N., *Biochem. J.*, **116**, 37P (1970)
86. HARRIS, E. J., *Biochem. J.*, **109**, 247 (1968)
87. FERGUSON, S. M., and WILLIAMS, G. R., *J. biol. Chem.*, **241**, 3696 (1966)
88. WOJTCZAK, A., *Biochim. biophys. Acta*, **172**, 52 (1969)
89. PAPA, S., *et al.*, *J. Bioenergetics*, **1**, 287 (1970)
90. LANOUE, K., NICKLAS, W. J., and WILLIAMSON, J. R., *J. biol. Chem.*, **245**, 102 (1970)
91. QUAGLIARIELLO, E., PALMIERI, F., and KLINGENBERG, M., *F.E.B.S. Letters*, **4**, 251 (1969)
92. HARRIS, E. J., and BERENT, C., *F.E.B.S. Letters*, **10**, 7 (1970)
93. HARRIS, E. J., *F.E.B.S. Letters*, **11**, 225 (1970)
94. TYLER, D., *Biochem. J.*, **107**, 121 (1968)
95. FONYÓ, A., *Biochem. biophys. Res. Commun.*, **32**, 624 (1968)
96. ROBINSON, B. H., and WILLIAMS, G. R., *F.E.B.S. Letters*, **5**, 301 (1969)
97. ROBINSON, B. H., WILLIAMS, G. R., HALPERIN, M. L., and LEZNOFF, C. C., *Eur. J. Biochem.*, **15**, 263 (1970)
98. MCGIVAN, J. D., and CHAPPELL, J. B., *Biochem. J.*, **116**, 37P (1970)
99. MCGIVAN, J. D., BRADFORD, N., and CHAPPELL, J. B., *F.E.B.S. Letters*, **4**, 247 (1969)
100. PAPA, S., D'ALOYA, R., MEIJER, A. J., TAGER, J. M., and QUAGLIARIELLO, E., in *C*, 159 (1969)
101. HARRIS, E. J., *J. Bioenergetics*, **2**, 93 (1971)
102. LOFRUMENTO, N., MEIJER, J., TAGER, J., PAPA, S., and QUAGLIARIELLO, E., *Biochim. biophys. Acta*, **197**, 204 (1970)
103. KLINGENBERG, M., HELDT, H. W., and PFAFF, E., in *C*, 237 (1969)
104. MITCHELL, P., and MOYLE, J., in *Electron Transport and Energy Conservation* (ed. J. TAGER *et al.*), Adriatica, Bari (1970)
105. HELDT, H. W., in *A*, 93 (1969)
106. MORET, V., LORINI, M. N., FOTIA, A., and SILIPRANDI, N., in *B*, 281 (1967)
107. SACCONE, C., GADALETA, M. N., and QUAGLIARIELLO, E., in *B*, 295 (1967)
108. MITCHELL, P., and MOYLE, J., *F.E.B.S. Letters*, **6**, 309 (1970)
109. PRESSMAN, B. C., in *Energy Linked Functions of Mitochondria* (ed. B. CHANCE), Academic, London (1963)
110. PRESSMAN, B. C., *J. biol. Chem.*, **238**, 401 (1963)

344

111. HENDERSON, P. J., and LARDY, H. A., *J. biol. Chem.*, **245**, 1319 (1970)
112. KLINGENBERG, M., in *Essays in Biochemistry*, vol. 6 (eds P. CAMPBELL and F. DICKENS), Academic, London, 119 (1970)
113. OUT, T. A., KEMP, A., and SOUVERIJN, J. A., *Biochim. biophys. Acta*, **245**, 299 (1971)
114. NISHIMURA, M., ITO, T., and CHANCE, B., *Biochim. biophys. Acta*, **59**, 177 (1962)
115. HARRIS, E. J., *J. Bioenergetics*, **2** 93 (1971)
116. MEISNER, H., and KLINGENBERG, M., *J. biol. Chem.*, **243**, 3631 (1968)
117. DUÉE, E. D., and VIGNAIS, P. V., *J. biol. Chem.*, **244**, 3932 (1969)
118. CHEN, R. F., *J. biol. Chem.*, **242**, 173 (1967)
119. VAN TOL, A., and HÜLSMANN, W. C., *Biochem. biophys. Acta*, **B.11**, 416 (1970)
119. VAN TOL, A., and HÜLSMANN, W. C., *Biochim. biophys. Acta*, **223**, 416 (1970)
120. CHANCE, B. C., and YOSHIOKA, T., *Arch. Biochem. Biophys.*, **117**, 451 (1966)
121. PURVIS, J. L., and LOWENSTEIN, J. M., *J. biol. Chem.*, **234**, 3044 (1963)

9

ION TRANSPORT IN LIVER AND KIDNEY

LINDA P. SOUTTER and J. D. JUDAH
*Department of Experimental Pathology, University College
Hospital Medical School, London*

INTRODUCTION

In this chapter we have not concerned ourselves with renal physiology in its widest aspects, e.g. renal clearance of dyes and other molecules, formation of urine, etc. For these we refer the reader to two excellent reviews by Forster[189] and by Thurau et al.[190].

IONIC COMPOSITION

The ionic composition of rat liver and of rabbit and guinea pig kidney is shown in Table 9.1 and that of rat serum in Table 9.2.

The apparent high concentrations of Na^+ and Cl^- in the tissues are due to contamination by extracellular fluid—the extracellular space of both liver and kidney cortex, measured by the distribution of inulin, is estimated to be between 24% and 34% of the total tissue fluid space[7-14]. Kidney medulla (of sheep) has been found to have a higher extracellular water content, 44.8%[11]. Other methods, such as the distribution of chloride or sulphate, have been used to determine extracellular space[8,9], but these are misleading in liver and kidney, as sulphate is found to be accumulated[9,15,16]; chloride also readily enters the cells and swelling of the cells under abnormal conditions is accompanied by uptake of isotonic extracellular fluid and by a proportional uptake of Na^+ and Cl^- (see reference 2 for example).

INTRACELLULAR pH

Walker et al.[19] found the mean pH of rat liver cells, measured *in*

Table 9.1 IONIC COMPOSITION OF LIVER AND KIDNEY

Tissue	No. of Samples	K^+	Na^+	Ca^{2+}	Mg^{2+}	Cl^-	Reference
Rat liver female	10	304 ± 15* (135)†	79 ± 10 (35)	6·26 ± 1·0 (2·8)	60·8 ± 2 27		1
Rat liver male	7	288 ± 9 (131)	82 ± 9 (37)	6·40 ± 1·0 (2·9)	56 ± 2 (25)		1
Rat liver	6	345 ± 6 (151)	114 ± 14 (50)			121 ± 4 (53)	2
Rabbit kidney cortex	17	313 ± 11	297 ± 36				3
Guinea pig kidney cortex	14	365 ± 8 (113)	350 ± 16 (109)			272 ± 7 (84)	2

* Top figure is content in mequiv per kg dry wt.
† Figure in brackets is concentration in mequiv per kg tissue H_2O.

vivo by the distribution of the weak acid 5,5-dimethyl-2,4-oxa-zolidinedione (DMO), to be 7·23, which was significantly more alkaline than other tissues studied (e.g. skeletal muscle pH 6·93, heart muscle pH 7·04). Measurements in liver slices agree well with this figure. It is questionable, however, whether such measurements have any real meaning. The liver cell is highly compartmented, but this method of pH measurement makes no allowance for this. Isolated mitochondria, for example, will take up K^+ with DMO as an accompanying anion. If this occurs in whole cells, measurement of pH by the distribution of DMO will be misleading.

Table 9.2 IONIC COMPOSITION OF RAT SERUM

Ion	No. of samples taken	Concn (mM)	Reference
Na^+	20	142·2 ± 0·4	4
K^+	20	4·0 ± 0·04	4
Ca^{2+}	9	2·37 ± 0·06	5
Mg^{2+}	4	1·33 ± 0·22	6
Cl^-	20	104·5 ± 0·4	4

348

METHODS OF INVESTIGATING TRANSPORT PHENOMENA

Tissue slice methods[20, 21] have been used successfully to demonstrate transport mechanisms in both liver and kidney cortex.

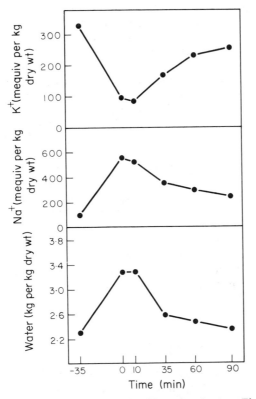

Figure 9.1. *K[+], Na[+], and water shifts of liver slices* in vitro. *The slices were cooled for 35 min at 0°C in 150 mM NaCl. At time 0, they were transferred to Ringer's at 37°C und –O₂ (From McLean[205]; courtesy* Biochem. J.)

Flink *et al.*[22] using liver, and Whittam and Davies[23] with kidney found that when slices were incubated in oxygenated medium at 38 °C (in the case of kidney with addition of a suitable substrate, see reference 173) an initial loss of K^+ and gain of Na^+ and water took place; however, this was reversed after about 15 min and almost normal ionic concentrations could then be maintained;

349

Mudge[24] described a convenient method of studying ion fluxes in slices that has been followed by many later authors. The slices are depleted of K^+ and loaded with Na^+ by incubation, in Ringer's or 154 mM NaCl, at 0 to 2 °C (see Figure 9.1). 60–70% of the tissue K^+ is lost in 30 min and is replaced by an equal amount of Na^+. In addition, the cells swell and take up Na^+ and Cl^- in isotonic solution. These changes are reversed on incubation in oxygenated medium at 30–38 °C, the cells shrinking while losing an isotonic solution of NaCl. For the first 10 min, little or no K^+ is gained, though the actual K^+ concentration within the cells may rise. Later, Na^+ is lost and K^+ gained in an approximately 1 : 1 ratio until the K^+ content of the tissue reaches about 300 mequiv per kg dry weight (i.e. almost its original value). The NaCl content, however, does not fall to its original value and the sum concentration of Na^+ and K^+ remains somewhat higher than normal (about 200 mM as opposed to approximately 160 mM). Cl^- movement is passive[25] and chloride in the incubation medium can be replaced (by, for example, bromide, nitrate, or sulphate) with little effect on the metabolism or K^+ content of kidney slices[3, 26]. In phosphate or bicarbonate Ringer's, the Cl^- content follows that of the Na^+, and is therefore at a lower concentration in the cells than in the Ringer's. When the fluoride in the medium is varied, the tissue chloride moves proportionately so as to maintain a constant $Cl_i : Cl_e$ ratio.

Most of the work on liver and kidney slices has been done on rat, guinea pig, or rabbit tissue, but some work has been done on kidney slices from other species such as frog[175, 181], snake, and chicken[177], and hibernating mammals[174, 176].

STUDY OF ISOLATED RENAL TUBULES

Separated renal tubules have also been used for the study of ion fluxes in kidney cells. Burg and Orloff[27–29] suggested that in slices the outermost tissue layers may act as a barrier between the bathing medium and the innermost cells, so instead they used a suspension of short lengths of proximal tubules, obtained by treating rabbit renal cortex with collagenase. Several differences were found between the behaviour of this preparation and that of slices. The tubules had a higher metabolic rate and faster exchange rate of Na^+ and K^+ than slices. It was possible to estimate unidirectional ion fluxes across the cell membranes more accurately. Other examples of the use of rabbit renal tubules are given in references 30–34. Murthy and Foulkes[35] compare the use of slices and tubules for the study of the properties of the kidney luminal membrane. Fish and amphibia have large, easily separated renal tubules, enabling the transport mechanisms of single nephrons to be studied; the flounder[36-41] and the amphibian *Necturus maculosus*[42-47] are most commonly used.

VOLUME REGULATION AND OSMOTIC CONTENT

Early experiments, in which the freezing points of liver and kidney tissue were measured, suggested that both had an osmotic pressure 50–100% greater than that of the extracellular fluid[48, 49]. This idea was thought to be supported by the fact that swelling was found to occur when the tissues were incubated in solutions isosmotic to the blood under conditions that inhibited metabolism, for example, anoxia, low temperature, or cyanide poisoning[50-55]. When the tissue metabolism was active, less swelling occurred[56, 57] and from this evidence it was assumed by some that the intracellular fluids were hypertonic to the extracellular and that this state was maintained by the active pumping of water from the cells[58-60]. However, other observers found that, on inhibition of metabolism, large quantities of Na^+ and Cl^- entered the tissues, in addition to Na^+ exchanging for the cellular K^+, and in fact the fluid entering was isosmotic with the extracellular fluid[2, 24, 55, 61-65]. On reactivation of the metabolism, the isosmotic solution of NaCl and water is expelled with consequent shrinking of the cells to their original size. This volume regulation is slowed but not abolished when the tissue is incubated in a 'K$^+$-free' (K^+ about 0.08 mM) medium or if ouabain is added[47, 66-69], and is energy-dependent, as it is prevented by conditions that cause lowering of the cell ATP levels[1]. Since the cell volume regulation is independent of Na^+–K^+ exchange[74], it seems likely that Na^+ ($+Cl^-$) is expelled by an Na^+ pump that is not necessarily linked to K^+ transport inwards[71]. It is probable that Na^+ is pumped out until the supply of available anion falls and the membrane potential is restored, K^+ then being taken up in exchange for Na^+. Measurements of membrane potentials in kidney slices support this idea[72]. Recently, Whittembury and coworkers suggested that in guinea pig kidney cortex two separate Na^+ pumps exist[73, 74], one of these being an exchange mechanism of intracellular Na^+ for extracellular K^+, with energy supplied from ATP by the ($Na^+ + K^+$) ATPase. Both Na^+–K^+ exchange and the ATPase are inhibited by similar concentrations of ouabain and are relatively insensitive to ethacrynic acid. The second pump is a volume-regulating mechanism, an electrogenic pump extruding

Na^+ and Cl^- that is relatively insensitive to ouabain, therefore not connected with the ATPase, but is inhibited by ethacrynic acid. However, MacKnight, working on rat kidney cortex slices[75], concluded that the effect of ethacrynic acid is more likely to be caused by interference with cell metabolism than by specific inhibition of a volume regulating mechanism.

Na^+–K^+ EXCHANGE

This may be demonstrated by the incubation of liver or kidney cortex slices. which have previously been loaded with Na^+ and leached of K^+ in the cold (in oxygenated K^+-free medium at 38 °C). The cells shrink, the total K^+ content (but not the concentration) of the slices may fall, and NaCl is lost until the slice water content falls to the normal value (about 2·3 kg H_2O per kg dry weight for liver). If K^+ is added to the incubation medium after this steady state has been reached, there is an immediate further loss of Na^+ and gain of K^+ in an approximately 1 : 1 ratio. This effect is inhibited both by metabolic inhibitors, such as DNP and cyanide, and by ouabain[47, 67, 76, 77] (unlike the volume regulation mechanism). There is also evidence that oligomycin has a direct inhibitory effect on cation transport in liver[178] and kidney[179], rather than just acting as an inhibitor of oxidative phosphorylation.

Studies on the rates of both influx and efflux of Na^+ and K^+ have been carried out with liver and kidney tissue, using radioisotopes. Van Rossum[78] found two main components of the $^{24}Na^+$ efflux from liver slices, the slower one apparently intracellular. About two thirds of the efflux of intracellular $^{24}Na^+$ was a passive exchange for ^{23}Na in the medium (i.e. it also took place anaerobically), while the remaining one third was active (inhibited by KCN and ouabain, and partly dependent on the present of K^+ in the medium). Whittam and Davies[79], using guinea pig kidney cortex slices, also found that Na^+ exchanged as at least two fractions but that all K^+ exchanged uniformly at 37 °C. However, at 0 °C there were at least two fractions of exchanging K^+. For other studies on flux rates see references 29, 80, 180.

LINKAGE OF Na^+ AND K^+ TRANSPORT

It is difficult to reconcile observations on Na^+ and K^+ movements in slices with the idea of a simple Na^+–K^+ exchange pump. In both liver and kidney, Na^+ transport can take place without K^+ movement. Willis[75], using kidney slices, suggests possible modifications to the pump theory:

1. There may be two parallel Na^+ pumps, one K^+-linked, one non-coupled (i.e. electrogenic).
2. The cells may contract mechanically when shrinking and extruding Na^+, Cl^-, and H_2O.
3. There may be an Na^+–K^+-coupled pump that can carry out Na^+–Na^+ exchange when no K^+ is available.
4. There may be a coupled Na^+–K^+ pump operating in contact with a restricted extracellular space between tubular cells or with infoldings of cell membrane so that, because of a long diffusion pathway, the K^+ concentration near the pump could be different from that in the medium.

Other authors have concluded that, for kidney tissue, Na^+ and K^+ transport have separate mechanisms[29, 80–83] and Kleinzeller and Cort[170–172] have found that K^+ (also Li^+ and Rb^+) moves passively into kidney cortex slices, following concentration and electro-chemical gradients only. The operation of a pure Na^+ pump, in which K^+ is merely a passive partner, demands that the distribution of K^+ depends on the membrane potential and that

$$\frac{[K]_i}{[K]_e} = \frac{[Cl]_e}{[Cl]_i}$$

where i and e refer to the intracellular and extracellular concentrations respectively. Whittembury[72] found that in kidney cells the K^+ distribution is no different from that required by the membrane potential when $[K]_e$ is 8 mM or more, but at a $[K]_e$ of 2 mM the intracellular K^+ concentration is higher than would be expected from measurement of the membrane potential and must therefore depend on active K^+ uptake (see Figures 9.2 and 9.3). It is possible that liver tissue exhibits similar behaviour, but measurements of the membrane potential of liver cells under appropriate conditions are not available. For results of measurements of cell membrane potentials in kidney see references 89–92; in liver, references 84–88.

UNCOUPLED Na^+ TRANSPORT

The expulsion of Na^+ unaccompanied by either the uptake of K^+ or the loss of Cl^- and H_2O has been reported for other tissues such as frog muscle, squid axon, and erythrocytes[93–95]. It is

N

possible, though unlikely, that an anion formed by metabolic processes accompanies the Na^{+} [96] or that it is accompanied by the uptake of H^{+} ions or their production within the cells.

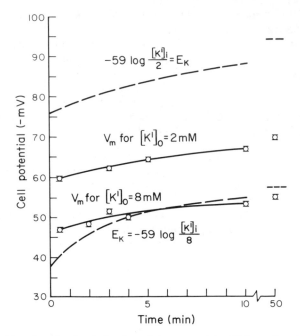

Figure 9.2. Actual and calculated membrane potentials of kidney slices in vitro. *The dashed lines are calculated for a passive distribution of* K^{+}; *the solid lines show the observed potentials (From Whittembury[72]; courtesy* J. gen. Physiol.)

Liver slices can be induced to conduct an apparently uncoupled Na^{+} transport by cooling to $0\,^{\circ}C$ and reincubation at $38\,^{\circ}C$ in Ca-free tris Cl-buffered medium. More Na^{+} is lost than can be accounted for by the loss of H_2O, and no K^{+} is gained, at the start of the reincubation. After a short time, the K^{+} content of the slice starts to rise and Na^{+} and K^{+} are exchanged in a 1 : 1 ratio. If the same procedure is carried out in a bicarbonate, instead of a tris Cl-buffered medium, no uncoupling is seen, Na^{+} and K^{+} being exchanged in a 1 : 1 ratio, both in the presence and in the absence of Ca^{2+}. One possible explanation of this result is that tris substitutes for K^{+} and is passively taken up by the cells as the membrane potential becomes more negative. Liver slices from female rats poisoned with ethionine, which have very low cell ATP levels, also seem to exhibit this uncoupled Na$^+$ transport.

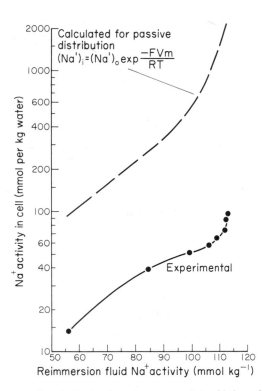

Figure 9.3. Actual and calculated membrane potentials of kidney slices in vitro. *The dashed lines show the calculated potentials for a passive distribution of Na$^+$; the solid lines show the observed potentials (From Whittembury[72]; courtesy J. gen. Physiol.)*

ENERGY RELATIONSHIPS OF ION TRANSPORT

In liver and kidney, as in other tissues investigated, it is probable that ATP supplies the energy for active transport. All conditions that lower the ATP content of the cells inhibit the expulsion of Na$^+$. However, ATP levels must fall very low before complete arrest of Na$^+$ movement takes place. For example, on treatment of female rats with ethionine *in vivo* (see Figures 9.4 and 9.5), the liver cell ATP levels drop by about 90%, which produces a fall of liver K$^+$ and gain of Na$^+$, whereas in male rats treated in the same way the

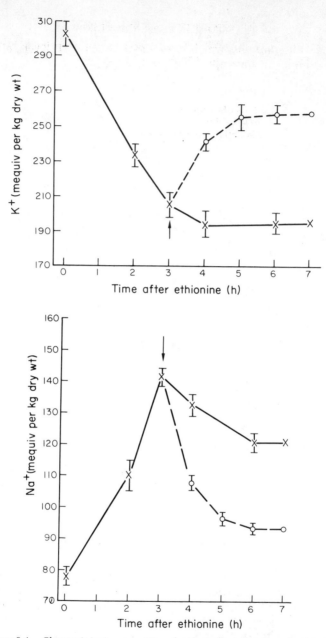

Figure 9.4.   Changes in ionic composition of rat liver after ethionine poisoning. Ethionine was administered at time 0. Each point shows the mean ± s.d. of five rats. At the arrow, the rats were given a dose of adenine which raises the level of liver ATP and reverses the changes (broken line) (From Judah et al.[1]; courtesy Lab. Invest.)

decrease in cell ATP levels is less (75–80%), and there is virtually no change in the ionic composition of the liver.

As expected, DNP blocks Na$^+$ transport in both liver and kidney[54, 66, 97, 98]. The effect of DNP on tissue slices may be rapidly reversed by the addition to the incubating medium of serum albumin that adsorbs the DNP. It has been assumed that active transport

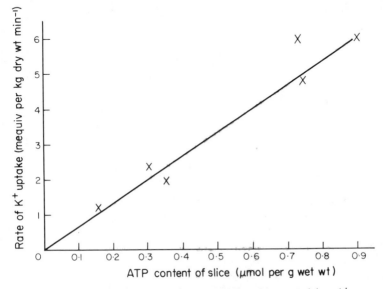

Figure 9.5. Liver ATP and rate of K$^+$ uptake compared (From Judah et al.[1]; courtesy Lab. Invest.)

was completely inhibited at 0 °C—the presence of a K$^+$ concentration gradient at the temperature was thought to indicate that part of the intracellular K$^+$ was bound[100], but Burg and Orloff[101], working on rabbit kidney tubules, found that at 0 °C, anoxia or the presence of ouabain or DNP reduced the K$^+$ concentration gradient further, suggesting that some active transport takes place at this temperature. It has been found[182–184] that foetal liver and kidney tissue can tolerate anoxia or lowering of temperature better than adult tissue, with less swelling and K$^+$ loss, because of its higher rate of anaerobic glycolysis. This ability (and the rate of anaerobic glycolysis) decreases rapidly after birth. Seidman and Cascarano[102] reported that they had observed anaerobic cation transport in adult liver slices when incubated with substrates such as oxaloacetate or pyruvate in an atmosphere of nitrogen. The ability of these slices to reaccummulate K^+ and extrude Na^+ was comparable to that of slices supplied with oxygen.

A different approach to the problem of the energy requirements of Na^+ transport was reported by Smyth[99], who calculated the critical energy barrier—the amount of energy required to just prevent the extrusion of Na^+ from kidney cortex slices loaded with Na^+. This was found to be between 5·9 and 7.5 J mequiv^{-1} for rat and guinea pig kidney.

$Na^+ + K^+$-ACTIVATED ATPases

Both liver and kidney possess these enzymes in substantial quantity[103–112]. The distribution of the $Na^+ + K^+$ ATPase in kidney cortex homogenates has been found to be mainly between the cell-debris and the microsomal fractions in equal amounts[113], whereas in liver it seems to occur mainly in the cell-debris fraction with very little in the microsomes[114]—in fact it has been suggested that it is absent from the microsomes[107]. The characteristics of the kidney ATPase have been more thoroughly investigated than those of the liver system. The kidney system is inhibited by ouabain, oligomycin[113, 115], organic mercurials[116], Ca^{2+} ions[113, 117], and other metals, e.g., Pb^{2+}, Zn^{2+}, Cu^{2+}, etc.[118]. It is not affected by sulphydryl-reacting compounds such as iodoacetate or by diuretics such as caffeine and theobromine[116]. The stimulation of the ATPase by K^+ and Na^+ depends partly on the concentrations of these ions and partly on the K^+–Na^+ ratio—a ratio of K^+ to Na^+ of less than one being most effective[119]. NH_4^+ has been found to substitute K^+ in the system[120]. Conflicting evidence is available on the effect of aldosterone on the ATPase—Landon et al.[121] reported that this produced a stimulatory effect, whereas Chignell and Titus[122] state that aldosterone has no effect. The liver Na, K, ATPase system is thought to have similar properties except that it is relatively insensitive to ouabain when compared with preparations from other tissues, for example brain ATPase has been found to be about 100 times more sensitive to ouabain[123]. On the other hand, it is inhibited by reagents such as chlorpromazine at concentrations that have little or no action on the brain system[123]. It is difficult to tell whether these differences indicate real differences in the enzyme systems or merely reflect minor differences in the methods of preparation used in the various studies.

THE RELATIONSHIP OF ION TRANSPORT TO RESPIRATION

Evidence has been obtained by Whittam and co-workers[124–127] that ion transport acts as a pacemaker of respiration. Decreases in the rate of K^+ uptake induced by means believed not to inhibit respiration directly (for example, the absence of external Na^+ or K^+ or a high Ca^{2+} concentration or the presence of ouabain) are nevertheless accompanied by parallel decreases in respiration. Therefore, the rate of respiration and active cation movements are interdependent. Blond and Whittam[127, 128] have shown that the respiration of rabbit kidney homogenate is controlled by the Na^+ + K^+ ATPase. A similar relationship in rat liver was suggested by Elshove and van Rossum[80].

MOVEMENTS OF DIVALENT CATIONS

The movements of calcium ions have not been extensively studied, except in isolated mitochondria[129, 130]. Preparations of liver cells *in vitro* tend to take up Ca^{2+}, especially when incubated under metabolically unfavourable conditions[191] or when they are dying[131, 132]. The Ca^{2+} content of the liver is low (see Tables 9.1 and 9.2) and it is probable that there is little or no free Ca^{2+} in the cell, because of the presence of substances such as ATP and Pi (which bind Ca^{2+}). It has been suggested that a mechanism exists for extruding Ca^{2+} from the cell. Van Rossum[17, 133] found that the extrusion of Ca^{2+} from rat liver slices was prevented by respiratory inhibitors but not by inhibitors of Na^+–K^+ transport (e.g. ouabain), and suggested that active Ca^{2+} extrusion occurs in this tissue. Judah and McLean (A.E.M.) (unpublished work) observed that in liver slices reduction of the external Na^+ concentration resulted in a gain of Ca^{2+}, while at the same time cell K^+ fell. This Ca^{2+} was lost if the cell K^+ content was made to rise, either by increasing the external Na^{2+} concentration (setting the Na^+ pump in action) or by increasing the external K^+ concentration so that K^+ entered passively. It was concluded from this that Ca^{2+} and K^+ compete for intracellular binding sites (see reference 18 for example) and that there is probably no active transport of Ca^{2+} as such. In kidney slices an energy-dependent accumulation of Ca^{2+} has been reported[134].

Still less work has been done on the transport and accumulation of magnesium. Van Rossum[17] reported that the Mg^{2+} content of rat liver slices remained much more constant under varying conditions than that of K^+, Na^+, or Ca^{2+} and much of it must therefore be either bound or retained within an impermeable membrane. However, the fact that there was some loss of Mg^{2+} from the cells on inhibition of respiration by cyanide and that this loss could be made good under certain circumstances, e.g. during incubation at 38 °C in the presence of octanoate and ethanol, suggested the possibility of an active accumulation of Mg^{2+}.

TRANSPORT OF ORGANIC COMPOUNDS

AMINO ACIDS

Active uptake of amino acids has been demonstrated in both liver and kidney, often using the non-metabolisable α-aminoisobutyric acid (AIB) as a model. Amino acid uptake in kidney has been found to depend on the external Na^+ and K^+ concentrations[135, 136], except that lysine and histidine transport were not completely abolished in Na^+-free media, so appear to have two separate uptake mechanisms[135]. The transport is inhibited by anoxia (or DNP)[137], and there is also competitive inhibition between various amino acids, indicating that the reabsorbtive mechanisms of these are closely related, perhaps even identical. Webber et al.[138] found mutual inhibition of tubular reabsorption of lysine, ornithine, arginine, and cystine (or cysteine), while Scriver and co-workers have found a renal tubular transport system with preference for proline, hydroxyproline and glycine in man and in the rat[139-141]. Amino acid transport in rat kidney cortex slices is inhibited by certain sugars. Thier et al.[142] found that glucose, galactose, and fructose specifically inhibit the uptake of glycine, valine, and cycloleucine, but not that of lysine, histidine or phenylalanine.

Recently, binding of L-proline by a membrane fraction prepared from isolated rabbit renal tubules had been demonstrated[147]. This binding is decreased by the removal of Na^+ or addition of cyanide but not by ouabain or DNP. Glycine and L-alanine competitively inhibit the binding of proline, but L-valine and phenylalanine do not. The authors suggest that this binding process is an initial step in the transtubular transport of L-proline. Amino acid transport in liver is apparently similar to that in kidney. AIB transport in rat liver slices is inhibited by anoxia, DNP, ouabain, and lack of Na^+, K^+ or Ca^{2+} [143]. Hydrocortisone and insulin appear to stimulate amino acid uptake in liver [144-146].

SUGAR TRANSPORT

Glucose and other monosaccharides are accumulated against a concentration gradient in kidney. Krane and Crane[148] used the non-metabolisable sugar D-galactose as a model for the uptake of glucose in rabbit kidney cortex slices and found that its active accumulation is inhibited by dinitrocresol and by the presence of glucose. Both active and passive uptake of D-galactose are inhibited by phlorizin. Kleinzeller and co-workers[149, 150], using the same model sugar, found that transport depended on Na^+ and K^+ and was inhibited by ouabain. Other sugars which the same authors have shown to be accumulated in rabbit kidney cortex are D-fructose, α and β methyl-D-glucosides, 2-deoxy-D-glucose, 2-deoxy-D-galactose, D-glucosamine and D-galactosamine [151, 152]. Kleinzeller et al. have also studied the kinetics and ion requirements of and effect of pH on sugar transport in kidney cortex[152–154]. Khuri et al.[155] showed that the rate constant for the absorption of glucose and the non-metabolised glucose analogue 3-O-methyl-D-glucose in Necturus kidney proximal tubules depends on the initial tubular concentration of NaCl; Li^+ or choline could not replace the Na^+. Crawhall and Segal[156] found no evidence for the active transport of sugars in rat liver slices.

TRANSPORT OF OTHER ORGANIC COMPOUNDS

A great variety of organic substances has been reported to be accumulated in kidney tissue, little information is available on liver. The transport of p-aminohippurate (PAH) has been extensively studied since Cross and Taggart[157] showed this to be accumulated by rabbit kidney slices from an oxygenated medium. Lactate, pyruvate, and acetate stimulate this uptake, while it is inhibited by α-ketoglutarate, succinate fumarate, and malonate, and by various fatty acids[157]. Transport is maximum at the normal tissue Na^+ and K^+ concentrations and is partially inhibited by replacing Cl^- by other anions[158]. In a K^+-free medium, there is no PAH accumulation in kidney cortex slices[34, 159] and it is also inhibited by substances thought specifically to inhibit K^+ transport, such as protamine and ouabain[159, 160]. Hypophysectomy has been found to reduce PAH uptake by rat kidney slices [163]. The kinetics of PAH transport has been studied by several authors[33, 164–168]. Recently, the metabolic requirements of PAH transport in rabbit kidney cortex slices were investigated by Maxild and Møller[169], who concluded that the metabolism of substances connected with carbohydrate metabolism,

361

(probably pyruvate and l-lactate), is necessary for active transport of PAH to occur.

Various other organic anions competitively inhibit the uptake of PAH in kidney and are thought to be actively transported by the same or a similar mechanism. These include phenol red, chlorphenol red, iodopyracet (diodrast)[36, 40, 161, 162] and urate[183–188]. It has also been reported by Huang and Lin[33] that DNP is accumulated in separated renal tubules by a mechanism similar to that of PAH uptake; the inhibitory effect of this substance on PAH uptake may therefore be partly due to a competitive effect.

Evidence for the active secretion of sixteen organic acids was found by Weiner et al.[192] in renal clearance experiments on dogs and Despopoulos[193] has demonstrated that transported weak acids have certain structural characteristics in common.

UREA

Active renal tubular secretion of urea has been demonstrated in frogs[194–196]. In the bullfrog, Forster[196] found that transport of urea was similar to the PAH transport system, being blocked by DNP and benemid and partially inhibited by PAH. Active reabsorption of urea from the renal collecting ducts is known to take place in Elasmobranch fishes[197, 198]. In the mammalian kidney, transport of urea has been thought to be purely passive, but recently evidence has been put forward that active movement does in fact occur. The mammalian renal medulla has the capacity to concentrate urea with an increasing gradient towards the tip of the papilla. This is thought to be a passive process, depending on the maintenance of the intramedullary electrolyte gradient that causes an osmotic flow of water from the collecting tubule, thereby creating a gradient down which urea might flow. Goldberg et al.[199] obliterated the medullary electrolyte gradient in dog kidneys by ethacrynic acid diuresis, but found that the urea gradient still persisted, the urinary urea concentration being always lower than the papillary water urea concentration. Several authors have found evidence that active transport of urea may occur in rat kidney[200–203]. Truniger and Schmidt-Nielsen[200] found that medullary concentrations of urea and two related substances methylurea and acetamide were higher than the urinary concentration in rats fed a low-protein diet and lower than the urinary concentration in rats fed a high-protein diet. They concluded from their results that urea was either actively transported against a concentration gradient or bound by the medullary tissue. Rabinowitz[204], however, found no evidence for active transport of urea in similar experiments using dog kidney.

REFERENCES

1. JUDAH, J. D., AHMED, K., MCLEAN, A. E. M., and CHRISTIE, G. S., *Lab. Invest.*, **15**, 167 (1966)
2. LEAF, A., *Biochem. J.*, **62**, 241 (1956)
3. MUDGE, G. H., *Am. J. Physiol.*, **165**, 113 (1951)
4. LEVITT, M. F., and TURNER, L. B., *J. clin. Invest.*, **35**, 99 (1956)
5. SMITH, P. K., and SMITH, A. H., *J. biol. Chem.*, **107**, 673 (1934)
6. EVELETH, D. F., *J. biol. Chem.*, **119**, 289 (1937)
7. WHITTAM, R., *J. Physiol.*, **131**, 542 (1956)
8. TRUAX, F. L., *Am. J. Physiol.*, **126**, 402 (1939)
9. PARSONS, D. S., and VAN ROSSUM, G. D. V., *Biochim. biophys. Acta*, **57**, 495 (1962)
10. ROSENBERG, L. E., DOWNING, S. J., and SEGAL, S., *Am. J. Physiol.*, **202**, 800 (1962)
11. GANS, J. H., BAILIE, M. D., and BIGGS, D. L., *Am. J. Physiol.*, **211**, 249 (1966)
12. ROBINSON, J. R., *Proc. Roy. Soc.* (B). **137**, 378 (1950)
13. MACKNIGHT, A. D. C., *Biochim. biophys. Acta*, **163**, 85 (1968)
14. KLEINZELLER, A., JANÁČEK, K., and KNOTKOVA, A., *Biochim. biophys. Acta*, **59**, 239 (1962)
15. BERGLUND, F., and DEYRUP, I. J., *Am. J. Physiol.*, **187**, 315 (1956)
16. DEYRUP, I. J., and USSING, H. H., *J. gen. Physiol.*, **38**, 599 (1954)
17. VAN ROSSUM, G. D. V., *J. gen. Physiol.*, **55**, 18 (1970)
18. COSMOS, E., and HARRIS, E. J., *J. gen. Physiol.*, **44**, 1121 (1961)
19. WALKER, W. D., GOODWIN, F. J., and COHEN, R. D., *Clin. Sci.*, **36**, 409 (1969)
20. DEUTCH, W., *J. Physiol.*, **87**, 56 (1936)
21. STADIE, W. C., and RIGGS, B. C., *J. biol. Chem.*, **154**, 687 (1944)
22. FLINK, E. B., HASTINGS, A. B., and LOWRY, J. K., *Am. J. Physiol.*, **163**, 598 (1950)
23. WHITTAM, R. W., and DAVIES, R. E., *Biochem. J.*, **55**, 880 (1953)
24. MUDGE, G. H., *Am. J. Physiol.*, **167**, 206 (1951)
25. GIEBISCH, G., and WINDHAGER, E. E., *Am. J. Physiol.*, **204**, 387 (1963)
26. GORDON, E. E., and MAEIR, D. M., *Am. J. Physiol.*, **207**, 71 (1964)
27. BURG, M. B., and ORLOFF, J., *Am. J. Physiol.*, **203**, 327 (1962)
28. BURG, M. B., *Fedn. Proc. Fedn. Am. Socs exp. Biol.*, **22**, 446 (1963)
29. BURG, M. B., GROLLMAN, E. F., and ORLOFF, J., *Am. J. Physiol.*, **206**, 483 (1964)
30. BURG, M. B., GRANTHAM, J., ABRAMOW, M., and ORLOFF, J., *Am. J. Physiol.*, **210**, 1293 (1966)
31. BURG, M. B., and ORLOFF, J., *Am. J. Physiol.*, **203**, 327 (1962)
32. BURG, M. B., and ORLOFF, J., *Am. J. Physiol.*, **217**, 1064 (1969)
33. HUANG, K. C., and LIN, D. S. T., *Am. J. Physiol.*, **208**, 391 (1965)
34. BOSÁČKOVÁ, *Biochim. biophys. Acta*, **71**, 345 (1963)
35. MURTHY, L., and FOULKES, E. C., *Nature, Lond.*, **213**, 180 (1967)
36. FORSTER, R. P., and TAGGART, J. V., *J. cell. comp. Physiol.*, **36**, 251 (1950)
37. PUCK, T., WASSERMAN, K., and FISHMAN, A., *J. cell comp. Physiol.*, **40**, 73 (1952)
38. FORSTER, R., and HONG, S., *J. cell. comp. Physiol.*, **51**, 259 (1958)
39. HONG, S., and FORSTER, R., *J. cell. comp. Physiol.*, **54**, 237 (1959)
40. KINTER, W., *Am. J. Physiol.*, **211**, 1152 (1966)
41. BURG, M. B., and WELLER, P. F., *Am. J. Physiol.*, **217**, 1053 (1969)
42. KINTER, W. B., *Am. J. Physiol.*, **196**, 1141 (1959)
43. OKEN, D. E., WHITTEMBURY, G., WINDHAGER, E. E., and SOLOMON, A. K., *Am. J. Physiol.*, **204**, 372 (1963)
44. OKEN, D. E., and SOLOMON. A. K., *Am. J. Physiol.*, **204**, 377 (1963)
45. SOLOMON, A. K., *Am. J. Physiol.*, **204**, 381 (1963)
46. TANNER, G. A., and KINTER, W. B., *Am. J. Physiol.*, **210**, 221 (1966)
47. WHITTEMBURY, G., *J. gen. Physiol.*, **51**, 303 (1968)
48. SABBATANI, L., *J. Physiol. Path. gén.*, **3**, 62 (1901)

49. OPIE, E. L., *J. exp. Med.,* **99**, 29 (1954)
50. SPERRY, W. M., and BRAND, F. C., *Proc. Soc. exp. Biol. Med.,* **42**, 147 (1939)
51. OPIE, E. L., *J. exp. Med.,* **87**, 425 (1948)
52. OPIE, E. L., *J. exp. Med.,* **89**, 185 (1949)
53. OPIE, E. L., *J. exp. Med.,* **91**, 285 (1950)
54. ROBINSON, J. R., *Nature, Lond.,* **166**, 989 (1950)
55. DEYRUP, I., *J. gen. Physiol.,* **36**, 739 (1953)
56. AEBI, H., *Helv. Physiol. Acta,* **10**, 184 (1952)
57. DEYRUP, I., *Am. J. Physiol.,* **175**, 349 (1953)
58. ROBINSON, J. R., and MCCANCE, R. A., *A. Rev. Physiol.,* **14**, 115 (1952)
59. ROBINSON, J. R., *Biol. Rev.,* **28**, 158 (1953)
60. BARTLEY, W., DAVIES, R. E., and KREBS, H. A., *Proc. Roy. Soc.* (B), **142**, 187 (1954)
61. HECKMAN, K. D., and PARSONS, D. S., *Biochim. biophys. Acta,* **36**, 203 (1959)
62. HECKMAN, K. D., and PARSONS, D. S., *Biochim. biophys. Acta,* **36**, 213 (1959)
63. SWAN, A. G., ELLINGTON, E. E., and MILLER, A. T., *Am. J. Physiol.,* **199**, 1227 (1960)
64. KLEINZELLER, A., *Biochim. biophys. Acta,* **43**, 41 (1960)
65. WHITTEMBURY, G., *J. gen. Physiol.,* **43**, 43 (1960)
66. KLEINZELLER, A., and KNOTKOVÁ, A., *J. Physiol.,* **175**, 172 (1964)
67. MACKNIGHT, A. D. C., *Biochim. biophys. Acta,* **150**, 263 (1968)
68. MACKNIGHT, A. D. C., *Biochim. biophys. Acta,* **163**, 500 (1968)
69. MACKNIGHT, A. D. C., *Biochim. biophys. Acta,* **163**, 557 (1968)
70. DEYRUP, I., *Am. J. Physiol.,* **188**, 125 (1957)
71. WILLIS, J. S., *Biochim. biophys. Acta,* **163**, 516 (1968)
72. WHITTEMBURY, G., *J. gen. Physiol.,* **48**, 699 (1965)
73. PROVERBIO, F., ROBINSON, J. W. L., and WHITTEMBURY, G., *Biochim. biophys. Acta,* **211**, 327 (1970)
74. WHITTEMBURY, G., and PROVERBIO, F., *Eur. J. Physiol.,* **316**, 1 (1970)
75. MACKNIGHT, A. D. C., *Biochim. biophys. Acta,* **173**, 223 (1969)
76. ELSHOVE, A., and VAN ROSSUM, G. D. V., *J. Physiol.,* **168**, 531 (1963)
77. VAN ROSSUM, G. D. V., *Biochim. biophys. Acta,* **122**, 323 (1966)
78. VAN ROSSUM, G. D. V., *Biochim. biophys. Acta,* **122**, 312 (1966)
79. WHITTAM, R., and DAVIES, R. E., *Biochem. J.,* **56**, 445 (1954)
80. MUDGE, G. H., *Am. J. Physiol.,* **173**, 511 (1953)
81. FOULKES, E. C., and MILLER, B. F., *Fedn Proc. Fedn Am. Socs exp. Biol.,* **15**, 254 (1956)
82. FOULKES, E. C., *Fedn Proc. Fedn Am. Socs exp. Biol.,* **16**, 181 (1957)
83. FOULKES, E. C., and MILLER, B. F., *Arch. biochim. Biophys.,* **73**, 327 (1958)
84. LI, C. L., and MCILWAIN, H., *J. Physiol.,* **139**, 178 (1957)
85. SCHANNE, O., and CORABOEUF, E., *Nature, Lond.,* **210**, 1390 (1966)
86. HUMPHREY, E. W., and MAENO, T., *Biochim. biophys. Acta,* **183**, 236 (1969)
87. HAYLETT, D. G., and JENKINSON, D. H., *Nature, Lond.,* **224**, 80 (1969)
88. CLARET, M., and CORABOEUF, E., *J. Physiol.,* **210**, 137 (1970)
89. GIEBISCH, G., *J. cell. comp. Physiol.,* **51**, 221 (1958)
90. GIEBISCH, G., *J. gen. Physiol.,* **44**, 659 (1961)
91. WHITTEMBURY, G., SUGINO, N., and SOLOMON, A. K., *J. gen. Physiol.,* **44**, 689 (1961)
92. GIEBISCH, G., *J. gen. Physiol.,* **51**, 3155 (1968)
93. KEYNES, R. D., and SWAN, R. C., *J. Physiol.,* **147**, 591 (1959)
94. FRUMENTO, A. S., and MULLINS, L. J., *Nature, Lond.,* **204**, 1312 (1964)
95. GARRAHAN, P. J., and GLYNN, I. M., *Nature, Lond.,* **207**, 1098 (1965)
96. BAKER, P. F., *Biochim. biophys. Acta,* **88**, 458 (1964)
97. MUDGE, G. H., and TAGGART, J. V., *Am. J. Physiol.,* **161**, 173 (1950)
98. SCHATZMAN, H. J., WINDHAGER, E. E., and SOLOMON, A. K., *Am. J. Physiol.,* **195**, 570 (1958)
99. SMYTH, H., *J. Physiol.,* **187**, 361 (1966)

364

100. FOULKES, E. C., *Am. J. Physiol.*, **203**, 655 (1962)
101. BURG, M. B., and ORLOFF, J., *Am. J. Physiol.*, **207**, 983 (1964)
102. SEIDMAN, I., and CASCARANO, J., *Am. J. Physiol.*, **211**, 1165 (1966)
103. SPATER, H. W., NOVIKOFF, A. B., and MASEK, B., *J. Biophys. Biochem. Cytol.*, **4**, 765 (1958)
104. BONTING, S. L., SIMON, K. A., and HAWKINS, N. M., *Arch. Biochem. Biophys.*, **95**, 416 (1961)
105. NOVIKOFF, A. B., DRUCKER, J., SHIN, W.-Y., and GOLDFISCHER, S., *J. Histochem. Cytochem.*, **9**, 434 (1961)
106. EMMELOT, P., and BOS, C. J., *Biochim. biophys. Acta*, **58**, 374 (1962)
107. BONTING, S. J., CARAVAGGIO, L. L., and HAWKINS, N. M., *Arch. Biochem. Biophys.*, **98**, 413 (1962)
108. MCCLURKIN, I. T., *J. Histochem. Cistochem.*, **12**, 654 (1964)
109. SKOU, J. C., *Biochim. biophys. Acta.*, **58**, 314 (1964)
110. EMMELOT, P., and BOS, C. J., *Biochim. biophys. Acta.*, **120**, 369 (1966)
111. BLAT, C., and HAREL, F., *Biochim. biophys. Acta*, **173**, 23 (1969)
112. EMMELOT, P., FELTKAMP, C. A., and VAZ DIAS, H., *Biochim. biophys. Acta*, **211**, 43 (1970)
113. WHEELER, K. P., and WHITTAM, R., *Biochem. J.*, **93**, 349 (1964)
114. AHMED, K., and JUDAH, J. D., *Biochim. biophys. Acta*, **93**, 603 (1964)
115. WHITTAM, R., WHEELER, K. P., and BLAKE, A., *Nature, Lond.*, **203**, 720 (1964)
116. TAYLOR, C. B., *Biochem. Pharmac.*, **12**, 539 (1963)
117. EPSTEIN, F. H., and WHITTAM, R., *Biochem. J.*, **99**, 232 (1966)
118. RUFKIN, R. J., *Proc. Soc. exp. Biol. Med.*, **120**, 802 (1965)
119. WHEELER, K. P., and WHITTAM, R., *Biochem. J.*, **85**, 495 (1962)
120. KINSOLVING, C. R., and POST, R. L., *Physiologist*, **3**, 94 (1960)
121. LANDON, E. J., JAZAB, N., and FORTE, L., *Am. J. Physiol.*, **211**, 1050 (1966)
122. CHIGNELL, C. F., and TITUS, E., *J. biol. Chem.*, **241**, 5083 (1966)
123. JUDAH, J. D., and AHMED, K., *J. cell. comp. Physiol.*, **64**, 355 (1964)
124. WHITTAM, R., *Nature, Lond.*, **191**, 603 (1961)
125. WHITTAM, R., *Biochem. J.*, **82**, 205 (1962)
126. WHITTAM, R., and WILLIS, J. S., *J. Physiol.*, **168**, 158 (1963)
127. BLOND, D. M., and WHITTAM, R., *Biochem. J.*, **92**, 158 (1964)
128. BLOND, D. M., and WHITTAM, R., *Biochem. J.*, **97**, 523 (1965)
129. CHANCE, B., *J. biol. Chem.*, **240**, 2729 (1965)
130. CARAFOLI, E., *J. gen. Physiol.*, **50**, 1849 (1967)
131. JUDAH, J. D., AHMED, K., and MCLEAN, A. E. M., *Ann. Med. exp. Fenn.*, **44**, 338 (1966)
132. WALLACH, S., REIZENSTEIN, D. L., and BELLAVIA, J. V., *J. gen. Physiol.*, **49**, 743 (1966)
133. VAN ROSSUM, G. D. V., *Nature, Lond.*, **225**, 637 (1970)
134. HOFER, M., and KLEINZELLER, A., *Physiologia bohemoslov*, **12**, 405 (1963)
135. THIER, S. O., FOX, M., ROSENBERG, L. E., and SEGAL, S., *Fedn Proc. Fedn Am. Socs exp. Biol.*, **22**, 166 (1963)
136. FOX. M., THIER. S., ROSENBERG, L., and SEGAL, S., *Biochim. biophys. Acta*, **79**, 167 (1964)
137. ROSENBERG, L. E., BLAIR, A., and SEGAL, S., *Biochim. biophys. Acta*, **54**, 479 (1961)
138. WEBBER, W. A., BROWN, J. L., and PITTS, R. F., *Am. J. Physiol.*, **200**, 380 (1961)
139. SCRIVER, C. R., EFRON, M. L., and SCHAFER, I. A., *J. clin. Invest.*, **43**, 374 (1964)
140. SCRIVER, C. R., and WILSON, O. H., *Nature, Lond.*, **202**, 92 (1964)
141. SCRIVER, C. R., and GOLDMAN, H., *J. clin. Invest.*, **45**, 1357 (1966)
142. THIER, S., FOX, M., ROSENBERG, L., and SEGAL, S., *Biochim. biophys. Acta*, **93**, 106 (1964)
143. TEWS, J. K., and HARPER, A. E., *Biochim. biophys. Acta*, **183**, 601 (1969)

144. SANDERS, R. B., and RIGGS, T. R., *Fedn Proc. Fedn Am. Socs exp. Biol.*, **22**, 417 (1963)
145. SANDERS, R. B., and RIGGS, T. R., *Fedn Proc. Fedn Am. Socs exp. Biol.*, **23**, 535 (1964)
146. CHAMBERS, J. W., GEORG, R. H., and BASS, A. D., *Molec. Pharmac.*, **1**, 66 (1965)
147. HILLMAN, R. E., and ROSENBERG, L. E., *Biochim. biophys. Acta*, **211**, 317 (1970)
148. KRANE, S. M., and CRANE, R. K., *J. Biol. Chem.*, **234**, 211 (1959)
149. KLEINZELLER, A., KOTYK, A., *Biochim. biophys. Acta*, **54**, 367 (1961)
150. KLEINZELLER, A., KOLÍNSKÁ, J., and BENÉŠ, I., *Biochem. J.*, **104**, 843 (1967)
151. KLEINZELLER, A., KOLÍNSKÁ, J., and BENÉŠ, I., *Biochem. J.*, **104**, 852 (1967)
152. KLEINZELLER, A., *Biochim. biophys. Acta.* **211**, 264 (1970)
153. KLEINZELLER, A., *Biochim. biophys. Acta.* **211**, 277 (1970)
154. KLEINZELLER, A., AUSIELLO, D. A., and ALMENDARES, J. A., *Biochim. biophys. Acta.*, **211**, 293 (1970)
155. KHURI, R. N., FLANIGAN, W. J., OKEN, D. E., and SOLOMON, A. K., *Fedn Proc. Fedn Am. Socs exp. Sci.*, **25**, 899 (1966)
156. CRAWHALL, J. C., and SEGAL, S., *Biochim. biophys. Acta*, **163**, 163 (1968)
157. CROSS, R. J., and TAGGART, J. V., *Am. J. Physiol.*, **161**, 181 (1950)
158. TAGGART, J. V., SILVERMAN, L., and TRAYNER, E. M., *Am. J. Physiol.*, **173**, 345 (1953)
159. FOULKES, E. C., *Fedn Proc. Fedn Am. Socs exp. Sci.*, **17**, 222 (1958)
160. BURG, M. B., and ORLOFF, J., *Am. J. Physiol.*, **202**, 565 (1962)
161. FORSTER, R. P., and COPENHAVER, J. H., *Am. J. Physiol.*, **186**, 167 (1956)
162. FOULKES, E. C., and MILLER, B. F., *Am. J. Physiol.*, **196**, 86 (1959)
163. FARAH, A., KODA, F., and FRAZER, M., *Endocrinol.*, **58**, 399 (1956)
164. FOULKES, E. C., *Am. J. Physiol.*, **205**, 1019 (1963)
165. ROSS, C. R., and FARAH, A., *J. Pharmacol.*, **151**, 159 (1966)
166. TUNE, B. M., BURG, M. B., and PATLAK, C. S., *Am. J. Physiol.*, **217**, 1057 (1969)
167. BURG, M. B., and ORLOFF, J., *Am. J. Physiol.*, **217**, 1064 (1969)
168. SHEIKH, M. I., and MØLLER, J. V., *Biochim. biophys. Acta.* **196**, 305 (1970)
169. MAXILD, J., and MØLLER, J. V., *Biochim. biophys. Acta*, **184**, 614 (1969)
170. CORT, J. H., and KLEINZELLER, A., *Biochim. biophys. Acta.*, **23**, 321 (1957)
171. KLEINZELLER, A., and CORT, J. H., *Nature, Lond.*, **180**, 1124 (1957)
172. CORT, J. H., and KLEINZELLER, A., *J. Physiol.*, **142**, 208 (1958)
173. BERNDT, W. O., and LE SHER, D. A., *Amer. J. Physiol.*, **200**, 1111 (1961)
174. WILLIS, J. S., *J. gen. Physiol.*, **49**, 1221 (1966)
175. PARSONS, D. S., and VAN ROSSUM, G. D. V., *Biochim. biophys. Acta*, **54**, 364 (1961)
176. WILLIS, J. S., *Biochim. biophys. Acta*, **163**, 506 (1968)
177. DANTZLER, W. H., *Am. J. Physiol.*, **217**, 1510 (1969)
178. VAN ROSSUM, G. D. V., *Biochim. biophys. Acta*, **82**, 556 (1964)
179. WHITTAM, R., WHEELER, K. P., and BLAKE, A., *Nature, Lond.*, **203**, 720 (1964)
180. WIGGINS, P. M., *Biochim. biophys. Acta*, **109**, 454 (1965)
181. VOGEL, G., and KROGER, W., *Pflügers Arch., ges. Physiol.*, **288**, 342 (1966)
182. VILLEE, C. A., and HAGERMAN, D. D., *Am. J. Physiol.*, **194**, 457 (1958)
183. WHITTAM, R., *Biochim. biophys. Acta*, **54**, 574 (1961)
184. VAN ROSSUM, G. D. V., *Biochim. biophys. Acta*, **74**, 1 (1963)
185. PLATTS, M. M., and MUDGE, G. H., *Am. J. Physiol.*, **200**, 387 (1961)
186. BERNDT, W. O., and BEECHWOOD, E. C., *Am. J. Physiol.*, **208**, 642 (1965)
187. MØLLER, J. V., *J. Physiol.*, **192**, 505 (1967)
188. MØLLER, J. V., *J. Physiol.*, **192**, 519 (1967)
189. FORSTER, R. P., *A. Rev. Physiol.*, **27**, 183 (1965)
190. THURAU, K., VALTIN, H., and SCHNERMANN, J. S., *A. Rev. Physiol.*, **30**, 441 (1968)
191. REYNOLDS, E. S., THIERS, R. E., and VALLEE, B. L., *J. biol. Chem.*, **237**, 3546 (1962)
192. WEINER, I. M., BLANCHARD, K. C., and MUDGE, G. H., *Am. J. Physiol.*, **207**, 953 (1964)

193. DESPOPOULOS, A., *J. theoret. Biol.*, **8**, 163 (1965)
194. MARSHALL, E. K., and CRANE, M. M., *Am. J. Physiol.*, **70**, 465 (1924)
195. WALKER, A. M., and HUDSON, C. L., *Am. J. Physiol.*, **118**, 153 (1937)
196. FORSTER, R. P., *Am. J. Physiol.*, **179**, 372 (1954)
197. SMITH, H. W., *Am. J. Physiol.*, **98**, 279 (1931)
198. SMITH, H. W., *Am. J. Physiol.*, **98**, 296 (1931)
199. GOLDBERG, M., WOJTCZAK, A. M., and RAMIREZ, M. A., *J. clin. Invest.*, **46**, 388 (1967)
200. TRUNIGER, B., and SCHMIDT-NIELSEN, B., *Am. J. Physiol.*, **207**, 971 (1964)
201. KLEINMAN, L. I., RADFORD, E. P., and TORELLI, G., *Am. J. Physiol.*, **208**, 578 (1965)
202. LASSITER, W. E., MYLLE, M., and GOTTSCHALK, C. W., *Am. J. Physiol.*, **210**, 965 (1966)
203. CLAPP, J. R., *Am. J. Physiol.*, **210**, 1304 (1966)
204. RABINOWITZ, L., *Am. J. Physiol.*, **209**, 188 (1965)
205. MCLEAN, A. E. M., *Biochem. J.*, **87**, 164 (1963)

10

PERIPHERAL NERVE

E. J. HARRIS*
Department of Biophysics, University College, London

INTRODUCTION

Nerve cells are unique in their ability to receive, conduct, and transmit information from one part of the body to another. The maintenance of normal function in nervous tissue depends both on mechanisms that are common to most animal cells and also on special properties of the nerve cell membrane. Common mechanisms include those responsible for maintaining the differences in ionic activity between the inside and outside of the cell, while the most important specialised feature of the nerve cell membrane is its variable permeability to ions.

MAINTENANCE OF THE ION GRADIENTS

GENERAL REMARKS: MEASUREMENT OF INTRACELLULAR ACTIVITIES

Table 10.1 compares the composition of two kinds of nerve, one of mammalian origin and the other from an invertebrate, with that of the solution that normally bathes them. In both cases, the interior of the nerve has an excess of K ions and a deficit of Na, Ca, and Cl ions. Two additional pieces of information are needed to determine whether this distribution is likely to involve the expenditure of energy or is a purely passive phenomenon. Measurements must be made of the transmembrane potential and also of the ion activities on the two sides of the membrane. The internal potential can be measured by inserting a microelectrode into the cell. It is always

* Based on the original chapter from the Second Edition and extensively revised by P. F. Baker, Physiological Laboratory, Cambridge.

found to be negative with respect to the outside. Activity coefficients for ions in the salt solution bathing the nerve can be measured by conventional methods; but the measurement of activity inside cells poses a difficult problem.

Table 10.1 CONCENTRATIONS (mmol PER kg WATER) OF SOME MAJOR IONS IN INVERTEBRATE (SQUID) AND VERTEBRATE (RABBIT NON-MYELINATED) NERVES

| Ion | Squid | | Rabbit | |
	axoplasm	blood	axon	blood
K^+	400	20	145	.5
Na^+	50	440	60	152
Ca^{2+}	0·5	10	—	2
Mg^{2+}	10	54	—	1
Cl^-	100	560	40	113

In the squid axon, which is up to 1 mm in diameter and many centimetres in length, it is possible to inject into the axoplasm a small radioactive sample of the ion under investigation and examine its mobility along the long axis of the fibre under the influence of an applied electric field. Using this technique, Hodgkin and Keynes[1, 2] were able to show that the mobilities of Na and K ions were only slightly less than (half to two thirds of) their values in free solution, whereas the mobility of Ca was very considerably reduced and a patch of Ca ions showed virtually no movement in an electric field.

These results have been confirmed and extended by other techniques. Na-selective and K-selective glass electrodes were used by Hinke[3] and Thomas[4] and chloride electrodes by Keynes[5] to show that the activity of these ions in squid axoplasm is not very different from that in free solution. Baker, Hodgkin, and Ridgway[6] used the Ca-sensitive protein aequorin to measure the ionised Ca in squid axoplasm. Aequorin reacts with Ca ions to produce light, and after injecting aequorin the axon emits light as a steady glow. Injection of various Ca–EGTA buffers can either enhance or depress the light emission. The buffer that caused no change in light output was assumed to have an ionised Ca concentration identical to that in axoplasm. The value obtained was 0·3 μM. The earlier mobility measurements of Hodgkin and Keynes[2] had suggested that the concentration of diffusible Ca was about 10 μM, so it would seem that of a total intracellular Ca concentration of 500 μM, only 10 μM is diffusible and of this only 0·3 μM is ionised Ca. The bulk of the Ca seems to be sequestered in intracellular organelles, probably mitochondria. This sequestration process requires energy, because metabolic inhibitors such as CN or DNP result in a rise in the free Ca concentration in the axoplasm, apparently through the release of Ca from intracellular stores.

In conclusion, although the available methods of measuring intracellular ion activities can all be criticised on the grounds that they involve some interference with the cell, there can be little doubt that the activities of Na, K, and Cl are not less than half their values in free solution, whereas the activity of Ca is very markedly less. As the interior of the cell is negative with respect to the outside, exclusion of the cations Na and Ca from the interior of the cell must involve the expenditure of energy. The same probably applies to Mg and H ions. The position for K and Cl is less clear. In some cells, separate K and Cl pumps seem to be required, whereas in others the distribution of these ions could be achieved by purely passive means.

THE Na–K EXCHANGE PUMP

Radioactive tracers can be used to investigate the processes responsible for maintaining the differences in Na ion activity across the membrane. In a squid axon at 20 °C, the influx and efflux of Na are both about 40 pmol cm^{-2} s^{-1}. For an ion moving across the membrane in a purely passive manner, provided there is no interaction between the ions, the ratio of influx to efflux can be calculated from the Ussing equation:

$$M_{in}/M_{out} = [C_{out} \exp (E_m F/RT)]/C_{in}$$

where C_{out} and C_{in} refer to ion activities outside and inside the nerve, and E_m is the membrane potential.

For Na ions, the negative internal potential and the higher activity of Na outside the cell should result in a flux ratio much greater than the observed value of unity. This provides further evidence that the efflux cannot be purely passive and must involve some form of active transport.

The properties of this active transport system have been investigated in the squid axon and are summarised below. They closely resemble the Na-transporting systems found in a variety of non-nervous tissues[7].

1. *Dependence on ATP.* The Na efflux, but not the Na influx, is reduced by poisoning with metabolic inhibitors such as CN or

371

DNP (Figure 10.1). Injection of ATP or the phosphagen arginine phosphate into these CN-poisoned axons restores the Na efflux (but see (7) below). It now seems that the primary source of energy for this Na efflux (commonly referred to as the sodium pump) is ATP, and substances such as arginine phosphate serve merely to regenerate ATP inside the axon. The best evidence for this comes from the work of Brinley and Mullins,[9] who inserted a porous

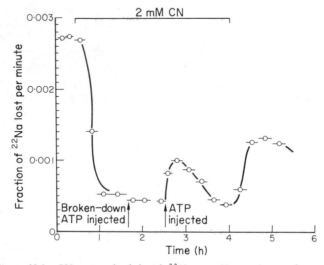

Figure 10.1.   780 μ axon loaded with [22]Na over 12 mm. At second arrow, 16·4 mM ATP was injected over the same 12 mm, giving a mean internal concentration of 2·9 (5·8 ∼ P). At the first arrow, an equal volume of the same ATP solution that had been hydrolysed by boiling was injected. At the end of the experiment the action potential was 99 mV, and the resting potential 56 mV (From Caldwell et al.[8]; courtesy J. Physiol)

capillary along the length of a squid axon and dialysed most of the small-molecular-weight materials from the axoplasm. In this preparation, the Na efflux falls to a low level and of the various energy-rich phosphate compounds tested only ATP, and to a lesser extent dATP, could restore a normal Na efflux.

The dependence of transport on ATP can be shown in another way. Most nervous tissue is aerobic, and the consumption of ATP by the pump leads to a rise in ADP and P_i, which in turn stimulate oxidative phosphorylation, leading to an increased uptake of oxygen. In some nerves, more than 50% of the resting oxygen consumption is associated with pumping[10].

2. *Inhibition by cardiac glycosides.* The energy-dependent extrusion of Na is very specifically inhibited by external application of such cardioactive glycosides as ouabain and strophanthidin. Injection of very high concentrations of these glycosides into the cell has no effect on the Na pump. This inhibitory action of the glycosides is antagonised by external K ions, apparently because K ions impede the binding of glycoside to the pumping sites[11]. The binding of labelled glycoside indicates that there may be as many as 1000 pumping sites per μm^2 axon membrane.

3. *Linkage of Na efflux to K influx.* The efflux of Na is reduced in the absence of external K ions (Figure 10.2). Hyperpolarisation of

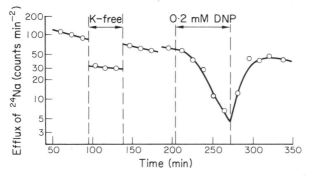

Figure 10.2.   Effects of potassium-free sea water and 0·2 mM DNP on sodium efflux from a Sepia *axon. When not in a test solution, the axon was in artificial sea water containing 10·4 mM K. A break has been left in the curve at 180 min because the excitability was tested at this point. Temperature 17°C (From Hodgkin and Keynes[12]; courtesy* J. Physiol.)

the membrane does not produce a comparable effect, which rules out the possibility that the apparent coupling between Na and K is electrical, the hyperpolarisation produced in K-free conditions leading to a reduction in Na efflux. It thus seems that some form of chemical coupling is more likely, and tracer experiments have established that Na efflux and K influx both require ATP and are both inhibited by ouabain. The parallel behaviour of these fluxes suggests they are linked, Na moving out of the cell in exchange for K moving in. High-speed measurements show removal of external K causes an almost immediate reduction in Na efflux, which suggests there is a rather tight linkage between Na and K movements[13].

The dependence of Na efflux on external K is shown in Figure 10.3. In this figure, the ouabain-sensitive Na efflux is plotted for a range of external Na concentrations as a function of the external K concentration. The points to note are that in the presence of external

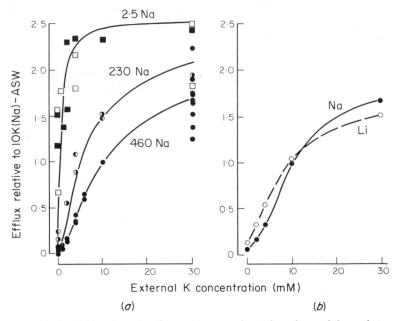

Figure 10.3. Collected results of experiments on the K dependence of the ouabain-sensitive Na efflux showing the effects of replacing external Na by choline or dextrose (a) or lithium (b). In both graphs the ordinate is the ouabain-sensitive Na efflux relative to that in 10-K(Na)-ASW and the abscissa is the external K concentration (mM). ● (460 Na)-ASW; ○ (460 Li)-ASW; ☐ (choline)-ASW; ■ (dextrose)-ASW; ◑ (230 Na-choline)-ASW; ◐ (230 Na-dextrose)-ASW. Temperature 17–19°C (After Baker et al.[13]; courtesy J. Physiol.)

Na the Na efflux is activated along a sigmoid curve that saturates at high K concentrations. Progressive replacement of external Na by choline or dextrose (a), but not Li (b), shifts the curve to the left without appreciably changing the maximum ouabain-sensitive Na efflux. In the absence of Na, the curves are not sigmoid. It has proved difficult experimentally to determine whether this change in shape is genuine or whether K ions leaking from the axon and Schwann cells into the small space immediately external to the axolemma maintain the pumping rate, even in nominally K-free conditions, above the range where sigmoid behaviour would be expected.

4. *Stoichiometry of the pump: evidence for electrogenic behaviour.* The ouabain-sensitive Na efflux is always greater than the ouabain-sensitive K influx. Ratios of about 2 Na : 1 K or 3 Na : 2 K seem most probable[13]. 3 Na : 2 K is preferred because the sigmoid shape

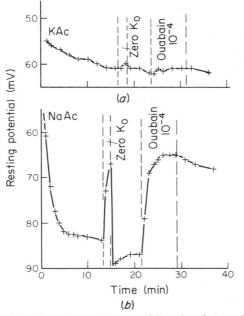

Figure 10.4. The effect of removing external K and ouabain on the resting potential of snail neurones. Removing external K or adding ouabain had little effect on the cell (a) when KAc diffused in, but both had a marked effect (b) on the hyperpolarisation induced by injection of NaAc. Note that reduction of external K, like ouabain, causes a fall in the hyperpolarisation (After Kerkut and Thomas[14]; courtesy Comp. Biochem. Physiol.)

of the K activation curve suggests that at least 2 K ions must co-operate in activating the Na efflux and measurements in both crab and squid axons suggest that 3 Na ions are extruded per energy-rich phosphate bond split[10]. The overall stoichiometry thus seems close to 3 Na pumped out : 2 K pumped in : 1 energy-rich phosphate bond split.

If more Na ions are pumped out than K ions reabsorbed, unless there is some other ouabain-sensitive ion movement that has not been detected, the Na pump should be electrogenic: that is, it should contribute directly to the membrane potential. The size of this contribution will, of course, depend on the rate of pumping and on the membrane resistance; but under favourable conditions it should be detectable experimentally. The squid axon is an unfavourable preparation in this respect—it can be calculated that if one third of the Na emerged as a current of Na ions, it would contribute only 1·8 mV to the resting potential. There is now much evidence for the electrogenic behaviour of the Na pump in nervous tissue. Perhaps the clearest demonstration is that of Thomas[4], working with a snail neurone. He showed that injection of Na (but not Li or K) resulted in hyperpolarisation of the cell—that is, the interior of the cell became more negative with respect to the outside. This effect required external K ions and was inhibited by application of ouabain (Figure 10.4). By voltage-clamping the neurone and simultaneously monitoring the internal Na concentration by a Na-selective glass electrode, Thomas was able to show that the hyperpolarisation that follows injection of Na was produced by an outward current of Na ions that amounted quantitatively to about one third of the total Na extruded from the cell. In other words, one third of the Na being pumped out of the cell emerged as a stream of Na ions. Coupled with the strong evidence for extrusion of 3 Na ions per energy-rich phosphate bond split, this is consistent with an overall stoichiometry of $3 \, Na : 2 \, K : 1 \sim P$.

The observation that the Na pump is electrogenic has considerable physiological significance because it means that, apart from the permeability mechanisms that will be discussed later in this chapter, in some nerve cells—or parts of nerve cells—changes in pumping rate may lead *directly* to alterations in membrane potential, as well as indirectly through changes in the ion gradients.

5. *Selectivity towards ions.* No ions have been found to replace completely internal Na in activating the Na pump. In most tissues, Li is a very poor substitute. At the external activation site, K can be replaced by a range of other cations. Of the ions tested, Tl^+, Rb, NH_4, Cs, and Li can all replace K to varying extents and their effectiveness is in the order listed, with Tl^+ slightly more effective than K, Rb about the same as K, and Cs much less effective. It is of interest that this sequence is the same as that for the effectiveness of different externally applied cations at depolarising nerve[10, 15]. The relation between these observations is not clear; but they suggest some similarity between the mechanisms that effect K uptake by the Na pump and passive movement of K down its electrochemical gradient.

6. *Enzymic basis of the Na pump.* The biochemical study of the Na pump was started in 1957 by Skou, who isolated from crab nerve an ATPase that required Na and K ions for maximum activity and was inhibited by ouabain. This enzyme is present in the membrane of intact crab nerve where it requires *internal* Na and *external* K for activation[10]. It is beyond the scope of this chapter to review the biochemical studies on this enzyme, except to point out that it seems that during each pumping cycle the membrane is alternately phosphorylated and dephosphorylated.

7. *Partial reactions.* The Na pump does not always exchange Na for K. Caldwell *et al.*[8] first noted that under certain conditions—for instance, after the injection of ATP into a CN-poisoned axon—the Na efflux was unaffected by removal of external K. In all these conditions the ATP–ADP ratio was lowered, suggesting that in such conditions internal Na may exchange with external Na. This has been confirmed by Baker *et al.*[13], who showed that in partially poisoned axons there is an approximately 1 : 1 exchange of internal Na for external Na by a ouabain-sensitive route. This mechanism serves no useful purpose; but needs to be borne in mind when interpreting flux data. It would seem that the mechanism that normally uses ATP to drive Na ions outwards across the membrane can be reversed if the ATP–ADP ratio falls to a low enough value; but as yet no one has shown that Na–Na exchange involves the alternate breakdown and resynthesis of ATP.

In red cells, it is also possible to find conditions that will allow exchange of internal K for external K by a ouabain-sensitive route; but this has not been looked for in nerve.

Ca PUMPS

If the movement of Ca across the cell membrane were purely passive, the intracellular concentration of ionised Ca should be about $0.1–1.0$ M, whereas in those cells in which measurements have proved possible the concentration inside the cell is about 10^{-7}–10^{-6} M. Again, some form of pump seems to be at work.

Schatzmann and Vincenzi[16] have demonstrated in red cell membranes an ATP-dependent Ca extrusion mechanism that has many features in common with the Ca uptake system of the sarcoplasmic reticulum (see p. 252). It is not clear whether a similar system exists in nerve; but if Ca extrusion from squid axons were solely dependent on ATP, application of a metabolic inhibitor such as cyanide should result in a fall in the Ca efflux. In practice, cyanide has a dual action: producing first a small fall and later a marked *rise* in the Ca efflux (Figure 10.5). The early fall may reflect cessation of an ATP-dependent extrusion mechanism; but the rise—which occurs when the ATP level is probably less than 100 µM—seems difficult to reconcile with a dependence of Ca extrusion on ATP. The explanation for the rise in Ca efflux seems to be the release of Ca from intracellular binding sites, probably mito-chondria. Energy for the efflux seems to be provided at least in part by the downhill movement of Na and Mg into the axon. Only part of the efflux is an exchange of internal Ca for external Ca.

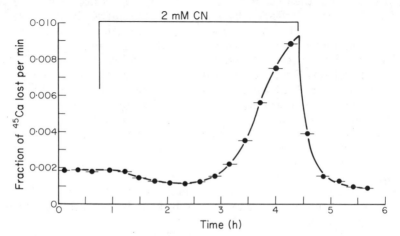

Figure 10.5. Effect of 2 mM CN on rate constant of Ca efflux. Abscissa: time. Ordinate: rate constant of Ca efflux. Axon diameter, 865 µm. The delay in the onset of the cyanide effect was 50–100% greater than average in this experiment. (After Blaustein and Hodgkin[17]; courtesy J. Physiol)

The suggestion that Na may exchange with Ca across the axon membrane is strengthened by the observation that reversing the Na gradient by replacing external Na by Li, choline, or dextrose results in an efflux of Na coupled to an influx of Ca[18]. Thus, depending on the magnitude and direction of the electrochemical gradient of Na, Ca movements can be either into or out of the axon.

Under physiological conditions, the electrochemical gradient for Na is maintained fairly constant by the operation of the Na pump, and it seems quite likely that the potential energy available in this gradient is used to maintain the low level of intracellular Ca. This hypothesis can be examined quantitatively[17]. If two Na ions were to move into the cell in exchange for one Ca ion moving out, the final intracellular Ca concentration would be given by

$$\frac{[Ca]_i}{[Ca]_o} = \frac{[Na]_i^2}{[Na]_o^2}$$

Thus, if the ratio $[Na]_i : [Na]_o = 1 : 10$, the ratio $[Ca]_i : [Ca]_o$ will be $1 : 100$. This is not adequate to account for the observed values; but if three Na ions moved in exchange for one Ca, with either one Na ion moving down the electrochemical gradient or in exchange for K, the intracellular Ca would be given by either

$$\frac{[Ca]_i}{[Ca]_o} = \frac{[Na]_i^3}{[Na]_o^3} \exp(E_m F/RT)$$

or

$$\frac{[Ca]_i}{[Ca]_o} = \frac{[Na]_i^3}{[Na]_o^3} \cdot \frac{[K]_o}{[K]_i}$$

both of which are capable of achieving the observed range of intracellular Ca concentrations.

A notable feature of Na-dependent Ca transport is that reducing the external Na concentration not only reduces the Ca efflux, but it also increases the Ca influx. It must therefore lead to a rise in intracellular Ca. In squid axons, these changes in intracellular Ca have been observed directly by introducing the Ca-sensitive protein aequorin into the cell[6]. Ca influx is also increased by raising the internal Na concentration. Of particular interest is the observation that the Ca influx increases at least as the square of the internal Na concentration. Thus a small rise in internal Na results in a large increase in Ca influx.

These observations have important physiological and pharmacological implications. For instance, the cardiac glycosides increase the force of contraction of the heart and also the resting and stimulus-dependent secretions of various nervous transmitters and hormones. Both actions require Na, and the glycosides are only effective at concentrations that effect partial or complete inhibition of the Na pump. It seems likely that the glycoside first produces a rise in internal Na, which leads in turn to a rise in internal Ca. Coupled with the markedly non-linear dependence of secretion and contraction on Ca, activation of Na–Ca exchange seems able to give a fairly complete account of the action of the cardiac glycosides on these cells[19].

379

OTHER TRANSPORT MECHANISMS

Little is known of the mechanisms responsible for maintaining the gradients of inorganic ions other than Na, K, and Ca. The mechanism of ion gradient coupling just discussed for Ca seems to be quite common, and there is evidence that many organic substances, for instance, nervous transmitters and amino acids are actively transported into nerves using the energy conserved in the gradients of Na and K ions.

PERMEABILITY PROPERTIES OF THE RESTING MEMBRANE

GENERAL FEATURES

The energy conserved in the ion gradients can also be used to generate a potential difference across the cell membrane. Each ion gradient represents a battery capable of generating a p.d.; but whether or not such a p.d. is generated depends on the other ion gradients present and the relative permeability of the membrane to these ions. Thus, if we consider only the relative permeability to Na and K ions, and the permeability to other ions is assumed to be very low, the potential difference V will be given by the Nernst equation

$$V = 58 \log \frac{[K]_o + b[Na]_o}{[K]_i + b[Na]_i}$$

where b is the selectivity factor P_{Na}/P_K and $[\ ]$ are activities.

If the membrane is completely impermeable to Na ions, the potential will be determined solely by K and as there is more K inside the cell than outside, the internal potential will be negative with respect to the outside. If, on the other hand, the membrane is impermeable to K ions, the potential will be determined solely by Na and because there is more Na outside the cell than inside the internal potential will be positive. These two examples represent the extremes of potential that can be obtained using the Na and K gradients alone. By varying the permeability ratio b, all intermediate potentials can be obtained. Under physiological conditions, the permeability to anions (usually chloride) must also be considered. In general, most animal cells are more permeable to K and Cl than to Na, and the internal potential of these cells is negative with respect to the external medium.

The dependence of the resting potential of nerve on K ions can be most readily demonstrated in the perfused squid axon. In this preparation, the axoplasm is squeezed out and replaced by a salt solution of known composition. Figure 10.6 shows the transmembrane potential—determined with an intracellular electrode—at a range of different internal and external K concentrations. When

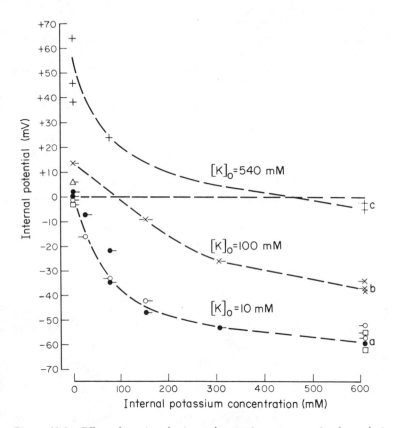

Figure 10.6. Effect of varying the internal potassium concentration by replacing isotonic KCl with isotonic NaCl in perfused axons surrounded by solutions containing 10 (a); 100 (b); or 540 mM, K (c). (a) :O, △, ▢, natural sea water with 10 mM K, 470 mM Na, and 550 mM Cl; each symbol represents a different axon. –O, after increasing internal potassium concentration. O–, after decreasing internal potassium concentration. ●, artificial sea water with 10 mM K, 526 mM Na, 633 mM Cl, 50 mM Ca. (b) × ; artificial sea water with 100 mM K, 436 mM Na, 633 mM Cl, 50 mM Ca: one axon in poor condition. (c) + ; high-K artificial sea water with 538 mM K. 635 mM Cl, 50 mM Ca, 6 different axons. Corrections for junction potential of 1 − 4 mV have been applied when necessary. The chloride concentration of the internal solutions was 560 mM (After Baker et al.[20]; courtesy J. Physiol.)

the K concentration inside exceeds that outside, the interior of the axon is negative; when the concentrations on both sides are the same, the p.d. is close to zero and, when there is more K outside than inside, the interior of the axon is positive. These results show quite clearly that the magnitude and size of the resting potential depends solely on the steepness and direction of the K ion gradient.

An interesting feature of Figure 10.6 is that as the K concentration gradient is increased the potential does not increase as much as would be predicted by the Nernst equation. The explanation seems to be that the potassium permeability falls as the resting potential becomes more negative, the potential ultimately reached being limited by the small inward leak of Na ions.

EVIDENCE FOR SINGLE FILE MOVEMENT

The fluxes of K through the resting squid axon membrane have been examined by Hodgkin and Keynes[12]. They inhibited the active influx of K by poisoning the axon with DNP, and then measured the influx and efflux of K at different values of membrane potential and external K concentration. Under these conditions the fluxes should be entirely passive, and provided there is no interaction between ions crossing the membrane, the flux ratio should be in accord with the Ussing equation.

$$M_{in}/M_{out} = [C_{out} \exp(E_m F/RT)]/C_{in}.$$

The relation can be rewritten after substituting $C_{in}/C_{out} = \exp(E_K F/RT)$, where E_K is the *equilibrium potential* for the potassium ion so that

$$M_{in}/M_{out} = \exp(E_m - E_K) F/RT$$

The difference $E_m - E_K$ has been called by Ussing the *driving potential*.

Hodgkin and Keynes's results departed by a large factor from the calculated ratio. For example, the ratio was 0.016 with a driving potential of $-45\,mV$, whereas the calculated ratio is 0.16; again, when the driving potential was $+27\,mV$ the ratio measured was 39, that calculated is 2.7. Therefore, the ratio depends much more steeply on driving potential than the equation predicts. Some sort of interaction between the fluxes is suggested by the fact that when the potential E_m was held at 30 mV (inside negative) and external K was raised by a factor of 10, the efflux was diminished by 2.6 times and the influx was increased thirty times. As E_m was kept constant in this experiment, the influx would have been expected to be proportional to the external concentration while the efflux should have remained constant. An explanation advanced by Hodgkin and Keynes is that the movement of K ions across the membrane takes place in single file through narrow pores that contain more than one K ion. If, say, three ions occupy the channel, for a radioactive ion added at one end of the channel to emerge at the other end, it must receive three successive collisions from the same side as it was added, without a single collision occurring at the other end. It follows that raising the K concentration on one side of the membrane will markedly increase the flux measured from that side and decrease that measured from the other side.

THE ACTION POTENTIAL: VOLTAGE-DEPENDENT CHANGES IN PERMEABILITY

Although under identical conditions a certain selectivity ratio $P_{Na} : P_K : P_{Cl}$ may be characteristic of a particular cell, it is not necessarily a fixed property of the cell membrane. Thus, the relative permeabilities may change as a function of membrane potential or following application of various chemicals such as transmitter substances or drugs. A particularly striking example of such changes in permeability is the action potential of excitable tissues. Conduction along nerve and muscle is essentially electrical in nature; but without some boosting mechanism a current fed in at one end of a nerve would be attenuated long before it reached its destination. The mechanism by which the flow of current is boosted is called the action potential. It results from three temporally distinct changes in permeability that occur in response to a sudden reduction in membrane potential (depolarisation). These are (1) a rapid increase in the permeability to Na ions, (2) a less rapid blocking or inactivation of the increase in permeability to Na ions, and (3) a slow maintained increase in the permeability to K ions. The upshot of these permeability changes is that the internal potential swings from negative to positive and back again to negative; the period of positivity serving to boost the flow of electric current along the fibre[21, 22].

The dependence of excitability on Na or Li ions was first described by Overton[23]; but it was Hodgkin and Katz[24] who showed that during the action potential the membrane becomes *selectively* permeable to Na ions. The potential at the crest of the spike can be calculated from the Nernst equation with a selectivity factor P_{Na}/P_K of about 10:1. Reducing the external Na concentration or raising the

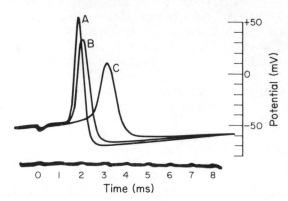

Figure 10.7. *Effect on the internally recorded action potential of a perfused squid axion of replacing internal potassium with sodium ions. A, isotonic potassium sulphate; B, one quarter of the potassium ions replaced by sodium ions; C, half the potassium ions replaced by sodium ions. The records were taken from a long experiment and were obtained in the order B, A, C (After Baker et al.[13]; courtesy J. Physiol.)*

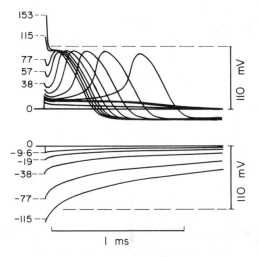

Figure 10.8. *Time course of membrane potential following a short shock at 23 °C. Depolarisations shown upwards. The numbers attached to the curves give the strength of shock in nC cm^{-2}. Shock strengths for unlabelled curves are 29, 23, 19·2, 17·3, 16·7, 15·3, 9·6 (After Hodgkin, Huxley, and Katz[26]; courtesy J. Physiol.)*

internal Na concentration both result in action potentials of smaller size (Figure 10.7) and ultimately block nervous conduction.

The nature of the permeability changes occurring during the action potential has been investigated using a technique known as the voltage clamp. In this method, current-passing and potential-measuring electrodes are inserted into the axon, thus enabling the membrane potential to be displaced to a new value and held there by electronic feedback. The current that flows through a definite area of membrane under the influence of the applied voltage is measured with a separate amplifier[25, 26].

Figure 10.8 shows the change in membrane potential following a brief displacement of the potential to either less negative or more negative values. In this experiment, feedback was not used to maintain a constant p.d. It will be seen that the immediate response of the membrane to a short shock is the attainment of a new and transient potential difference, which corresponds to the charging of a condenser by the stimulus current. From the ratio (charge passed)/(change of p.d.) the capacity of the nerve membrane was found to be $0.9\,\mu F\,cm^{-2}$. Although the initial charging process is linear, the subsequent behaviour of the membrane potential varies with the strength of shock. If the potential is increased (hyperpolarised) or decreased (depolarised) by not more than 15 mV, the p.d. decays back to the original value in a few msec. The time course of the decay is much slower than for the initial capacity surge, because now the current is not flowing through the wires of the stimulating circuit but is carried by ions through the membrane.

For depolarisations much greater than 15 mV, an action potential is produced. The displacement of the potential becomes regenerative and the membrane continues to depolarise, until the internal potential is about $+40\,mV$. The potential does not stay at this value but slowly declines to its resting level.

The voltage clamp technique provides information on the currents flowing through the membrane in response to different impressed voltages. Figure 10.9 shows the membrane currents for depolarisations of 91 to 143 mV. For depolarisations of 20 to 110 mV, the ionic current consists of two phases: an early inward current and a later outward current. The early inward current must represent either an inward flow of positive ions or an outward flow of negative ions. In practice, it depends on Na ions and is abolished or reversed in solutions in which the bulk of the Na is replaced by choline or dextrose (Figure 10.10). Li ions can substitute for Na. Further evidence that the early inward current is carried by Na ions is that when the potential is made equal to that of the Na battery (as predicted by the Nernst equation) there is no early current and only the delayed (potassium) current is seen. For voltage displacements beyond 110 mV, both components of the membrane current are outward. The internal potential at which there is no early current is known as the 'Na equilibrium potential' and is normally about 40–50 mV inside positive (i.e. a total displacement of membrane potential of 100–110 mV from a resting value of about 60 mV inside negative). Changing Na_o or Na_i alters the equilibrium potential in accord with the Nernst equation. Apart from the electrical evidence, there is also much tracer data showing a net entry of Na during the action potential. The main conclusion from these observations is that the early current is carried by Na ions and that the magnitude and direction of the current depends on the concentration gradient for this ion and the potential difference across the membrane.

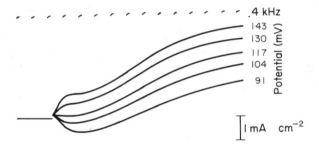

Figure 10.9. Membrane currents for different displacements of the membrane potential at a temperature of $3.5\,°C$; outward current upwards. The figures at the right give the change in internal potential (After Hodgkins, Huxley, and Katz[26]; courtesy J. Physiol.)

Hodgkin and Huxley[27–30] made use of the observation that the early current depends on external Na to separate the membrane current into its two components. Measurements were made of membrane current for the same displacement of membrane potential, first in the presence of Na ions, then with most of the Na replaced by choline, and finally in Na again. On the assumption that choline does not affect the time courses of either current, they were able to obtain separate curves for the Na and K currents. Nowadays, tetrodotoxin (TTX) is preferred to choline as a means of separating Na and K currents (see p. 397).

Currents obtained in these ways do not immediately provide information about the permeability of the membrane; for instance, at the Na equilibrium potential there is no sodium current; but the membrane is permeable to Na ions. The current carried by a particular ion depends on two factors: (1) the driving force on the ion that is itself the resultant of the concentration difference of the ion on the two sides of the membrane and the p.d. across the membrane, and (2) the permeability of the membrane to the ion in

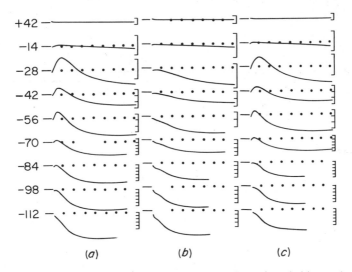

Figure 10.10. Records of membrane current during 'voltage clamps': (a) axon in sea water; (b) axon in choline sea water; (c) after replacing sea water. Displacement of membrane potential indicated in mV. Axon no. 21, temperature 8·5°C. Vertical scale: 1 division is 0.5 mA cm⁻². Horizontal scale: interval between dots is 1 ms. Note that outward current is downwards (After Hodgkin and Huxley[27]; courtesy J. Physiol.)

question. A measure of the driving force on an ion is the difference between the membrane potential (V) and the equilibrium potential (V_E) for the ion, ($V - V_E$). The permeability of the membrane to the ion can be expressed as a conductance (current divided by p.d.). The Na conductance, g_{Na} is measured by $I_{Na}/(V - V_{Na})$ and the K conductance, g_K, by $I_K/(V - V_K)$. Figure 10.11 shows how g_{Na} and g_K vary with time following a step depolarisation of 56 mV. It illustrates a number of important features:

1. Both g_{Na} and g_K rise along sigmoid curves, g_K rising 10–30 times more slowly than g_{Na}. The maximum conductance for both ions is about 20–30 mmho cm⁻².

387

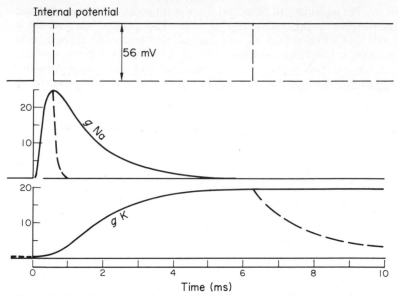

Figure 10.11. *Time course of sodium conductance (g_{Na}) and potassium conductance (g_K) associated with depolarisation of 56 mV; vertical scale in mmho cm^{-2}. The continuous curves are for a maintained depolarisation; broken curves give the effect of repolarising the membrane after 0.6 or 6.3 ms (After Hodgkin[31] based on Hodgkin and Huxley[27, 28]; courtesy Proc. R. Soc. and J. Physiol.)*

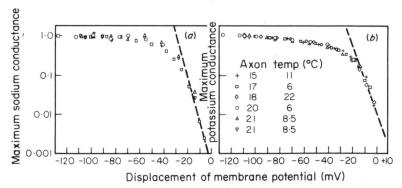

Figure 10.12. *(a) Maximum sodium conductance reached during a voltage clamp. Ordinate: peak conductance relative to value reached with depolarisation of 100 mV, logarithmic scale. Abscissa: displacement of membrane potential from resting value (depolarisation negative). (b) Maximum potassium conductance reached during a voltage clamp. Ordinate: maximum conductance relative to value reached with depolarisation of 100 mV, logarithmic scale. Abscissa: displacement of membrane potential from resting value (depolarisation negative) (After Hodgkin and Huxley[27]; courtesy J. Physiol.)*

2. The increase in g_{Na} is transient. In less than 1 ms it begins to decline exponentially. This decline is called inactivation.

3. The increase in g_K does not show any appreciable inactivation.

4. Returning the membrane potential to its resting level cuts off both currents exponentially. The resting value of g_K is greater than that of g_{Na}.

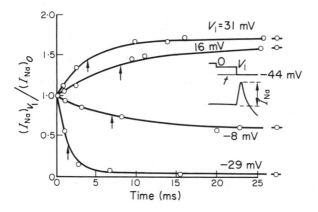

Figure 10.13. Time course of inactivation at four different membrane potentials. Abscissa: duration of conditioning step. Ordinate: circles, sodium current (measured at inset) relative to normal sodium current; smooth curve, $y = y_\infty - (y_\infty - 1)$ exp $(-t/\tau_h)$, where y_∞ is the ordinate at $t = \infty$ and τ_h is the time constant (shown by arrows) (After Hodgkin and Huxley[29]; courtesy J. Physiol.)

To this list should be added three further observations not illustrated in Figure 10.11.

5. The rate of change of both g_{Na} and g_K increases with increasing depolarisation, but the maximum conductance attained remains fairly constant for depolarisations greater than about 50 mV (Figure 10.12).

6. Temperature markedly alters the rates of change of conductance; but has rather little effect on their final values. Cooling a nerve greatly prolongs the action potential.

7. The size of g_{Na} in response to a large depolarisation is very dependent on the potential from which the depolarisation is made. This is illustrated in Figures 10.13 and 10.14. A small

maintained depolarisation markedly reduces (inactivates) the maximum Na current that can be elicited, whereas a small hyperpolarisation allows a larger Na current. The influence of membrane potential on inactivation in the steady state is shown in Figure 10.15.

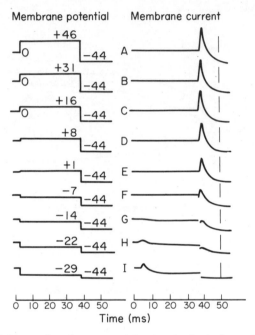

Figure 10.14. Influence of membrane potential on inactivation in the steady state. The vertical lines show the sodium current with no conditioning step; they vary between $0.74\,mA\,cm^{-2}$ in A and $0.70\,mA\,cm^{-2}$ in I (After Hodgkin and Huxley[29]; courtesy J. Physiol.)

At this stage, it is perhaps worth considering some physical models that have been put forward to explain the above data. The steep dependence of permeability on potential suggests that the permeability changes must arise from some effect of the electric field on the orientation of charged particles within the membrane. The striking influence of potential on g_{Na} and g_K is shown by the observation that g_K increases e-fold when the p.d. is reduced by 5–6 mV and g_{Na} is increased e-fold by a reduction of 4 mV.

The proportion (p) of Na^+ carriers at the interior of the membrane may be related to the proportion outside $(1 - p)$ by considering the mean difference in potential energy possessed by each one on the respective sides. This difference is zeE, where E is the p.d., and ze is the charge on the carrier. If work w is needed to cross the barrier in addition to the electrical work, then the proportion of carriers on the inside will be approximately

$$p = \{1 + \exp[-(w + zeE)/kT]\}^{-1}$$

and when z is negative and E high, $p = \text{constant} \times \exp(zeE/kT)$.

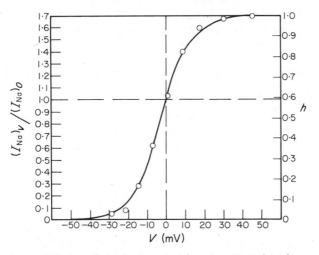

Figure 10.15.   Influence of membrane potential on 'inactivation' in the steady state. Abscissa; displacement of membrane potential from its resting value during conditioning step. Ordinate: circles, sodium current during test step relative to sodium current in unconditioned test step (left-hand scale) or relative to maximum sodium current (right-hand scale) (After Hodgkin and Huxley[27]; courtesy J. Physiol.)

To fit the observed results for Na^+, the carrier will need -6 electronic charges. A similar effect is obtained if a special arrangement of six separate molecules bearing single negative charges is involved. The actual processes involved are not known; but various ideas have been put forward.

1. *Movement of a charged carrier to which the ions are attached.* If the ion-carrier complex bears a number of negative charges, in the resting state it will be attracted to the outside of the membrane.

Depolarisation will allow the carrier to move and the extent of the redistribution will depend on the total negative charge of the complex. Attractive as this hypothesis is, it is rendered very unlikely, as it predicts that there should be an initial inward movement of negative charge, i.e. an outward current, carried by the carrier molecules. Only when Na^+ ions are released into the axon and the carriers return to the outside will the current become inward. No clearly outward current has been observed immediately preceding the rise in inward Na^+ current.

2. *Movement of a charged substrate or enzyme activator.* This remains a possibility; but one piece of evidence that argues against it is the observation that axons perfused only with simple salt solutions will conduct as many as 5×10^5 impulses[20]. If conduction involves the breakdown of a substrate, it seems unlikely that enough substrate for 5×10^5 impulses could be stored in the membrane. However, some form of cyclic breakdown and re-synthesis using the energy stored in the ion gradients cannot be ruled out. It has been suggested in the past that acetylcholine might play some part in the permeability changes of the action potential, just as it does at cholinergic synapses and at the neuromuscular junction. Careful experiments on squid axons have failed to reveal any evidence in support of this suggestion.

3. *Movement of charged particles that do not act as carriers in the usual sense, but that allow ions to cross the membrane when they occupy particular sites.* One can think of the particles as creating a channel across the membrane. This model predicts that the rate at which the particles move will determine the rate at which the ionic current approaches its maximum; but it should have little or no effect on the maximum current. This is in agreement with experiment. Temperature has a large effect on the rate at which the Na^+ current increases, but has little effect on its maximum value. Further, many particles may be required to form a channel; but the removal of only one will block it. This is consistent with the sigmoid rise in Na^+-ion and K^+-ion permeabilities after depolarisation, and their exponential decline after repolarisation. The transient nature of the Na^+ current can be explained by supposing either that, after forming a channel, the activating particles undergo some change that leads to blockage of the channel, or that there is a slowly moving inhibitory particle that blocks the Na^+-ion channel when it reaches a certain position. This model has many points in its favour and it is perhaps relevant to note that the empirical equations used by Hodgkin and Huxley to reconstruct the conducted action potential can be given a physical basis in terms of a model of this kind.

DESCRIPTION OF g_K

In order to describe the changes in g_K, Hodgkin and Huxley[30] assumed that a path for K ions is formed when four similar charged particles occupy a certain region of the membrane. If n is the probability that a single particle is in the right place for forming a K channel, then $g_K = \overline{g_K} n^4$, where $\overline{g_K}$ is the maximum potassium conductance. The value of n depends on the membrane potential and can be expressed in terms of the respective rate constants for the movement of the charged particles into and out of the region of the K channel. The value of n is given by

$$\frac{dn}{dt} = \alpha_n(1 - n) - \beta_n n$$

where α_n is the rate constant for the movement of a particle into the channel and β_n the rate constant for its removal. If the particle has a negative charge, α_n should increase and β_n decrease when the membrane is depolarised (Figure 10.16).

The rise of g_K is described by $[1 - \exp(-t/\tau)]^4$, while the fall is described by $\exp(-4t/\tau)$. These equations fit the observation that the rise has a delay while the fall is purely exponential.

For an experiment in which the p.d. is changed, making g_K change from an initial value of $g_{K,0}$ towards one of $g_{K,\infty}$, the value at time t is given by

$$g_K = \{(g_{K,\infty})^{\frac{1}{4}} - [(g_{K,\infty})^{\frac{1}{4}} - (g_{K,0})^{\frac{1}{4}}] \exp(-t/\tau_n)\}^4$$

where $\tau_n = 1/(\alpha_n + \beta_n)$. This equation was fitted to curves of g_K like that illustrated in Figure 10.17. The values for g_K and τ were chosen for best fit, and differ, of course, during the depolarised time from the value holding after restoration of polarisation. From the relations

$$n = n_\infty - (n_\infty - n_0) \exp(-t/\tau_n)$$

with

$$n_0 = \alpha_{n,0}/(\alpha_{n,0} + \beta_{n,0})$$

where $\alpha_{n,0}, \beta_{n,0}$ are values for the resting level, and

$$n_\infty = \frac{\alpha_n}{\alpha_n + \beta_n} = \left(\frac{g_{K,\infty}}{g_K}\right)^{\frac{1}{4}}, \qquad \alpha_n = \frac{n_\infty}{\tau_n}, \qquad \beta_n = \frac{1 - n_\infty}{\tau_n}$$

the various constants were evaluated, with the assumption that the

393

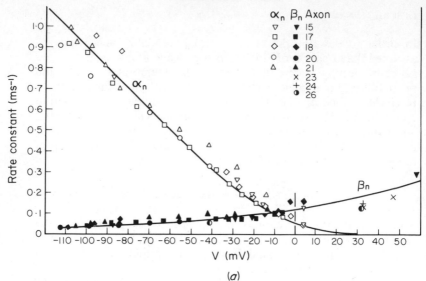

(a)

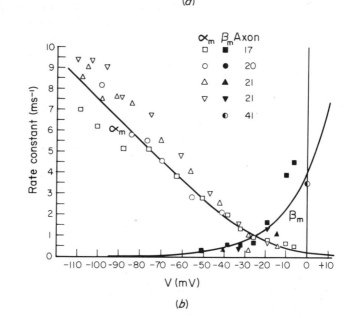

(b)

Figure 10.16. (a) *Abscissa: membrane potential minus resting potential in sea water. Ordinate: rate constants determining rise (α_n) or fall (β_n) of potassium conductance at 6°C.* (b) *Abscissa: membrane potential minus resting potential in sea water. Ordinate: rate constants (α_m and β_m) determining initial changes in sodium conductance at 6°C. (After Hodgkin and Huxley[27]; courtesy J. Physiol.)*

394

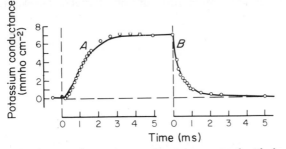

Figure 10.17. A, *rise of potassium conductance associated with depolarisation of 25 mV;* B, *fall of potassium conductance associated with repolarisation to the resting potential. The last A is the same as the first point in* B. *Axon 18, 21 °C, in choline sea water. The smooth curve is drawn with the following parameters.*

	Curve A	Curve B
	(V = − 25mV)	(V = 0)
$G_{K, 0}$	0·09 mmho cm⁻²	7·06 mmho cm⁻²
$G_{K, \infty}$	7·06 mmho cm⁻²	0·09 mmho cm⁻²
τ_n	0·75 ms	1·1 ms

(After Hodgkin and Huxley[28, 30], courtesy J. Physiol.)

limiting value of $g_{K, \infty}$ is 5% more than the value reached at a depolarisation of 100 mV.

The values of α_n and β_n for various states of membrane p.d. can be expressed by use of two empirical equations, that resemble in form the solutions obtained by the constant field assumption for the ionic fluxes through a membrane.

V = displacement of p.d. from resting value in mV

$$\alpha_n = 0·01 \, (V + 10)/\{\exp\left[(V + 10)/10\right] − 1\}$$

$$\beta_n = 0·125 \exp (V/80)$$

There is reasonable agreement between theoretical and experimental curves, except that the latter show more delay. Better agreement might be obtained with a power higher than four.

Description of g_{Na}

For the Na channel, Hodgkin and Huxley assumed that three simultaneous events, each of probability m, open the channel to Na, and a single event of probability $(1 − h)$ blocks it. The probability of there being three activating particles and no blocking particle is therefore m^3h. Hence

$$g_{Na} = \overline{g_{Na}}m^3h$$

395

where $\overline{g_{Na}}$ is the maximum sodium conductance. The values of m and h are given by

$$\frac{dm}{dt} = \alpha_m (1 - m) - \beta_m m$$

and

$$\frac{dh}{dt} = \alpha_h (1 - h) - \beta_h h$$

where the α's and β's are rate constants, α_m and β_h increasing and β_m and α_h decreasing with depolarisation.

The solutions with time boundary conditions shown by subscripts 0 and ∞ are:

$$m = m_\infty - (m_\infty - m_0) \exp(-t/\tau_m)$$
$$h = h_\infty - (h_\infty - h_0) \exp(-t/\tau_h)$$
$$m_\infty = \alpha_m/(\alpha_m + \beta_m)$$
$$\tau_m = 1/(\alpha_m + \beta_m)$$
$$h_\infty = \alpha_h/(\alpha_h + \beta_h)$$
$$\tau_h = 1/(\alpha_h + \beta_h)$$

m_0 is negligible compared with values attained in depolarisations exceeding 30 mV, and inactivation is nearly complete under these conditions, so h_∞ is negligible, then

$$g_{Na} = \overline{g_{Na}}\, m_\infty^3 h_0 \left[1 - \exp(-t/\tau_m)\right]^3 \exp(-t/\tau_h)$$

This expression was fitted to the appropriate records of g_{Na} for depolarisation > 30 mV. For smaller depolarisations, the full equations were employed. In this way, Hodgkin and Huxley obtained the respective rate constants α and β for h and for m as functions of the p.d. These could be expressed as

$$\alpha_m = 0{\cdot}1\, (V + 25)/\{\exp[(V + 25)/10] - 1\}$$

(displacement from resting potential $= V$ mV)

$$\beta_m = 4 \exp(V/18)$$
$$\alpha_h = 0{\cdot}07 \exp(V/20)$$
$$\beta_h = 1/\{\exp[(V + 30)/10] + 1\}$$

The complete set of equations that relate the total current to the membrane potential can be used to calculate the current resulting from a small displacement of potential. The total ionic current I_i is the sum of I_{Na}, I_K, and I_L, where I_L stands for leakage current which does not change with the voltage. The expression for total ionic current is then $I_i = I_{Na} + I_K + I_L = g_{Na}(V - V_{Na}) + g_K(V - V_K) + g_L(v - V_L)$. When the inward sodium current just exceeds the outward potassium current, the depolarisation of the membrane becomes regenerative. In the absence of feedback this results in the appearance of the action potential, the shape of which is determined by the transitory high g_{Na}, followed by the restorative effect of the delayed high g_K which appears later owing to its longer time constant. Hodgkin and Huxley made the calculation of current for a number of conditions of p.d. and showed that the equations they had derived from the constant p.d. experiments would fit the observed p.d. change during the action potential. The variation of impedance during the action potential could also be found, and this agreed with the measurements of Curtis and Cole[32]. The actual net quantities of Na^+ and K^+ moving into, and out of, the nerve during the action potential were found to agree fairly well with the values that have been measured.

An essentially similar analysis has been applied successfully to the action potentials of amphibian myelinated nerve[33], amphibian skeletal muscle[34] and cardiac muscle[35].

AGENTS THAT BLOCK OR MODIFY THE PERMEABILITY CHANGES

Although the equations of Hodgkin and Huxley were formulated on an entirely empirical basis, there is now much evidence in their support. Perhaps the clearest evidence for the existence of separate Na and K channels is the discovery of agents that block g_{Na} without affecting g_K, and vice versa (see Table 10.2 and reviews by Baker[36], Hille[37]).

The puffer fish poison tetrodotoxin (TTX) is the best example of a substance that blocks g_{Na} without affecting g_K[38]. TTX, which is only effective when applied to the outer face of the membrane, does not interfere with the kinetics of either opening or closing the Na channels; it seems simply to block up the Na channels as soon as they are formed. The interaction between TTX and the membrane is so powerful and specific that it has proved possible to use the binding of TTX to estimate the number of Na ion channels in the axon membrane[39]. Only 13 molecules of TTX were bound per μm^2 of lobster nerve membrane. If it is assumed that one TTX binds per Na channel, this gives an upper limit of 13 channels per μm^2 of membrane. Although there are a number of errors inherent in an estimate of this kind, the result emphasises how thinly scattered

Table 10.2 COMPARISON OF THE PROPERTIES OF THE Na^+-ION-SELECTIVE AND K^+-ION-SELECTIVE CHANNELS, WHICH APPEAR IN RESPONSE TO A STEP DEPOLARISATION OF AN AXON MEMBRANE

Type of experiment	Na^+ ion channels	K^+ ion channels
Kinetics		
In response to depolarisation from -60 mV to 0 mV	Open along a sigmoid time-course, but rapidly become blocked (inactivated) along an exponential time-course	Open along a sigmoid time-course, but more slowly than the Na^+-ion channels; remain open as long as the depolarisation persists
In response to repolarisation from 0 mV to -60 mV	If open, the channels will close along an exponential time-course	Close along an exponential time-course
Selectivity (in squid)		
At the inner end	The ease with which ions enter is: $Na^+ : Li^+ : K^+ : Rb^+ : Cs^+$, $1 : 1\cdot1 : 0\cdot08 : 0\cdot025 ; 0\cdot016$. These ions do not block	Blocked by Cs^+ ions and to a lesser extent by Rb^+ ions
At the outer end	Similar to the inner end; other ions which can pass are: NH_4^+, $NH_2NH_3^+$, NH_3^+OH, guanidinium and Ca^{2+}	Rb^+ ions behave much like K^+ ions; Cs^+ ions do not enter or block
Blocking agents		
Tetrodotoxin	Blocks externally; no effect internally	No effect
Procaine	Blocks	Produces some block in most nerves
Xylocaine (at the node)	Blocks	No effect
Tetraethylammonium	No effect	Blocks; but mechanism shows species differences
Condylactic Toxin (CTX)	Blocks inactivation of the channel	Little or no effect
Modifying agents		
Veratridine	Both interfere with the closing of the channels	No effect
DDT		No effect

the Na channels are throughout the membrane material. It underlines the problems that must be faced in any attempt to isolate the molecules concerned. Measurement of the number of Na pumping sites gives about 1000 per μm^2, so the pumping sites are 50–100 times as numerous as the Na channels.

While TTX is the best example of a chemical that blocks specifically the increase in Na^+-ion permeability, it is not the only one. The paralytic shellfish poison saxitoxin, the anaesthetics procaine, xylocaine, and urethane, and the tranquilliser Compazine (prochlorperazine) all act in a similar way although, in many tissues, they also exert some effect on the K^+-ion-permeability mechanism.

Any process tending to block specifically the K^+-ion channels would be expected to slow down the repolarisation phase of the action potential. Long-lasting action potentials are seen to some extent when the K^+ ions inside a perfused axon are replaced by Rb^+ ions and very dramatically when they are partially replaced by Cs^+ ions. Voltage clamp experiments show that internal, but not external, Cs^+ ions block the delayed currents normally carried by K^+ ions, without much affecting the transient currents carried by Na^+ ions[40].

In a number of excitable tissues, including the node of Ranvier, the action potential can be prolonged by inclusion of tetraethylammonium (TEA) in the medium. In the squid, a similar prolongation is seen only when TEA is applied internally. In both cases the results are consistent with the idea that the TEA blocks the K^+ ion channels without affecting the transient rise in Na^+ ion permeability[41, 42]. Inclusion of TEA and TTX in the medium blocks both the Na^+ and K^+ currents. In squid, but not at the node, TEA blocks only outward currents flowing through the K^+ ion channel. These results suggest that internal TEA might be swept into the membrane by outward K^+ currents producing a block and be flushed out again by a current of K^+ ions flowing in the opposite direction. At the node of Ranvier, the results suggest that the binding of one TEA molecule blocks the flow of ions in either direction through one K^+ ion channel. The binding of TEA is not affected by the presence of TTX: this suggests that these drugs act at different sites on the membrane.

Once the Na^+ ion channels are open, they normally become closed by the process of inactivation. Anything that keeps the Na^+ ion channels open, either by slowing the process of inactivation or by acting in some other way, will tend to prolong the action potential and depolarise the membrane.

A toxin that seems to act by blocking inactivation specifically has been isolated from the sea anemone *Condylactis gigantea*. In the presence of condylactis toxin (CTX) the action potential is greatly prolonged and under voltage clamp conditions depolarisation of the membrane produces a maintained Na current that can be turned off only by repolarising the membrane to the resting potential. This toxin has a molecular weight of 10000 to 15000 and it is interesting that it affects inactivation specifically with little or no effect on the processes turning on g_{Na} and g_K. A component of scorpion venom also has an action very like CTX.

The Na channel can also be held open in other ways. The alkaloid veratridine and the insecticide DDT both interfere with the closure of the Na channels without much affecting the kinetics of activation of either g_{Na} or g_K. These substances seem not to work solely by slowing the process of inactivation. They also block the closure of the Na channel that normally follows repolarisation to the resting potential. Their mode of action is therefore to stabilise the Na channel once it is formed.

NATURE OF PERMEABILITY CHANGES

Despite the success of Hodgkin and Huxley in elucidating the events that underlie the action potential, the biochemical basis of permeability changes is still a mystery. Indeed, the membrane remains a black-box, the behaviour of which is well understood but the essential components of which are unknown. This is one of the central problems for those interested in the basis of excitability, and there are a number of experimental approaches that may lead to a solution. A few of these are:

1. *Role of calcium ions.* Calcium ions seem to play at least two roles in nervous conduction. (a) They have a stabilising action. A reduction in external Ca is equivalent to depolarising the membrane in that a smaller depolarisation is needed to turn on a particular g_{Na} or g_K. A fivefold reduction in Ca is in many respects similar to a depolarisation of 10–15 mV[43]. (b) They are essential for the permeability changes. In the complete absence of Ca, it is no longer possible to obtain the normal conductance changes. Mg ions can replace Ca in stabilising the membrane, but seem unable to replace it in the permeability changes.

Two kinds of hypothesis have been advanced to account for the stabilising action of Ca: (a) that Ca ions adsorb to the outer face of the membrane and thereby create an electric field which adds to that provided by the resting potential (a rather similar explanation involving the appearance of negative charges at the inner face of the membrane has been advanced to explain the ability of axons perfused with solutions of low ionic strength to conduct impulses when the measured resting potential is close to zero[44]), and (b) that Na cross the membrane through Na-selective channels that are blocked when occupied by Ca. According to this view, depolarisation increases g_{Na} by first removing Ca from combination with a sodium 'carrier'. Data relevant to this latter possibility has recently been described by Baker, Hodgkin, and Ridgway[6]. They used internally injected aequorin to follow the entry of Ca into squid axons. The Ca entry in response to a depolarisation occurred in two phases (Figure 10.18), an early phase that was blocked by TTX and reached a peak at about the same time as the sodium conductance,

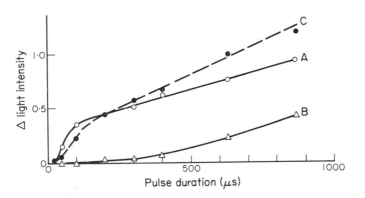

Figure 10.18. Evidence of Ca entry in response to depolarisation of a squid axon. The axon was injected with aequorin. Ordinate: increment in light intensity after 500 pulses relative to increment after 500 spikes, both pulses and spikes at $200\ s^{-2}$; A, before applying tetrodotoxin (TTX). B, in 0·8 μm TTX and C, after removing TTX. Abscissa: pulse duration. Pulse amplitude 80 mV; 22°C; 112 mM Ca and 400 mM Na in external solution. From tracer data (After Hodgkin and Keynes[2]) 1 unit in A, B, C corresponds to ca 0·08 pmol cm^{-2} (After Baker et al.[6]; courtesy J. Physiol.)

and a later phase that was unaffected by TTX and internal TEA, but was blocked by Mn^{2+} and La^{3+} ions. The early Ca entry can be largely attributed to Ca entering through the Na channel and there is no evidence for Ca entry associated with the increase in g_{Na}. The late entry resembles the Ca-entry mechanisms involved in transmitter release (see reference 45).

2. *Chemical modification of the membrane.* Mention has already

401

been made of the use of various toxins and drugs to dissect out the components of the permeability changes. Two other approaches are: (a) application of various enzymes either externally or internally, in the hope of achieving controlled modifications in permeability; a particularly striking observation is that perfusion with the proteolytic enzyme pronase seems to abolish inactivation fairly selectively[46]; (b) alterations in external or internal pH in an attempt to characterise the groups involved in the permeability changes[37].

3. *Heat production.* Associated with the action potential there is a rapid phase of heat production followed by a period during which most of the heat is reabsorbed[47]. There is a rough correspondence between the phase of positive heat production and the rising phase of the action potential and the phase of negative heat production and the falling phase of the action potential. It seems likely that at least three factors contribute to this heat production: (a) the heats of mixing of the ions which interchange across the membrane, (b) the discharging and recharging of the membrane capacity, and (c) changes in the entropy of the nerve membrane when it is depolarised and repolarised.

4. *Optical changes.* A number of workers have described changes in the optical properties of nerves during the passage of a nerve impulse. In a detailed analysis, Cohen, Hille, and Keynes[48] were unable to relate the birefringence changes in squid axons to the permeability processes underlying the action potential. The changes seemed rather to result from the alterations in p.d. across the membrane that occur during the action potential. Despite the failure of this approach, there are a number of other optical methods that may yet provide useful information on the mechanisms underlying permeability changes.

5. *Use of artificial membranes.* This is a very powerful method for studying the genesis of selective permeability mechanisms. It is, however, a model system and great caution must be exercised in applying results obtained by this approach to living nerves. Perhaps the most striking discovery is that various cyclic antibiotics and related substances can act as carriers for ions. Examples of these ionophores are valinomycin and alamethicin. It is not difficult to imagine that such carriers might represent the particles postulated by Hodgkin and Huxley. For instance, if the ionophores were negatively charged, depolarisation might allow a number of particles to form into a column across the membrane, thus providing a channel for the passage of ions. As yet, substances of this kind have not been isolated from nervous tissue.

GENERATOR POTENTIALS AND TRANSMITTER RELEASE

At the start of this chapter, it was pointed out that nerves are specialised both to receive and to transmit information. The bulk of this chapter has been devoted to the mechanism of nervous conduction; some mention will now be made of the membrane events involved in reception and transmission.

GENERATOR POTENTIALS

The final common mechanism of all receptors is a variation in the membrane potential of a nerve terminal. Depolarisation of this terminal leads to the discharge of a train of action potentials, whereas hyperpolarisation reduces this discharge. These changes in potential are brought about in the main by alterations in the permeability of the nerve terminal, either directly, for instance by deformation, or indirectly via the release of chemicals such as transmitter substances (see p. 397).

TRANSMITTER RELEASE

Although it is still too early to be certain, the release of nervous transmitters seems to involve a basically similar mechanism in all nerves—the only variable being the transmitter released. Transmission has been most studied at the neuromuscular junction and the squid giant synapse and has recently been reviewed by Katz[45]. The following are a few important features:

1. The transmitter substance appears to be packaged in small vesicles. In the resting nerve there is a slow random release of groups or quanta of transmitter molecules that is presumed to reflect the loss of the contents of individual vesicles. The rate of release is greatly increased for a brief period following on action potential or electrical depolarisation of the terminal.

2. Calcium ions are essential for transmitter release. Mg ions cannot replace Ca and they antagonise the actions of Ca. Dodge and Rahamimoff[50] have analysed the quantitative dependence of transmitter release at the neuromuscular junction on Ca ions. At low Ca concentrations, there is a sigmoid relation between the endplate potential (taken as a measure of transmitter release) and the external Ca concentration. At higher Ca concentrations, the relation approaches a limiting value. When plotted on double logarithmic

coordinates, the initial part of this relation gives a straight line with a slope of nearly 4. Addition of a constant amount of Mg reduces the endplate potential, but does not alter the slope of the log endplate potential–log Ca relation. These results can be explained quantitatively, on the hypothesis that Ca ions bind to a specific site X in the nerve membrane to form CaX and that the number of packets of transmitter released is proportional to $[CaX]^4$. Thus, if at the terminal there are two parallel reactions

$$Ca + X \rightleftharpoons CaX \text{ (with dissociation constant } K_1)$$

$$Mg + X \rightleftharpoons MgX \text{ (with dissociation constant } K_2)$$

where CaX is the complex necessary for quantal release, while MgX is ineffective.

$$[CaX] = \frac{W[Ca]}{1 + \dfrac{[Ca]}{K_1} + \dfrac{[Mg]}{K_2}}$$

where W is a constant. If transmitter release is proportional to $[CaX]^4$,

$$\text{end-plate potential} = K[CaX]^4 = K\left(\frac{W[Ca]}{1 + \dfrac{[Ca]}{K_1} + \dfrac{[Mg]}{K_2}}\right)^4$$

where K is a proportionality constant.

This analysis suggests that a cooperative action of about 4 Ca ions is necessary for the release of each packet of transmitter by the nerve impulse. At the squid giant synapse, Katz and Miledi[50] have found that transmitter release involves co-operation between 2 to 4 Ca ions. It is not known where these Ca ions are required; but it is an attractive idea that they may be required to facilitate fusion of a vesicle with the surface membrane.

3. By analogy with experiments on squid axons, it seems most likely that the Ca ions enter the nerve terminal; but this is still not proved. Katz and Miledi[49] have investigated the effects of various blocking agents on transmitter release at the squid giant synapse. They found that transmitter release in response to depolarising pulses applied to the presynaptic terminal is unaffected by TTX, and internally applied TEA; but is blocked by external Mn ions. These properties are strikingly similar to those of the second phase of Ca entry during the action potential of squid axons (p. 401) and it seems likely that Ca entering the terminal by this route leads to the release of transmitter. Two further pieces of evidence favour Ca entry

triggering transmitter release: (a) when the currents of Na and K have been reduced to low levels by application of TTX and TEA, depolarisation of a terminal may lead to a calcium-based action potential—implying an inward current carried by Ca ions, and (b) if Ca ions are entering the terminal down their electrochemical gradient, by making the internal potential very positive, it should be possible to stop Ca entry and so block transmitter release. This has been achieved at the squid giant synapse. When the internal potential is pulsed to about + 130 mV, there is no transmitter release during the pulse. An equilibrium potential of + 130 mV is consistent with an internal Ca concentration of 4×10^{-7} M, which is very close to that measured by Baker et al.[6].

REFERENCES

1. HODGKIN, A. L., and KEYNES, R. D., *J. Physiol.*, **119**, 513 (1953)
2. HODGKIN, A. L., and KEYNES, R. D., *J. Physiol.*, **138**, 253 (1957)
3. HINKE, J. A. M., *J. Physiol.*, **156**, 314 (1961)
4. THOMAS, R. C., *J. Physiol.*, **201**, 495 (1969)
5. KEYNES, R. D., *J. Physiol.*, **169**, 690 (1962)
6. BAKER, P. F., HODGKIN, A. L., and RIDGWAY, E. B., *J. Physiol.*, **218**, 709 (1971)
7. BAKER, P. F., *Endeavour*, **25**, 166 (1966)
8. CALDWELL, P. C., HODGKIN, A. L., KEYNES, R. D., and SHAW, T. I., *J. Physiol.*, **152**, 561. 591 (1960)
9. BRINLEY, F. J., and MULLINS, L. J., *J. gen. Physiol.*, **52**, 181 (1968)
10. BAKER, P. F., *J. Physiol.*, **180**, 383 (1965)
11. BAKER, P. F., and WILLIS, J. S., *Nature, Lond.*, **226**, 521 (1970)
12. HODGKIN, A. L., and KEYNES, R. D., *J. Physiol.*, **128**. 61 (1955)
13. BAKER, P. F., BLAUSTEIN, M. P., KEYNES, R. D., MANIL, J., SHAW, T. I., and STEINHARDT, R. A., *J. Physiol.*, **200**, 459 (1969)
14. KERKUT, G. A., and THOMAS, R. C., *Comp. Biochem. Physiol.*, **14**, 167 (1965)
15. RANG, H. P., and RITCHIE, J. M., *J. Physiol.*, **196**, 163, 183 (1968)
16. SCHATZMANN, H. J., and VINCENZI, F. J., *J. Physiol.*, **201**, 369 (1969)
17. BLAUSTEIN, M. P., and HODGKIN, A. L., *J. Physiol.*, **200**, 497 (1969)
18. BAKER, P. F., BLAUSTEIN, M. P., HODGKIN, A. L., and STEINHARDT, R. A., *J. Physiol.*, **200**, 431 (1969a)
19. BAKER, P. F., 'Transport and Metabolism of Calcium ions in Nerve', *Progr. Biophys. biophys. Chem.*, in press
20. BAKER, P. F., HODGKIN, A. L., and SHAW, T. I., *J. Physiol.*, **164**, 330 (1962)
21. HODGKIN, A. L., *Conduction of the Nervous Impulse*, Liverpool University Press (1964)
22. KATZ, B., *Nerve, Muscle and Synapse*, McGraw-Hill, London (1966)
23. OVERTON, E., *Arch.für. Physiol.*, **92**, 115 346 (1902)
24. HODGKIN, A. L., and KATZ, B., *J. Physiol.*, **108**, 37 (1949)
25. COLE, K. S., *Arch. Sci. physiol.*, **3**, 253 (1949)
26. HODGKIN, A. L., HUXLEY, A. F., and KATZ, B., *J. Physiol.*, **116**, 424 (1952)
27. HODGKIN, A. L., and HUXLEY, A. F., *J. Physiol.*, **116**, 449 (1952)
28. HODGKIN, A. L., and HUXLEY, A. F., *J. Physiol.*, **116**, 473 (1952)
29. HODGKIN, A. L., and HUXLEY, A. F., *J. Physiol.*, **116**, 497 (1952)
30. HODGKIN, A. L., and HUXLEY, A. F., *J. Physiol.*, **117**, 500 (1952)
31. HODGKIN, A. L., *Proc. R. Soc.* (B), **148**, 1 (1958)
32. CURTIS, H. J., and COLE, K. S., *J. cell. comp. Physiol.*, **19**, 135 (1942)

33. FRANKENHAEUSER, B., *Prog. Biophys.*, **18**, 97 (1968)
34. ADRIAN, R. H., CHANDLER, W. K., and HODGKIN, A. L., *J. Physiol.*, **208**, 607 (1970)
35. NOBLE, D., *Physiol. Rev.*, **46**, 1 (1966)
36. BAKER, P. F., *Brit. med. Bull.*, **24**, 179 (1968)
37. HILLE, B., *Prog. Biophys.*, **21**, 1 (1970)
38. KAO, C. T., *Pharm. Rev.*, **18**, 997 (1966)
39. MOORE, J. W., NARAHASHI, T., and SHAW, T. I., *J. Physiol.*, **188**, 99 (1967)
40. CHANDLER, W. K., and MEVES, H., *J. Physiol.*, **180**, 788 (1965)
41. ARMSTRONG, C. M., and BINSTOCK, L., *J. gen. Physiol.*, **48**, 859 (1965)
42. HILLE, B., *J. gen. Physiol.*, **50**, 1287 (1967)
43. FRANKENHAEUSER, B., and HODGKIN, A. L., *J. Physiol.*, **137**, 217 (1957)
44. CHANDLER, W. K., HODGKIN, A. L., and MEVES, H., *J. Physiol.*, **180**, 821 (1965)
45. KATZ, B., *The Release of Neurotransmitter Substances*, Liverpool University Press (1969)
46. ROJAS, E., *Nature, Lond.*, **225**, 747 (1970)
47. HOWARTH, J. V., KEYNES, R. D., and RITCHIE, J. M., *J. Physiol.*, **194**, 745 (1969)
48. COHEN, L., HILLE, B., and KEYNES. R. D., *Nature, Lond.*, **218**, 438 (1958)
49. KATZ, B., and MILEDI, R., *J. Physiol.*, **203**, 419 (1969)
49. DODGE, F., and RAHAMIMOFF, R., *J. Physiol.*, **193**, 419 (1967)
50. KATZ, B., and MILEDI, R., *J. Physiol.*, **207**, 789 (1970)

11

ION TRANSPORT AND RELATED PHENOMENA IN YEAST AND OTHER MICRO-ORGANISMS

W. McD. ARMSTRONG
Department of Physiology, Indiana University School of Medicine, Indianapolis

INTRODUCTION

Micro-organisms offer certain advantages to the investigator of membrane transport. Among these is the fact that they are discrete cellular entities which are easily obtained in large numbers by suitable culture methods, and readily maintained in a fully viable condition. It is in fact a simple matter to produce homogeneous populations which are the progeny of a single cell and to reproduce, within a relatively short time, many successive generations of the same population. For this reason micro-organisms are 'ideal' systems for the study of such problems as the genetic control of membrane function[1].

Because they are discrete cells, without the attachments (connective tissue, blood vessels, etc.) which are unavoidable complications in the study of many animal tissues, the distinction between 'cell' and 'environment' can usually be made with more precision when dealing with micro-organisms. Consequently, the experimental methods used in the study of a given transport process and the interpretation of the results obtained are frequently much simpler with micro-organisms than with other cell species[2].

Thus, micro-organisms are attractive systems for the study of 'statistical' or 'population' aspects of membrane transport. However, from another viewpoint they have serious limitations. The interpretation of membrane phenomena often requires data, e.g. membrane potentials, intracellular activities of individual molecules or ions, etc., which can only be obtained by measurements on single cells. Because of their small size (and, in some cases, the fact that the cell proper is surrounded by a tough, relatively thick wall) micro-organisms do not lend themselves readily to such measurements and the parameters in question must be estimated by indirect methods.

To date, membrane function has been extensively studied in only a few species of micro-organisms. Because of space limitations, this review will be restricted largely to two types, yeast (as exemplified by baker's yeast, *Saccharomyces cerevisiae*) and bacteria (particularly *Escherichia coli*) and will be concerned mainly with the transport of inorganic ions and related problems. Useful additional material can be found in a number of reviews[1, 3–7]. Surveys of the earlier literature are given by Conway[8], Rothstein[9, 10] and Mitchell and Moyle[11].

INTRACELLULAR OSMOLALITY OF YEAST AND BACTERIA

Unlike the cells of higher animals which normally exist in a tightly controlled osmotic environment, micro-organisms (with the exceptions of marine forms) must frequently adapt to wide variations in the osmolality of their environment. They must and can, therefore, withstand large differences in osmotic pressure between the intracellular fluid and the external medium.* In walled forms the osmolality of the cell fluid is frequently much higher than that of the surrounding medium, the volume of the cell and its response to changes in internal or external osmolality being largely determined by the mechanical properties of the cell wall[12, 13].

There is abundant evidence that internal osmolalities up to 1 molal or thereabouts are quite common in walled forms of bacteria. The existence of these high internal solute concentrations can be inferred from chemical analysis of the low-molecular-weight constituents of the cytoplasm (Table 11.1) on the assumption that the greater part of these solutes is in 'free' or osmotically active form.† On the basis of this assumption, Mitchell and Moyle[11]

---

* Mitchell and Moyle[11] have found that the ability of certain bacteria to withstand large transmural osmotic pressure gradients is correlated with the 'age' of the cells, i.e. their rate of growth at the time of harvesting. The more rapid their rate of growth at this time, the less able they are to withstand prolonged exposure to media which differ widely in osmotic activity from the cell contents.

† This assumption is very questionable in the case of certain halophilic bacteria (e.g. *Halobacterium salinarium*; Table 11.1). In extreme halophiles[15] fantastically high electrolyte 'concentrations' are observed. These are sometimes greater than can be realised in aqueous solution (see p. 437 for further discussion).

calculated the internal osmotic pressure of several bacterial species. Their values ranged from about 5–6 atm for *Bacterium coli* to 20–30 atm for *Staphylococcus aureus*. The reality of these estimates was experimentally confirmed. Determination of the plasmolysis threshold yielded a value of 2–3 atm for *B. coli* and the use of a vapour pressure equilibrium method gave an internal osmotic pressure of 20–25 atm (equivalent to an effective internal solute activity of about 1 molal) for the unplasmolysable *S. aureus*[11]. The existence,

Table 11.1 SOLUTE CONCENTRATIONS (mM) IN FOUR BACTERIAL SPECIES

| Bacteria | Medium | | Cells | | | | |
	Na$^+$	K$^+$	Na$^+$	K$^+$	Amino acids	Cl$^-$	PO$_4^=$
Salmonella oranienburg	150	25	130	240	110	<5	50
Staphylococcus aureus	150	25	100	680	440	8	80
Vibrio costicolus	1000	4	680	220	330	140	70
Halobacterium salinarium	4000	32	1370	4600	210	3600	90

Data from Christian and Waltho[14] as modified by Epstein and Schultz[7]; courtesy *Biochim. biophys. Acta* and Williams and Wilkins.

in walled bacterial cells, of high activities of osmotically active solutes (averaging about 0.5 molal for a number of species) is strongly indicated by a variety of other experimental techniques. For example, treatment of these cells by a number of chemical or physical agents such as detergents, organic solutes, heat, or osmotic shock[16, 17] releases large amounts of low-molecular-weight solutes to the external medium. Studies with broken cells have failed to reveal extensive binding of low molecular weight substances to macromolecular cell constituents which would be required to explain accumulation and retention of the former on the basis of an 'adsorption' model[17]. Finally, the cell wall of many bacteria can be removed by lysozymes[18]. If this is done in media containing a nonpenetrating sugar in approximately 0.5 M concentration, intact protoplasts are obtained which are stable as long as the external osmolarity is maintained. In hypotonic media protoplasts undergo rapid lysis. In hypertonic media they behave as osmometers. Similar evidence has been found for the existence of a high internal osmolality in yeast cells. Eddy and Williamson[19] found that stable protoplasts of yeast can be obtained by treating whole cells with the intestinal juice of the snail *Helix pomatia* in a medium containing 0.5 M rhamnose. Like bacterial protoplasts, these tend to swell and burst in media containing less than 0·5 M sugar and to shrink in solutions containing more than 0·5 M.

Conway and Armstrong[13] determined the total intracellular solute activity of baker's yeast directly by a microcryoscopic method.

Table 11.2 AVERAGE INTRACELLULAR CONTENTS OF RESTING BAKER'S YEAST AS DETERMINED BY CHEMICAL ANALYSIS

Substance	mmol per kg resting yeast	No. of observations	mmol per litre cell water
Potassium	128	34	261·8
Amino acids (B)	110	3	224·0
Orthophosphate	13·2	5	26·9
Succinic acid (B)	13·2	3	26·9
Magnesium	7·1	8	14·5
Hydroxy acids (ether-soluble) (B)	5·2	3	10·6
Hexose esters	4·7	5	9·6
Calcium	4·7	8	9·6
Citric acid (B)	3·5	3	7·1
Sodium	2·7	8	5·5
Metaphosphate	2·6	5	5·3
Keto acid (B)	0·4	3	0·8

Data from Conway and Armstrong[13] courtesy *Biochem. J.* (B) indicates data obtained from Professor T. G. Brady (personal communication); reproduced with permission.

Packed cells were alternatively frozen and thawed in liquid oxygen and freezing-point measurements were made on the resulting fluid mass. An average value of 0·59 M was found for the total solute activity of washed resting cells. This value was confirmed in experiments in which samples of the fluid material obtained by freezing and thawing the cells were dialysed to equilibrium against measured volumes of distilled water and freezing-point determinations were made on the dialysate. Marked increases in the total internal osmolality were found in cells which had been fermented in solutions containing 5% glucose together with KCl, NaCl or $MgCl_2$. The rather extensive chemical analysis of the same commercial strain of yeast shown in Table 11.2 indicates that most of the low-molecular-weight constituents of the cytoplasm must be in free solution to account for the observed osmalality of the cell contents, though one may infer[3] that the bulk of the cellular Ca^{2+} and Mg^{2+} are not.

IONIC COMPOSITION OF YEAST AND BACTERIA

Table 11.1 summarises the concentrations of the major low-molecular-weight constituents found in four species of bacteria. Disregarding for the moment the two halophilic species included, it is evident that the composition of bacterial cells with respect to the monovalent cations Na^+ and K^+ is basically similar to that of animal cells, i.e. the K^+ content of the cells is high relative to their Na^+ content even though the opposite may be true in the external medium. A similar situation exists in yeast cells, especially when cultured in low Na^+ media (Table 11.2; see Rothstein[3] for a similar analysis of another strain of baker's yeast). It is evident that, under these conditions, K^+ is by far the most abundant cation in the cell interior. The question of the regulation of the K^+ content of the cells is therefore of major concern in the analysis of ionic movements in yeast and bacteria.

THE TRANSPORT OF IONS AND OTHER SOLUTES IN BACTERIA AND YEAST

ION TRANSPORT IN YEAST

The transport of a number of ionic species, including H^+, the alkali metal cations, Ca^{2+}, Mg^{2+}, Mn^{2+} and the anion $H_2PO_4^-$ has been rather extensively studied in resting (i.e. nonproliferating) yeast cells. The transport of K^+ which, as already mentioned, is normally the major cationic constituent of the cytoplasm, has been the subject of particularly intensive investigation.

PASSIVE PERMEABILITY OF THE YEAST CELL MEMBRANE

At the outset, it can be said that the passive or 'leak' permeability of yeast cells to ions is comparatively low. In fact, in this sense, the cells are virtually impermeable to most anions including chloride, sulphate, citrate, $H_2PO_4^-$ and succinate[10].* Similarly, passive permeability to cations in the absence of exogenous substrate is also low, for example, although the K^+ concentration in resting

---

* In certain situations (see pp. 413 and 426) special mechanisms are brought into play which enable succinate or $H_2PO_4^-$ to cross the cell membrane freely. These movements are, however, associated with active movements of cations (specifically K^+) and are quite distinct from the type of permeability process under present discussion.

cells is usually about 0.25 M[3, 13], the rate of loss of this ion from cells suspended in distilled water is only about 1% per hour[10].

FACTORS AFFECTING THE PASSIVE PERMEABILITY OF YEAST

Gross changes in the permeability of the yeast cell membrane, as judged by the onset of rapid loss of cellular K^+ and anions, are induced by a number of agents in concentrations of 1 mM or less. These include cationic detergents[20], cationic dyes[21], cationic redox dyes[22] and certain heavy metal ions[23], e.g. Hg^{2+}. The detailed mechanisms of these effects have not in all cases been worked out but certain conclusions can be drawn, at least tentatively.

It seems likely that the primary site of action of cationic detergents and cationic dyes (including redox dyes) is an anionic ligand in the cell membrane since the cytolytic effects of cationic detergents and the binding of cationic redox dyes by yeast cells are inhibited by inorganic cations[24, 25]. The release of cell K^+ by cationic redox dyes is an all-or-none effect[22]. It has been suggested that the anionic ligands involved in the interaction of these agents with the cell membrane are the phosphoryl and carboxyl groups which have been implicated in the binding of inorganic cations to the yeast cell surface (see p. 431)[26], particularly the former[20], and that the cytolytic action of cationic redox dyes may involve an interaction with membrane —SH groups following attachment to an anionic binding site[22]. Passow and Rothstein[23] also suggest that —SH groups are implicated in the interaction of Hg^{2+} with the yeast cell surface, although they are careful to point out that no simple explanation of the effects of heavy metals in general based on interaction with a single ligand species seems adequate to explain the experimental findings with such ions as Ag^+, Cu^{2+}, Pb^{2+} and Zn^{2+}.

Anionic detergents also cause disruption of membrane function and inhibition of metabolism in fermenting yeast[27, 28]. In unbuffered suspensions, considerably higher concentrations are necessary than with cationic detergents of similar structure. The mechanism of action of these agents appears to be quite different from that involved in the interaction of cationic detergents with the yeast cell membrane. With long-chain alkyl sulphates inhibition of metabolism and membrane damage increase rapidly over a concentration range which corresponds closely to the critical concentration range for micelle formation. With a given detergent, metal cations lower the concentration range over which inhibitory and cytolytic effects are observed in a manner which parallels their lowering effect on the critical micelle concentration[28]. These findings strongly suggest

412

that the interaction of anionic detergents with the yeast cell membrane involves detergent micelles rather than single detergent molecules. A similar mechanism has been invoked to explain the effect of bile salts on the permeability of the yeast cell membrane[29, 30].

ACTIVE UPTAKE OF IONS BY YEAST

When yeast cells are provided with an exogenous source of metabolic energy, e.g. glucose, several major pathways for the uptake of ions by the cells can be identified. These mechanisms are highly active and can effect the rapid movement of relatively large amounts of ions into the cells. Three such uptake systems will be discussed in detail. Following Rothstein's[4] classification, these systems will be identified in terms of their specificities as (a) the K^+–H^+ system, (b) the $H_2PO_4^-$ system, and (c) the phosphate–Mg^{2+} system.

THE K^+–H^+ SYSTEM

In the absence of exogenous substrate net movement of K^+ across the yeast cell membrane is usually negligible, though K^+–K^+ exchange across the membrane occurs to some extent, particularly in the presence of oxygen[31, 32]. When glucose is added to the external medium there is a rapid uptake of K^+ by the cells under both aerobic and anaerobic conditions[33, 34]. If the supply of external K^+ is adequate, the cells under suitable conditions virtually double their K^+ content in 10–15 minutes and net K^+ uptake can be demonstrated from media in which the K^+ concentration is as low as 1×10^{-4} M (ref. 35). Under aerobic conditions, K^+ uptake can be induced by a number of oxidisable substrates other than glucose[36].

The mechanism of the glucose-induced uptake of K^+ by yeast has been the subject of much detailed study. The initial impetus for this stemmed from the discovery[33, 34] that K^+ uptake occurs by a 1 : 1 exchange for H^+. The amount of H^+ secreted may be quite large. In concentrated unbuffered suspensions of 'starved' cells (see below) with high K^+ concentrations in the medium, sufficient H^+ may be secreted to lower the external pH from about 6 to well below 2^{37}. Although these H^+ ions are derived from metabolism there is no accumulation of H^+ within the cells. On the contrary, during fermentation and concomitant K^+ uptake the pH of the cell interior was found to increase from about 5·8 to 6·35, although that of a special outer region fell from 5·9 to below 4·2[38]. In freshly harvested and washed yeast cells much of the H^+ secreted is accompanied by organic anions, notably succinate, and considerable H^+ secretion can occur even in the absence of K^+ (Figure 11.1). In 'starved' cells, i.e. cells depleted of endogenous carbohydrate reserves by prolonged oxygenation in substrate-free media (e.g. tap water), very little secretion of H^+ occurs in the absence of K^+. In the presence of K^+ a 1:1 secretion of free H^+ ions in exchange for K^+ taken up by the cells is observed (Figure 11.1a).

ENERGY REQUIREMENTS FOR K^+–H^+ EXCHANGE

It is clear from the foregoing that K^+–H^+ exchange in yeast involves net movement of both ionic species against large concentration gradients. Direct estimates of intracellular pH^{38} show that H^+ is moved out of the cells against an activity gradient of several orders of magnitude.* Unless one assumes that the bulk of the cell K^+ is 'bound' or 'sequestered' in one or more states of extremely low thermodynamic activity a similar conclusion must be drawn concerning the inward movement of K^+. 'Binding' of the greater part of the cellular K^+ is difficult to reconcile with the freezing-point measurements of Conway and Armstrong[13] and with subsequent experiments of a similar kind[40] in which it was shown directly that some 92% of the cytoplasmic K^+ can be recovered in a freely diffusible state from disintegrated cells. In the absence of direct estimates of the membrane potential in yeast cells, it cannot be stated categorically that net movements of both K^+ and H^+ under these conditions are examples of active transport according to Rosenberg's[41] definition. However, since these ions are moving in opposite directions, it is clear that at least one is being transferred against an electrochemical gradient and, unless the membrane potential is unusually high, the inferential evidence in favour of active movement of both is quite strong.

As one would anticipate from its dependence on a supply of exogenous substrate, the K^+–H^+ exchange system is sensitive to a number of metabolic inhibitors including azide, 2,4-dinitrophenol,

---

* Somewhat different results for cell pH in yeast were reported by Kotyk[39] who found that the intracellular pH tends to remain close to 5·8 when the external pH is low and rises to a second 'plateau' at 7·6 when the external pH is increased above 5. During fermentation without external K^+, internal pH fell to 6·7. With K^+ in the fermentation medium. intracellular pH fell to 6·3. These discrepancies in detail between the results of Kotyk and those of Conway and Downey[38] do not affect the major point at issue in the present discussion, the ability of the yeast cell to excrete H^+ against a high activity gradient.

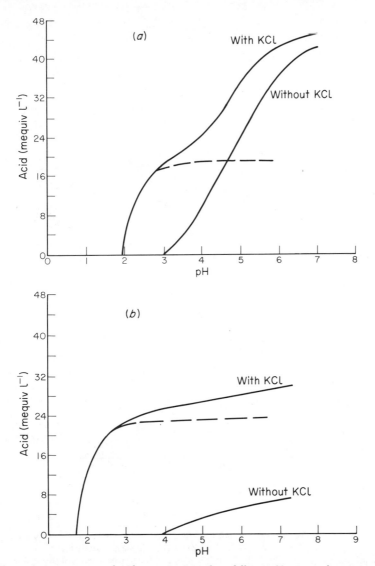

Figure 11.1.   Titration of acid appearing in medium following 30 minutes fermentation of yeast (1·0 g centrifuged cells per 0·6 ml suspension medium) in 5% glucose with and without 0·1 M KCl. (a) Fresh washed cells. (b) Cells washed and suspended following 30 hours oxygenation in water (After Conway and Brady[37]; courtesy Biochem. J.)

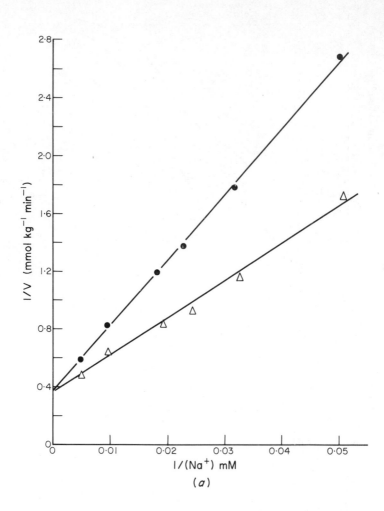

Figure 11.2   Competition between alkali metal cations for uptake by fermenting yeast.
(a) *Competitive effect of K^+ on Na^+ uptake (After Conway and Duggan[43]; courtesy* Biochem. J.). (b) *Effect of Rb^+ on K^+ uptake (After Armstrong and Rothstein[48];* *courtesy* J. gen. Physiol.)

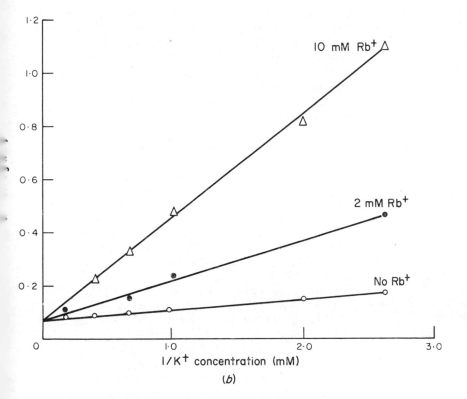

(b)

and monoiodoacetate[8]. In the absence of external glucose and inorganic phosphate, K^+ uptake is blocked by inhibitors of respiration, e.g. cyanide[42]. In the presence of glucose and phosphate, cyanide has no effect on K^+ uptake.

KINETICS OF K^+ UPTAKE: COMPETITION BETWEEN OTHER CATIONS AND K^+

At constant external pH the relationship between the initial rate of entry of K^+ and the external K^+ concentration follows simple Michaelis–Menten kinetics[35, 43] and can be fitted by an equation of the form $V/V_{max} = (S_e)/(S_e + K_m)$ where S_e is the external K^+ concentration and K_m is the Michaelis constant. This kind of kinetic behaviour is usually interpreted as indicating that combination between the transported species and a membrane 'carrier' is an integral part of the transport process. At neutral pH, each of the alkali metal cations (and also Mg^{2+} and Ca^{2+}) can, if present alone in the external medium, exchange for cellular H^+ by a similar saturable mechanism but with considerably less facility than K^+[43, 44]. The ability of other ions to be taken up, albeit relatively slowly, in this fashion has been utilised in the preparation of yeasts in which the major part of the intracellular K^+ has been replaced by another cation, e.g. Na^+[45, 46].

Evidence that K^+ and other cations (including H^+, Rb^+, Cs^+, Na^+, Li^+, Mg^{2+}, Ca^{2+}, NH_4^+, and organic cations such as basic amino acids and substituted ammonium ions) share a common carrier has emerged from competition studies[43, 44, 47, 48]. The interactions between the uptake of K^+ and that of H^+, the other alkali metal cations, Mg^{2+}, and Ca^{2+} have been rather extensively studied. It has been found that K^+ competitively inhibits the uptake of the other alkali metal cations, Mg^{2+} and Ca^{2+}[43, 48]. Similarly, Rb^+ is a competitive inhibitor of K^+ uptake[48]. These relationships are exemplified in Figure 11.2a and b. On the basis of such studies and their interpretation in terms of competition for a single carrier site, the relative affinities of these ions for this site have been determined in terms of their apparent Michaelis constants (Table 11.3).

As mentioned above, the interactions between K^+ and Rb^+ appear to be completely symmetrical, i.e. each ion acts as a simple competitive inhibitor of the uptake of the other. Thus a carrier model involving competition for a single receptor site[43] in the transport system would seem to describe adequately the kinetics of uptake of these two ions. The interactions between K^+ and the other alkali metal cations, K^+ and H^+, K^+ and Mg^{2+}, and K^+ and Ca^{2+}

are more complicated however and appear to require a different explanation[48].

External H^+ has marked effects on the transport of alkali metal cations by yeast. As the external pH is reduced the net rate of K^+ uptake decreases and may even become negative at pH values below 2^{44}. Similar effects of pH on net uptake of other cations have been reported[49]. Efflux of K^+ from yeast cells into a K^+-free medium

Table 11.3 KINETIC CONSTANTS FOR CATION UPTAKE IN YEAST

	K_m (mM)	K^+ modifier (mM)	β^* H^+ against cations	β^* cations against K^+
H^+	0·2	0·01–0·03	—	0·5–0·7
Li^+	27	19	0·4	0·7
Na^+	16	14·4	0·3	0·46–0·51
K^+	0·5	1·6	0·67	1·0
Rb^+	1·0	—	0·5	1·0
Cs^+	7·0	1·3	0·15	0·7
Mg^{2+}	500	4·0	—	0·7
Cs^{2+}	600	1·5	—	0·6

From Armstrong and Rothstein[48]; courtesy *J. gen. Physiol.*
* See p. 424.

increases regularly with decreasing external pH[50]. Examination of the effect of external Na^+ on K^+ efflux at different external pH levels showed that, above pH 4·5, K^+ efflux was markedly enhanced by Na^+ ions which were taken up by the cells in equivalent quantity to the K^+ lost. Below pH 4·5 external Na^+ did not induce any increase in K^+ efflux[50]. These findings are consistent with the observation[43, 50] that cells suspended at near neutral pH values in media containing Na^+ ions plus added substrate readily take up large amounts of Na^+, whereas at low pH, Na^+ uptake is minimal.

Armstrong and Rothstein[44] studied the kinetics of cation uptake in cells exposed to the chlorides of individual alkali metal cations at constant pH values ranging from 3 to 8. Figure 11.3 which shows the data for K^+ uptake is typical of the results obtained. The effects of pH on the uptake of alkali metal cations can be conveniently analysed in terms of the parameters K_m and V_{max} of the Michaelis–Menten equation, these parameters being estimated from plots such as those illustrated in Figure 11.3. The results of such an analysis are illustrated in Figure 11.4a and b. It is evident from these figures that three zones of behaviour can be distinguished in the

419

effect of H^+ on the uptake of alkali metal cations. At pH values below 4, the apparent K_m for each ion is increased with comparatively little change in V_{max}, i.e. the effect of external H^+ on alkali metal uptake has the kinetic characteristics of a predominantly competitive inhibition (compare the curves for K^+ uptake at pH 3 and pH 3·5 in Figure 11.3; also the very similar data obtained by

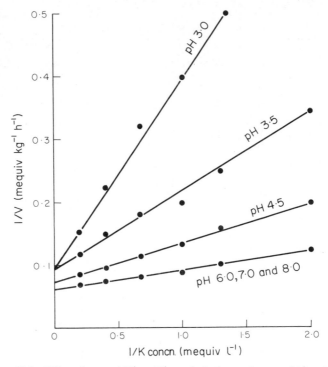

Figure 11.3   Effect of external H^+ on K^+ uptake by fermenting yeast (After Armstrong and Rothstein[44]; courtesy J. gen. Physiol.)

Conway et al.[47]). In the pH range 4 to 6, V_{max} increased markedly with relatively little change in the apparent K_m. This is kinetically equivalent to a partial but predominantly non-competitive effect of H^{+51}. Above pH 6, virtually no change in V_{max} or K_m was observed.

These two kinetic effects of H^+ on K^+ uptake suggested the existence in the cation transporting system of two distinct proton binding sites (or classes of proton binding sites) with different affinities for

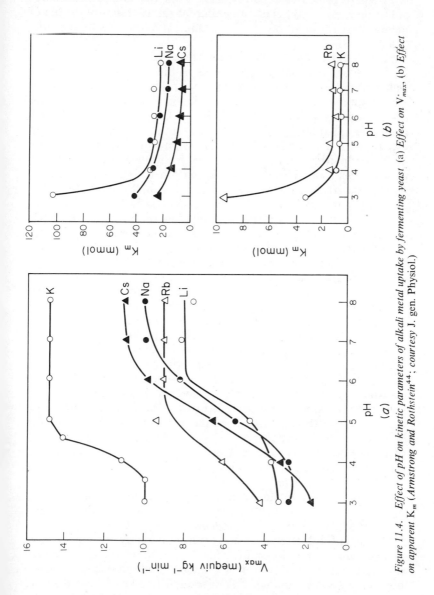

Figure 11.4. *Effect of pH on kinetic parameters of alkali metal uptake by fermenting yeast* (a) *Effect on* V_{max}, (b) *Effect on apparent* K_m *(Armstrong and Rothstein[44]; courtesy J. gen. Physiol.)*

H^+. One of these (which is responsible for the 'competitive' effect of H^+ on cation uptake) has an apparent pK' (in terms of the pH of the external medium) of 3·7. The second site has an apparent pK' of 4·5[44].

Subsequent studies[44, 48] were concerned with the further characterisation of these two sites in terms of their affinities for various transported cations and with the interpretation of their role in cation uptake by yeast cells. It was found that K^+ and the other alkali metal cations in sufficiently high concentrations could overcome the apparently 'non-competitive' inhibitory effect of H^+. In experiments run at pH 8 (to eliminate any significant interaction of H^+ with either site) K^+, as already mentioned, always behaved as a competitive inhibitor towards the other alkali metal cations. Also, as stated above, Rb^+ was a competitive inhibitor of K^+ uptake. Li^+, Na^+ and Cs^+ behaved like H^+, i.e. at relatively low concentrations they behaved as apparent partial non-competitive inhibitors[51] of K^+ uptake. At higher concentrations they competitively inhibited the remaining moiety of K^+ transport. With Li^+ and Na^+ there was considerable 'overlap' between these two inhibitory functions (i.e. significant 'competitive' inhibition appeared before maximal 'non-competitive' inhibition was observed). With Cs^+ a clear-cut experimental separation between the two inhibitory effects could be achieved. In all cases the apparently non-competitive inhibition of K^+ uptake could be overcome by sufficiently high concentrations of K^+. It thus appeared that the two sites differed strikingly in their *relative* affinities for K^+ and the other cations studied. Further study showed that the site with the higher relative affinity for K^+ is the transporting or 'carrier' site. The second site (designated the 'modifier' site) is capable, when occupied by H^+ or another alkali metal cation (except Rb^+), of modifying (i.e. reducing) the rate of uptake of K^+ and, when occupied by H^+, of modifying the rate of uptake of all the alkali metal cations. That it is not itself a transporting site is indicated by the following observations[44, 48]. The concentrations of K^+ required to overcome the 'modifying' effects of other cations were always much greater than those required to 'saturate' the K^+ uptake system when this ion was present alone in the external medium. Ca^{2+} and Mg^{2+}, in relatively low concentrations (1 to 2 mM), partially inhibited K^+ uptake in an apparently non-competitive manner, although their affinities for the transport site are extremely low relative to that of K^+[8, 43]. Competitive inhibition of K^+ transport is always accompanied by appreciable uptake of the competing cation whereas, with Cs^+ and Ca^{2+}, maximal 'non-competitive' inhibition of K^+ uptake could be elicited under conditions where there was virtually no uptake of the inhibiting cation.

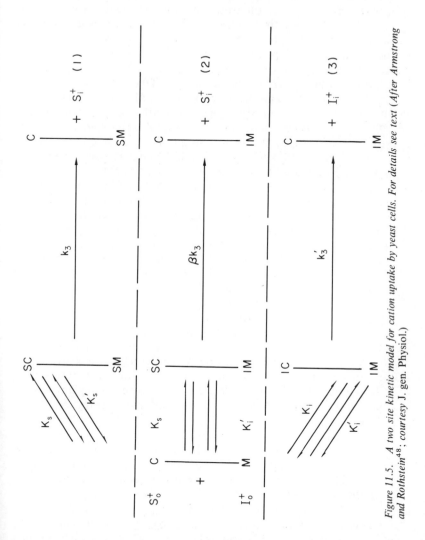

Figure 11.5. A two site kinetic model for cation uptake by yeast cells. For details see text (After Armstrong and Rothstein[48], courtesy J. gen. Physiol.)

A kinetic model for cation uptake in yeast based on these studies is shown in Figure 11.5, and Table 11.3. illustrates the relative affinities of the two sites for the various cations studied (based on their apparent K_m values for the transport site and the dissociation constants for binding to the modifier site). The essential features of the model shown in Figure 11.5 can be summarised as follows: (1) C and M are two independent combining sites for cations, C being the carrier or transporting site and M a modifier site which is not a transporting site. (2) C and M have different affinity arrays for cations. (3) H^+, K^+, and the other alkali metal cations can combine with both C and M. (4) Mg^{2+} and especially Ca^{2+}, except at very high concentrations, combine only with M. (5) When C alone or both C and M are occupied by a single transportable ionic species species (S or I), uptake follows simple Michaelis–Menten kinetics (pathway 1 or 3). (6) Occupation of C by a competing ion (I) results in competitive inhibition of the uptake of S with I being transported at a rate proportional to the fraction of C sites it occupies (path 3). (7) Occupation of M by I usually results in a reduction of V_{max} for S. This reduction occurs without transport of I and can be expressed in terms of an empirical constant β, where $V_{max} = k_3 C_t$ (C_t being the total number of transport sites available) is the maximal rate of transport of S in the absence of I, and $V_{max} = \beta k_3 C_t$ is its maximal transport rate when M is occupied by I (pathway 2). (8) β can assume any value between 0 and 1. If $\beta = 1$, no inhibition results from the occupation of M by I. If $\beta = 0$, inhibition is complete. This is the classical non-competitive inhibition of enzyme kinetics. If $0 < \beta < 1$, a partial inhibition occurs which is kinetically of the 'partially non-competitive' type discussed by Webb[51]. (9) Since it is postulated that all the cations considered herein have finite affinities for M, it is possible to displace I from M and reverse the apparently non-competitive effects of I by increasing the concentration of S to a sufficiently high level.

The model illustrated in Figure 11.5 provides a satisfactory qualitative framework into which the known interactions of a number of cations with the inward facing or uptake component of the K^+–H^+ exchange system can be fitted. Its quantitative evaluation however requires further work.*

* Professor W. Wilbrandt (personal communication) has pointed out that at least one other model might be used to explain these data. This does not involve a change in k_3 (Figure 11.5), but rather a reduction in the steady state amount of carrier at the outer surface of the membrane. However this model seems to imply that the parameter β (Table 11.3) should be the same for all transported cations. This is clearly not the case for the 'modifying' action of H^+. For further discussion see Armstrong and Rothstein[48].

An important consequence of the interaction of H^+ with the modifier site is that the reduction in V_{max} (Figure 11.4a and Table 11.3) is not the same for all cations. For example, at pH 8 (with virtually no H^+ associated with the modifier site) the ratio of the V_{max} for K^+ to that for Na^+ is about 1·5, whereas at pH 3·5 (where the modifier site is virtually saturated with H^+) this ratio is 3·3. At low pH therefore the discrimination of the carrier in favour of K^+ over Na^+ (and other cations) is much greater than would be predicted from the K_m values for the transport site.

THE AMOUNT AND NATURE OF THE K^+–H^+ EXCHANGE CARRIER

Foulkes[42], using azide binding and Conway and Duggan[43], by selective displacement of K^{42} from surface binding sites with Mg^{2+} and Rb^+ at low temperature (4–5 °C), estimated the amount of the K^+ carrier (site C in Figure 11.5) to be 0·1–0·2 mM per kg yeast. On this basis Rothstein[4] has calculated that the turnover rate of the carrier is some 10 to 50 times per minute, depending on the temperature of the system. As yet the amount of 'modifier' sites has not been estimated, nor has the chemical nature of either the C or the M site (Figure 11.5), apart from the fact that they are probably fixed anions, been determined.

There seems to be little evidence for the direct participation of a Na^+–K^+ specific ATPase in ion transport and accumulation in yeast and other micro-organisms[6]. Conway[8] favours a mechanism in which the energy requirement for transport is linked directly to electron transport. In support of this contention it was found[52] that redox dyes at a concentration of 0·1 mM, at which no gross disruption of membrane function is observed[22], increase K^+ uptake and H^+ secretion above control levels when the E_h is above 150 to 200 mV. Conversely, dyes with E_h values below 150 mV inhibit K^+ uptake and H^+ loss. Similar results were obtained with inorganic redox systems. More recently Conway et al.[47] showed that the effect of pH on the redox potential of yeast fermenting anaerobically in the absence of inhibitors is virtually negligible in the pH range of 4 to 6 ($\Delta E/\Delta pH$ 3 to 4 mV). In the absence of glucose or in the presence of inhibitors ($HgCl_2$, azide, or 2,4-dinitrophenol) $\Delta E/\Delta pH$ was much higher (about 21 to 27 mV). These authors interpreted these findings as indicating 'domination of the redox potential by a metal system during fermentation'. Clearly, further work on the nature of the carrier mediating K^+–H^+ exchange in fermenting yeast is desirable.

THE $H_2PO_4^-$ SYSTEM

This system has been studied in particular detail by Rothstein and his collaborators (see, for example, reference 4). Resting yeast contains considerable amounts of osmotically active phosphate, in the form of both inorganic orthophosphate and soluble phosphate esters (hexose phosphates, triose phosphates, ATP, etc., see Table 11.2 and reference 3). Nevertheless, when yeast is suspended in distilled water, leakage of phosphate from the cells to the medium is negligible and, in the absence of exogenous substrate, ^{32}P-labelled phosphate added to the medium exchanges very slowly with intracellular phosphate. When glucose is added to the medium, phosphate is rapidly absorbed by the cells. During absorption of ^{32}P-labelled phosphate, the specific activity of the medium remains unchanged, indicating that phosphate movement is essentially 'one-way' and does not involve any appreciable exchange with cellular phosphate[53]. Phosphate absorption under these conditions takes place against considerable concentration gradients, e.g. absorption takes place from external concentrations as low as 0.1 mM, displays saturation kinetics (K_m, 0.4 mM) and is inhibited by concentrations of azide and dinitrophenol which do not inhibit cellular metabolism[4, 54]. During phosphate absorption, the intracellular concentration of orthophosphate remains virtually constant, most of the extra phosphate taken up appearing as metaphosphate, though the uptake system appears[4] to be specific for the ionic species $H_2PO_4^-$.

INTERACTIONS OF K^+ WITH THE $H_2PO_4^-$ SYSTEM

When phosphate is presented to the cells as KH_2PO_4 much more phosphate is absorbed than when it is added to the medium as NaH_2PO_4. In the presence of KH_2PO_4, both K^+ and phosphate are absorbed[55]. Some absorption of phosphate occurs without absorption of its accompanying cation, e.g. from triethylamine phosphate. During such absorption the medium becomes more alkaline and the cell interior more acid, a factor which may limit phosphate uptake[4].

Although K^+ and phosphate are both absorbed, the rates of absorption are quite different. When small amounts of KH_2PO_4 are present initially in the medium, the K^+ may be virtually all absorbed before phosphate absorption begins. Even more strikingly, cells enriched in K^+ by previous fermentation in media containing K^+ and glucose can subsequently absorb phosphate at an increased rate from K^+-free media[56]. Evidently, the stimulating effect of K^+ on phosphate uptake is not due to coupled transfer of these two ions. In fact the K^+ content of the cells, rather than that of the medium, seems to be the dominant factor in phosphate absorption. K^+-rich cells absorb phosphate readily, whereas cells depleted of K^+ do not. Rothstein[4] believes that the acid–base balance of the cell interior may be the controlling mechanism in the regulation of K^+-dependent phosphate absorption.

THE PHOSPHATE–Mg^{2+} SYSTEM

As already mentioned, Mg^{2+} and other cations can enter the yeast cell via the inwardly directed component of the K^+–H^+ carrier[49, 57]. Such entry however is slow and the affinity of Mg^{2+} for the carrier (judged on the basis of its apparent K_m value) is extremely low, so that care must be taken, in studying Mg^{2+} uptake by this route, to eliminate competing cations (especially K^+) from the bathing medium[57]. In media containing K^+, phosphate and Mg^{2+} together with glucose, a second mechanism for Mg^{2+} absorption is brought into play[55]. This mechanism is also capable of transporting other bivalent cations into the cell, the specificity array being Mg^{2+}, Co^{2+}, $> Zn^{2+} > Mn^{2+} > Ni^{2+} > Ca^{2+} > Sr^{2+}$ [58, 59]. Uptake of Ni^{2+} is associated with a partial inhibition of fermentation due to specific interaction of this ion with the alcohol dehydrogenase system of the yeast cell[60]. The uptake system is saturable with respect to the concentration of transported ions, and competition studies have shown that these ions compete for a common carrier. The apparent K_m for Mg^{2+} and for Co^{2+} is <0.01 mM. That for Ni^{2+} is about 0.5 mM[59]. Much information concerning this system has been obtained with Mn^{2+} (a suitable isotope of which, ^{54}Mn, is readily available). Thus it has been shown isotopically that, once inside the cell, absorbed ions do not readily exchange with ions in the external medium[4].

This system shows an interesting dependence on K^+ and phosphate (Figure 11.6). As shown in this figure, Mn^{2+} absorption is an increasing function of external phosphate concentration over the range studied. In the absence of phosphate, no Mn^{2+} absorption occurs. With fixed amounts of Mn^{2+} and phosphate, K^+, in relatively low concentrations, stimulates Mn^{2+} absorption. Higher concentrations of K^+ are somewhat inhibitory. Further, Mn^{2+} absorption is not absolutely dependent on K^+, considerable absorption occurring from K^+-free phosphate media. Figure 11.6 also shows that, as already mentioned, Mn^{2+} absorption is a

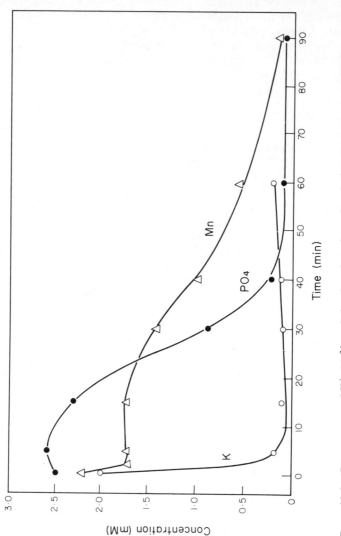

Figure 11.6. Disappearance of K^+, Mn^{2+}, and phosphate from the medium bathing respiring yeast cells. Suspension contained 100 mg yeast per ml of 0.1 M glucose. K^+, Mn^{2+}, and phosphate all added at zero time (After Rothstein et al.[56]; courtesy J. gen. Physiol.)

saturable function of external Mn^{2+} concentration when K^+ and phosphate are held constant.

The time course of uptake of K^+, Mn^{2+}, and phosphate from media containing all three ions is quite different. K^+ is absorbed first and, in solutions containing little K^+, may virtually disappear from the medium before the onset of phosphate or Mn^{2+} absorption. The onset of Mn^{2+} absorption in turn lags behind phosphate absorption. It thus appears that, as in the case of phosphate-dependent K^+ absorption (p. 427) it is the cellular content of K^+ and phosphate, rather than their concentrations in the external medium, which determine Mn^{2+} uptake. This was confirmed in experiments in which cells were first allowed to accumulate K^+ from a KCl–glucose medium, washed, and transferred to a medium containing glucose plus triethylamine phosphate (from which they absorbed phosphate faster than did a control suspension), and finally washed again and transferred to a medium containing glucose and $MnCl_2$. Under these conditions Mn^{2+} absorption was as fast as in suspensions exposed simultaneously to K, phosphate and Mn^{2+}.

The Mg^{2+} transport system appears to be quite labile, e.g. it is completely lost following 8 hours aeration in the absence of substrate and even faster (2 hours) during fermentation with glucose in phosphate free media[4]. Thus the Mg^{2+} carrier appears to be specifically activated by phosphate uptake. Rothstein[4] has suggested that the carrier may be or may at least depend upon a phosphorylated component of the cell membrane.

During uptake of bivalent cations by this mechanism approximately $2 K^+$ are excreted from the cells for each bivalent cation absorbed. In Na^+-rich yeast (see p. 430) 2 Na^+ are excreted. Uptake rates are similar under aerobic or anaerobic conditions and uptake is reduced below pH 5^{59}, though no direct exchange of bivalent cations for cellular H^+ has been found.

INTERACTIONS OF ARSENATE WITH THE $H_2PO_4^-$ AND PHOSPHATE-Mg^{3+} SYSTEMS

Arsenate competes with phosphate for entry into the yeast cell[61]. The two ions involved (primarily $—H_2AsO_4^-$ and $—H_2PO_4^-$ under the experimental conditions employed) have approximately equal affinities for the carrier system but the V_{max} (initial rate) for phosphate is about twice that for arsenate. In addition to competing with phosphate for entry into the cells, arsenate irreversibly inactivates the phosphate transport system. Kinetically, inactivation can be fitted by a first-order rate equation, but it is not directly related to the amount of arsenate absorbed. In the presence of phosphate inactivation is delayed but develops when less arsenate has been absorbed than is the case in the absence of phosphate. Once developed, the inhibition is not reversed by phosphate and does not appear to be related to the known effects of arsenate on phosphate metabolism, e.g. its inhibitory effects on glyceraldehyde dehydrogenase or phosphorylase[62, 63].

Arsenate inhibits in similar fashion the phosphate-induced uptake of Mg^{2+} and other bivalent cations; both the rate of uptake and the total amount of cation absorbed during prolonged fermentation are reduced[64].

OUTWARD CATION TRANSPORT BY YEAST: THE Na^+ CARRIER

For at least one obvious reason—the relative inaccessibility of the cell interior to experimental manipulation—outwardly directed cation transport in yeast has been much less extensively investigated than cation uptake. From the foregoing discussion it is evident that absorption of cations, unless associated with special mechanisms such as those involved in phosphate transfer, is an exchange process, the amount of positive charge taken up being balanced by excretion of an equivalent amount of positive charge. In normal yeast, grown in low-Na^+ media and containing trivial amounts only of Na^+ ions, H^+ excretion, as already discussed, is the usual mechanism by which the uptake of K^+ and other cations is electrically balanced. In the phosphate–Mg^{2+} system, K^+ rather than H^+ is the preferred excretory product, except in cells artificially enriched in Na^+ by suitable means[45, 46].

The excretion of Na^+ from such 'Na yeasts' has some interesting properties[4]. When such cells are suspended in water they lose Na^+ ions, together with bicarbonate and organic acid anions. Na^+ excretion into 0·1 M KCl is about three times as fast as into water and appears to involve a 1 : 1 exchange of Na^+ for K^+. The rate of Na^+ secretion into solutions containing 0·1 M NaCl together with various concentrations of KCl is maximal when the external KCl concentration is 0·04 M. With media containing 0·1 KCl and various concentrations of NaCl there is no inhibition of Na^+ secretion until the external NaCl concentration reaches 0·2 M. At 0·43 M NaCl, Na^+ secretion is 50% inhibited. Na^+ secretion is inhibited by heavy metal ions.

There has been some disagreement concerning the nature of the

Na$^+$ carrier. Conway and Moore[45] found that in substrate-free media, Na$^+$ secretion was inhibited by anoxia and by 2 mM cyanide but not by 2 mM azide or 2,4-dinitrophenol. On the other hand, the K$^+$ carrier (involved in K$^+$ uptake, or Na$^+$ uptake in the absence of external K$^+$) was inhibited by anoxia, cyanide, azide, and dinitrophenol. On the basis of these findings, the opposite effects of redox dyes on K$^+$ uptake and Na$^+$ secretion[52], and their different specificities for Na$^+$, Li$^+$ and K$^+$, Conway[45, 8] considered the Na$^+$ carrier to be separate and distinct from the inwardly directed K$^+$ carrier. These views were challenged by Foulkes[42] who reported the presence in yeast of two K$^+$ carriers, one sensitive to azide and anoxia, the other (activated by the presence of glucose or glucose plus phosphate in the medium) insensitive to these inhibitors. This author contended that since the experiments of Conway and Kernan[52] on the effects of redox dyes on K$^+$ uptake were performed under conditions where the azide insensitive K$^+$ transport mechanism was dominant they are not germane to the properties of an azide sensitive carrier. Foulkes further reported that, contrary to the findings of Conway and Moore[45], Na$^+$ extrusion into a K$^+$-free medium was not inhibited by cyanide or anoxia, that in all cases examined the effect of inhibitors on K$^+$ uptake and Na$^+$ extrusion were identical, and that NH$_4^+$ was apparently exchanged for K$^+$ by a NH$_4^+$-enriched yeast in a manner very similar to Na$^+$–K$^+$ exchange in Na$^+$ yeast. These findings, in Foulkes's opinion, make the assumption of a separate Na$^+$ carrier in yeast unnecessary. Further complications in the excretion of Na$^+$ by Na$^+$ yeast have emerged from a study by Kotyk and Kleinzeller[46]. These authors found that Na$^+$ extrusion into K$^+$-free media is markedly increased by 0·28 M glucose but that a similar effect is obtained with non-metabolised sugars. Under these conditions net Na$^+$ extrusion is accompanied by loss of water from the cells. In the presence of external K$^+$, Na$^+$ extrusion is markedly increased and is accompanied by an equivalent uptake of K$^+$.

In view of its possible analogies to the almost universal mechanism for active Na$^+$ transport which exists in animal cells, further studies on Na$^+$ excretion by yeast cells seem desirable.

METAL ION BINDING BY YEAST AND ITS IMPLICATIONS IN SUGAR TRANSPORT

In addition to the cation uptake mechanisms already discussed, yeast cells have the capacity to bind small quantities of metal ions to anionic groups in the cell surface[26]. Binding is easily distinguished from absorption by several criteria, among which the following may be listed: (1) it is rapid, reaching completion in 2 minutes or less; (2) the amounts of cations involved in surface binding are small, about 3 mM per kg yeast in all; (3) binding, unlike absorption, is readily reversible, i.e. surface-bound cations can exchange readily with ions in the bathing medium; (4) binding is little affected by such factors as temperature and metabolism; (5) the specificity of binding is quite different from the specificity of the cation transporting systems discussed above.

NATURE AND SPECIFICITY OF CATION BINDING

The binding of cations to the yeast cell surface can be expressed in terms of simple mass law equations, assuming a one-to-one inter-action between the metal ion and a membrane ligand[26]. In the case of Mn^{2+} a Scatchard plot[65] reveals the existence of two kinds of binding site, higher affinity sites with a total concentration of about 1 mM per kg cells and a mass law dissociation constant of 4.7×10^{-5} M, and a group of sites with considerably lower affinity and a total concentration of about 2 mM per kg cells. Utilising the ability of other cations to displace ^{54}Mn from the higher affinity sites, Rothstein and Hayes were able to assess the relative binding affinities for these sites of a number of metal ions. Setting the affinity for Mn^{2+} at unity, the affinity for UO_2^{2+} was 300, that for Ba^{2+} was 3.0 and those for the other bivalent cations tested (Zn^{2+}, Co^{2+}, Mg^{2+}, Ca^{2+}, Sr^{2+}, Cu^{2+}, and Hg^{2+}) ranged from 1.0 to 1.5. The relative binding affinities for the alkali metal cations were much lower (0.005 to 0.01) and those for organic cations (diethylamine and tris-hydroxymethyl-aminomethane) were vanishingly small (about 0.001).

Particular interest attaches to the binding of UO_2^{2+} by yeast cells[66, 67]. This ion is a potent and highly specific inhibitor of a number of biological systems, e.g. proximal renal tubules[68], intestinal epithelia[69], skeletal muscle[70, 71] and cardiac muscle[72]. In the yeast cell, two inhibitory effects of UO_2^{2+} have been described, an inhibition of exogenous sugar metabolism by low concentrations ($< 10^{-5}$ M) and an inhibition of surface bound invertase which requires concentrations of UO_2^{2+} about ten times higher[73]. External UO_2^{2+} has no effect on the metabolism of endogenous carbohydrate or on the metabolism of exogenous 2-carbon substrates[67]. Inhibition of fermentation is a consequence of blockage by UO_2^{2+} of sugar transfer across the cell membrane[74] and is associated with binding of this ion to the high-affinity ligands[67] described above for Mn^{2+}.

UO_2^{2+} has also provided evidence concerning the chemical nature of the ligands involved in cation binding. There is strong inferential evidence that the high-affinity ligands are phosphoryl groups in a chelating array. This was inferred from studies of the effect of exogenous polyphosphates on the distribution of UO_2^{2+} between yeast cells and the external medium[75] and supported by studies of UO_2^{2+} binding and sugar fermentation in phosphate depleted cells[76, 77]. Phosphate depletion reduces high-affinity UO_2^{2+} binding and at the same time reduces the rate of uptake of glucose by the cells. There is a 1 : 1 correlation between the loss of these two functions[78].

Similarly, it can be inferred that the low affinity ligands are carboxyl groups. This conclusion is based on the fact that inhibition by UO_2^{2+} of invertase activity in intact cells is very similar to UO_2^{2+} inhibition of the isolated enzyme[67] and that, in the latter case, inhibition involves the formation of a UO_2^{2+} carboxyl complex[79]. Invertase activity does not involve phosphoryl groups.

Some further interactions of UO_2^{2+} and other cations with sugar transport in yeast are of interest in the present context.* Both the phosphoryl and carboxyl groups in the cell membrane appear to be implicated in the inhibitory effects of UO_2^{2+} on glucose uptake[67]. Uptake of glucose in unpoisoned cells and galactose uptake in galactose-induced cells appear to depend mainly on the former. These processes are inhibited by relatively low concentrations of UO_2^{2+}, are characterised by a reduction in Ni^{2+} binding, with a low K_m and high V_{max}. Glucose uptake by iodoacetate-poisoned cells and sorbose and galactose uptake by starved cells do not involve reduction in Ni^{2+} or Co^{2+} binding, and are not inhibited by Ni^{2+}. They are inhibited by relatively high concentration of UO_2^{2+} and thus appear to involve carboxyl groups in the cell surface[82]. It has been suggested[80, 82] that the first of these pathways involves an active transport mechanism while the second is a facilitated diffusion of the equilibrating type[83].

ION TRANSPORT AND ACCUMULATION IN BACTERIA: NET MOVEMENTS OF Na⁺ AND K⁺

Earlier investigations of the Na^+ and K^+ content of several bacterial species[84–86], suggested that the intracellular Na^+ did not differ greatly from that of the culture medium and led to the conclusion that the cell membrane is freely permeable to Na^+, and that

---

* For fuller accounts of the general topic of sugar transport in yeast and other micro-organisms see references 80, 81, 66, 67.

this ion is essentially in equilibrium across it. Similarly, it was proposed[84] that the relatively high K^+ content of bacterial cells could be explained by supposing that it consists of two fractions, one in free solution at a concentration equal to that of the external medium and another attached to non-diffusible intracellular binding sites. It was further proposed[84] that these sites were formed during carbohydrate metabolism and that their number (and hence the total intracellular K^+ content) was under metabolic control. Apart from the difficulty of reconciling this mechanism of K^+ accumulation with the high internal osmotic pressure of bacterial cells (p. 408), more recent investigations, notably those of Schultz and his collaborators[87-90], have led to quite different conclusions.

Schultz and Solomon[87] found that the intracellular Na^+ and K^+ concentrations of E. coli during growth are a function of extracellular cation concentrations and of the age of the culture. During the early logarithmic phase of growth the intracellular K^+ concentration greatly exceeds that of the medium, whereas intracellular Na^+ is lower than medium Na^+. As the culture ages, intracellular K^+ decreases and intracellular Na^+ increases so that during the stationary phase the intracellular Na^+ and K^+ levels approach those of the bathing medium. The marked increase in intracellular K^+ during early growth is observed over a wide range of medium K^+ concentrations (0·05 to 120 mM) and the maximum intracellular K^+ concentration achieved may be as much as 3600 times greater than the extracellular K^+ concentration. The decline in intracellular K^+ and the increase in intracellular Na^+ during the late logarithmic and stationary phases of the growth cycle are accompanied by a decrease in medium pH and an accumulation, in the medium, of metabolic products such as formate. Evidence that these phenomena are interdependent emerges from the observation that when non-growing cells from a 48 hour culture were resuspended in fresh medium they rapidly accumulated K^+ and extruded Na^+. K^+ accumulation was inhibited when the pH of the fresh medium, instead of 7, was 4·5.* However, inhibition of Na^+ extrusion required either that the cells be inoculated into an old medium at pH 4·5 or that, in addition to an increased H^+ content, the medium contain organic anions such as formate. That K^+ uptake following inoculation was not a consequence of renewed cell growth and division was shown by the fact that uptake was not

---

* Eddy and Hinshelwood[91] observed a similar effect of pH on ^{42}K uptake by B. lactis aerogenes.

accompanied by any significant increase in the optical density of the cultures and was not inhibited by omission of a source of nitrogen from the medium or by addition of chloramphenicol, an inhibitor of bacterial protein synthesis[92].

Since the membrane potential of E. coli cannot be measured directly, the nature of these Na^+–K^+ exchanges is subject to the same uncertainties already noted (p. 414) for ionic transfer processes in yeast. However, as in yeast, there is strong inferential evidence that at least one of the processes involved, K^+ uptake or Na^+ extrusion, must take place against an electrochemical gradient. Thus, both processes are markedly reduced during glucose deprivation, and are inhibited by 1 mM iodoacetate and 0·1 mM dinitrophenol but not by 1 mM cyanide[87]. Of interest is the finding that 3×10^{-2} M NaF completely inhibits Na^+ extrusion but does not appear to affect K^+ uptake[87]. This suggests that the two processes are not tightly coupled[6]. The differential effects of medium pH mentioned above support this conclusion.

Returning to the question of the electrochemical gradient, it is evident[87] that since the two net flows occur in opposite directions they cannot both be driven by an electrical potential. An unusually high E_K (210 mV, inside negative) would be required for passive inward K^+ movement at the maximum intracellular–extracellular concentration ratio. Alternatively, the corresponding E_{Na} (20 mV, inside positive) would impose a considerable electrochemical barrier to K^+ entry. If one assumes[88] that Cl^- is in electrochemical equilibrium across the cell membrane, the calculated values of E_{Cl} (29 mV during the logarithmic phase, and 3 mV during the stationary phase; both inside negative) require that both Na^+ and K^+ be actively transported.

In later experiments[90] it was found that during uptake of K^+ and simultaneous extrusion of Na^+, K^+ uptake exceeded net Na^+ loss under all conditions studied. Examination of concomitant net movements of other ions showed that Mg^{2+} uptake was accompanied by an equivalent amount of phosphate. Similarly, net Cl^- movement also failed to account for this discrepancy. It was concluded that excess K^+ uptake is balanced by extrusion of cellular H^+ similar to the K^+–H^+ exchange in yeast already discussed.* Measurement of H^+ secretion during glucose uptake at pH 7 showed an average of 2·4 mol H^+ secreted per mol glucose consumed, this ratio being independent of the cation content of the

---

* This conclusion is in agreement with the acidification of the medium that occurs during fermentation and K^+ uptake[87] and with the fact that K^+ influx is inhibited by low external pH[90].

cells over a fairly wide range (K^+ 37 to 185, Na^+ 61 to 152 mmol per litre cell water). These observations and the fact that the steady-state efflux of H^+ was some 16 times the rate of K^+ efflux[88] suggest that only a small part of H^+ efflux is obligatorily coupled to K^+ uptake[89], the remainder being accompanied by organic anions. Under steady-state conditions the kinetics K^+ efflux (determined from the rate of loss of ^{42}K from the cells to an initially unlabelled medium) is consistent with the existence of a single intracellular compartment and a flux rate of 1 pmol cm^{-2} s^{-1}). The flux rate is independent of external K^+ concentration over the range 1·3 to 5·2 mM. Analysis of the kinetics of K^+ influx following transfer of K^+ poor cells to a fresh stationary phase medium[90] suggests the existence of two exchange processes, a K^+–H^+ exchange and a K^+–Na^+ exchange. These processes appear to have widely different time constants and to be separable by suitable manipulation of the external conditions. Thus at pH 7, K^+–Na^+ exchange accounts for about 40% of the total K^+ taken up during 18 minutes exposure to fresh medium. At pH 5·15 to 5·50, 80% of the K^+ taken up is exchanged for cellular Na^+. Further, during the first few minutes of exposure K^+ uptake greatly exceeds Na^+ loss. Subsequently, these two ions exchange on an approximately 1:1 basis. The K^+–H^+ exchange process has a K_m with respect to external K^+ of 4·5 mM[90].

In later experiments Weiden et al.[93] described a K^+-dependent phosphate uptake mechanism in E. coli which is reminiscent of the K^+-linked phosphate system in yeast (p. 426). When E. coli is grown in media containing limiting amounts of K^+, growth proceeds normally until virtually all the extracellular K^+ is taken up. Thereafter, growth rate, glucose utilisation, and O_2 consumption decrease progressively and the K^+ and phosphate contents of the 'K$^+$-limited' cells fall. All these changes are reversed by exogenous K^+. The metabolic effects of K^+ limitation are not related to intracellular K^+ or Na^+ levels but appear to be dependent on extracellular K^+ per se. Phosphate uptake is seriously impaired in the absence of external K^+ and Weiden et al.[93] consider the inhibition of phosphate uptake the cause of the metabolic deficiencies observed in the K^+-limited condition. Following long periods of K^+ limitation phosphate uptake and growth rate are only partly restored by exogenous K^+. Unlike the situation in yeast (p. 426), markedly enhanced phosphate uptake cannot be induced by prior enrichment of the cells with K^+. It seems likely that the dependence on external K^+ of phosphate uptake in E. coli may involve a direct coupling between transmembrane movements of these two ions.

CELL K$^+$ AND PROTEIN SYNTHESIS IN *E. Coli*

Cells in which, unlike the K$^+$-limited cells of Weiden *et al.*[93], *intracellular* K$^+$ is markedly depleted, show severe disturbances in growth and protein synthesis. Investigation of these phenomena was greatly facilitated by the isolation[94] of a mutant strain which lacked the ability to concentrate K$^+$ from a low external concentration. In these mutants, cell K$^+$ concentrations are readily controlled by adjusting the external K$^+$ level. Loss of cell K$^+$ is apparently compensated for by a gain in Na$^+$, other cellular constituents being little changed. Cells in which the K$^+$ content had been decreased from the normal level (about 200 mM) to about 10 mM synthesised considerable amounts of RNA but no protein[95]. This uncoupling of RNA and protein synthesis is similar to that induced in growing cells by chloramphenicol and suggested that blockage occurred at a late stage in protein synthesis. Experiments with the cell-free poly-U system[96] yielded results consistent with this hypothesis. The most K$^+$ sensitive step in this system was the transfer of amino acid from amino acetyl-S-RNA to polypeptide[96, 97].

SOME OTHER EFFECTS OF K$^+$ ON BACTERIAL GROWTH AND FUNCTION

Many intracellular enzymes including a number found in yeast and bacteria are strongly dependent on K$^+$. Among these are enzymes involved in the glycolytic pathway and in purine biosynthesis which are vital to the growth of K$^+$-deficient *E. coli* cells growing in minimal media[98]. The overall effects of K$^+$ depletion on bacterial cell growth can therefore be expected to be the result of a complex array of causes.

A link between the uptake of K$^+$ and that of amino acids by *Salmonella oranienburg* has recently been demonstrated. Certain amino acids (e.g. proline) stimulate K$^+$ uptake from high-NaCl media, but the effect appears to be relatively non-specific[99].

IONIC REGULATIONS IN HALOBACTERIA

Halophilic bacteria present an interesting case of ionic accumulation and regulation under conditions of extreme salinity. In media rich in NaCl these cells accumulate large amounts of K$^+$ (Table 11.1) and in the stationary state can apparently sustain large inwardly directed concentration gradients for Na$^+$ and outwardly directed

K^+ concentration gradients which are fantastically high. In addition, the total intracellular ionic content frequently exceeds that of the medium by amounts which, if osmotically active, would give rise to osmotic pressure differences of several hundred atmospheres during the logarithmic phase of growth[15]. The question of the thermodynamic activity of intracellular ions, particularly K^+, is therefore of paramount importance in unravelling the mechanisms underlying ionic regulation in halobacteria.

It is not possible at present to determine directly ionic activities in these cells. However, several lines of evidence suggest that the intracellular K^+ activity is significantly reduced. A considerable reduction in intracellular Na^+ activity in animal cells, probably through complex formation with intracellular polyanions of very restricted diffusivity, seems now to be well established by a variety of techniques[100]. Reduction through a similar mechanism of intracellular K^+ activity, though strongly advocated by a number of workers[101, 102] does not as yet have a wide basis of direct experimental support, though recent studies strongly indicate that there is some reduction in cytoplasmic K^+ activity, at least in muscle[100, 103].

Returning to the question of the state of intracellular K^+ in halobacteria, one notes that the freezing-point measurements of Christian and Ingram[104] indicate an internal osmolality only a little higher than that of the medium. Further, as pointed out by Ginzburg et al.[15], successful resistance to a high osmotic pressure would require a cell wall of high mechanical strength. There seems to be fairly general agreement that the chemical composition of the cell wall of halobacteria would not lead one to expect a high degree of rigidity[105, 106]. These findings, together with the fact that the membranes of some halophiles are permeable to fairly large molecules such as insulin and serum albumin[107], point rather strongly to the conclusion that there must be a considerable 'sequestration' of cytoplasmic ions. Because of the tremendously high $[K_i]/[K_o]$ ratios in these organisms (Table 11.1) and the fact that these gradients can be successfully maintained under conditions unfavourable to 'active' transport (e.g. abolition of O_2 uptake), the conclusion[15] that considerable amounts of cell K^+ are sequestered seems quite reasonable.

SUGAR TRANSPORT BY ISOLATED BACTERIAL MEMBRANES

Although their exact relationship to the major topic of this discussion is not yet apparent, some recent studies on sugar transport by isolated bacterial membranes will be reviewed briefly because of the insights they provide into the biochemical and molecular basis of membrane transport. A much fuller account of these studies is given by Kaback[108].

Kundig et al.[109] reported the isolation of a bacterial phosphotransferase which catalysed the transfer of phosphate from phosphoenolpyruvate (PEP) to several carbohydrates according to the reaction scheme:

$$PEP + HPr \quad \xrightarrow{\text{Enzyme I + Mg}^{2+}} \quad \text{pyruvate} + \text{P–HPr}$$

$$P\text{–HPr} + \text{sugar} \quad \xrightarrow{\text{Enzyme II, Mg}^{2+}} \quad \text{sugar–P} + \text{HPr}$$

The system involves two enzymes (I and II) plus a heat stable, low-molecular-weight protein (HPr) which functions as a phosphate carrier[110]. Enzyme I and HPr are soluble; Enzyme II is membrane-bound.

Following its initial description, much biochemical and genetic evidence was found which implicated the phosphotransferase system in sugar transport by both gram-positive and gram-negative bacteria. Thus E. coli subjected to cold-shock treatment[111] lost their capacity for uptake of α-methyl glucoside and TMG (methyl-β-D galactoside). This effect was completely reversed by addition of HPr[110]. Pleiotropic mutants of Aerobacter aerogenes and E. coli which failed to grow on a number of carbohydrates also lacked HPr and Enzyme I activities when tested in vitro[112, 113]. In a mutant of Salmonella typhimurium which was unable to utilise nine carbohydrates for growth, the physiological lesion was an inability to transport carbohydrates[114]. The mutant appeared to be the result of a single genetic event involving loss of Enzyme I activity.

The transport of carbohydrates by membrane-bounded vesicles about 1 μm in diameter prepared from a number of bacterial species[115, 116] has been extensively studied by Kaback and his associates[108]. In membranes prepared from cells containing the complete phosphotransferase system, uptake and phosphorylation of α-methyl glucoside were markedly dependent on the PEP concentration in the medium. With membranes prepared from Enzyme I or HPr-deficient cells no stimulating effect of PEP on α-methylglucoside uptake or phosphorylation was observed. The stimulating effect was specific for PEP and showed an absolute requirement for bivalent cations[115]. In further experiments[108] it was shown that PEP acts as a phosphate donor for α-methyl-glucoside and that the PEP–P–transferase system acts primarily in promoting the passage of sugars through the membrane and their accumulation as phosphorylated derivatives in the intramembranal space. HPr and Enzyme I markedly increase the stimulatory effects of low concentrations of PEP (5×10^{-4} M). At higher PEP concentrations (4×10^{-4} M or above) these agents had no effect. The stimulatory effects of HPr and Enzyme I on sugar transport were completely abolished by 2 mM NaF, although phosphorylation still continued under these conditions (sugar phosphates appearing in the external medium rather than in the intramembrane space).

Using these isolated vesicles it has also been possible to separate the general osmotic or 'barrier' permeability of the membrane from the specific energy-requiring transport function associated with the PEP–P–transferase system. By comparing the effect of temperature on the initial rates of uptake and phosphorylation of α-methyl-glucoside and the steady-state level of its phosphorylated derivative within the cells, it was found that the steady-state level was optimal at 40 °C whereas the optimal temperature for the initial rates was 46 °C. This discrepancy was found to result from the onset of leakage from the cells of α-methyl glucoside phosphate at 40 °C, the rate of leakage increasing rapidly with temperature to reach a maximum at 55 °C. Between 40 and 46 °C leakage was a reversible function of temperature. Kaback[108] suggests that these experiments show that membrane transport of sugars and the ability of the membrane to retain sugar phosphate are discrete, separable functions, that the latter function is determined largely by the membrane phospholipids, and that, between 40 and 46 °C, these may undergo reversible phase changes similar to those reported for aqueous suspensions of phospholipids[117]. Tentative support for this hypothesis has been obtained in experiments in which the effects of intestinal phospholipase, nonionic detergents, and alteration in the phospholipid composition of the membranes (achieved by growing cells on different media) on α-methyl glucoside uptake and phosphorylation and leakage of α-methyl glucoside phosphate were compared[108].

PHYSIOLOGICAL IMPLICATIONS OF IONIC ACCUMULATION IN YEAST AND BACTERIA

In conclusion, it seems appropriate to consider briefly the possible role in the general cellular economy of micro-organisms of the complex array of ionic accumulation processes discussed above. This topic is discussed in more detail by Rothstein[1, 5] and by Epstein and Schultz[7].

In walled cells, the volume is determined primarily by the cell wall and by the limited amount of stretching or shrinking permitted by the relative rigidity of this structure[13]. Thus, walled cells do not need the exquisitely sensitive osmotic mechanisms for volume regulation required by naked cells such as those of animal tissues[5]. Nevertheless, some form of osmoregulation appears to be a necessary condition for normal growth and division in many walled species. Most bacteria maintain an osmolality somewhat higher than that of the medium in which they grow. The resulting osmotic imbalance across the cell membrane is called turgor pressure. That the maintenance of a positive turgor pressure is essential for growth and division is shown by the following observations. Plasmolysed cells do not grow[118]. During rapid growth the turgor pressure of *E. coli* is several times greater than that of stationary phase cells[119]. A sudden decrease in turgor pressure by increasing medium osmolality (osmotic upshock) causes a temporary stoppage of nucleic acid and protein synthesis which is resumed as the turgor pressure is re-established. Bacteria appear to maintain a more or less constant turgor pressure over a wide range of medium osmolality, e.g. plasmolysis measurements on growing *E. coli* cells indicated that turgor pressure changed by only 20% over a tenfold change in medium osmolality (reference 119; compare also the freezing-point measurements of Christian and Ingram[103] on cells growing in media of widely different osmolalities).

The existence of osmoregulatory mechanisms in bacteria such as *E. coli* which normally exist in media containing moderately high concentrations of dissolved solutes is revealed by the imposition of osmotic stress, i.e. upshock or the converse downshock[7]. In the presence of external K^+ and substrate, plasmolysis of *E. coli* produced by upshock is rapidly reversed. This reversal is caused by a rapid uptake of K^+ which is blocked by metabolic inhibitors[120]. Uptake of K^+ appears to take place largely in exchange for cellular H^+ and is not associated with appreciable uptake of anions from the medium[17]. This suggests a simple mechanism by which the cells can respond to increases in medium osmolality and re-establish turgor pressure by a corresponding increase in intracellular osmolality. The essentials of this model are as follows[7]. Following osmotic upshock in the presence of K^+ and substrate, substrate is taken up by the cells. During substrate utilisation considerable amounts of organic acids (mainly univalent acids) are produced. In the absence of osmotic stress these acids are largely secreted into the medium[87]. Following upshock an osmotically sensitive K^+–H^+ pump comes into play by which H^+ is exchanged for external K^+, the acid anions remaining in the cell. These, together with the K^+ taken up provide the increase in intracellular osmolality necessary to re-establish turgor pressure[7].

At the other end of the scale, cells like *E. coli* respond to sudden increases in turgor pressure (downshock) by an increased leakiness to small ions and molecules, which results in a decrease in intracellular osmolality. K^+, sugars, amino acids, nucleotides and other low molecular weight phosphate compounds have all been shown to leak out rapidly following downshock[7]. Moderate downshock is reversible. Severe downshock can result in extrusion of protoplasts and destruction of the cell.

The response of micro-organisms to osmotic downshock is variable, some species (e.g., *S. aureus*) being much less sensitive than others (e.g. *E. coli*). In essentially fresh-water forms such as yeast, response to osmotic stress as well as steady-state leak permeability is minimal and, unlike *E. coli*, active transport systems continue to operate in the stationary phase. Osmoregulation, in the strict sense, does not appear to exist. As discussed above the cell has the potential to accumulate relatively large amounts of electrolytes, thus markedly increasing its turgor pressure[13], but the factors which determine the final steady-state levels of K^+, Mg^{2+}, $H_2PO_4^-$, etc., are not osmoregulatory but are in part environmental (e.g. amounts of electrolytes in the medium, substrate supply) and in part cellular (e.g. cell-to-medium pH ratio). In the limit (i.e. with adequate amounts of substrate and electrolytes in the medium) accumulation appears to be limited by the acid–base balance of the cell interior[5]. During the stationary phase, the capacity of the active transport systems in yeast declines. Following reinoculation into fresh growth medium this capacity increases, reaching a maximum during the logarithmic phase of growth. The cation content of the cells increases in a concomitant fashion[5]. During the same time a marked increase in discrimination in favour of K^+ develops[120].

Rothstein[1, 5] has summarised the role of ionic accumulation and turgor pressure in growth and division in walled cells (e.g. yeast) as follows: At a certain critical point in the growth cycle a local softening of part of the cell wall occurs. Because of this and the positive turgor pressure of the cells, water enters and the wall is stretched. As stretching proceeds new wall material is synthesised and, when the appropriate size is reached, the wall re-hardens, a new wall or septum is formed, and the daughter cell splits off. The process of wall softening and hardening in yeast is complex and appears to involve enzymatic breaking and re-forming of —S—S— bridges in a protein component of the wall as well as synthesis of glucan and mannan[1]. In terms of this model of growth and division, the marked accumulation of electrolytes which precedes the increase in total volume may be somewhat fancifully envisaged as the laying down of an ionic dowry for the eventual daughter cell.

ION TRANSPORT IN YEAST AND OTHER MICRO-ORGANISMS

REFERENCES

1. ROTHSTEIN, A., in *The Cellular Functions of Membrane Transport* (ed. J. F. HOFFMAN), Prentice-Hall, Englewood Cliffs, N.J., 23 (1964)
2. ARMSTRONG, W. MCD., and ROTHSTEIN, A., in *Laboratory Techniques in Membrane Biophysics* (eds H. PASSOW and R. STAMPFLI), Springer Verlag, New York, 34 (1969)
3. ROTHSTEIN, A., in *Regulation of the Inorganic Ion Content of Cells*, Ciba Fdn Study Group No. 5 (eds G. E. W. WOLSTENHOLME and C. M. O'CONNOR), Churchill, London, 53 (1960)
4. ROTHSTEIN, A., in *Membrane Transport and Metabolism* (eds A. KLEINZELLER and A. KOTYK), Academic, London, 270 (1961)
5. ROTHSTEIN, A., in *Microbial Protoplasts, Spheroplasts and L-Forms* (ed. L. B. GUZE), Williams Wilkins, Baltimore, Md, 174 (1968)
6. CIRILLO, V. P., *Bact. Rev.*, **30**, 68 (1966)
7. EPSTEIN, W., and SCHULTZ, S. G., in *Microbial Protoplasts, Spheroplasts and L-Forms* (ed. L. B. GUZE), Williams and Wilkins, Baltimore, Maryland, 186 (1968)
8. CONWAY, E. J., *Int. Rev. Cytol.*, **4**, 377 (1955)
9. ROTHSTEIN, A., *Discussion of Faraday Society*, **21**, Membrane Phenomena, 229 (1956)
10. ROTHSTEIN, A., *Bacteriol. Rev.*, **23**, 175 (1959)
11. MITCHELL, P., and MOYLE, J., Bact. anatomy, in *6th Symp. Soc. Gen. Microbiol.*, 150 (1956)
12. CONWAY, E. J., and DOWNEY, M., *Biochem. J.*, **47**, 347 (1950)
13. CONWAY, E. J., and ARMSTRONG, W. MCD., *Biochem. J.*, **81**, 631 (1961)
14. CHRISTIAN, J. H. B., and WALTHO, J. A., *Biochim. biophys. Acta*, **65**, 506 (1962)
15. GINZBURG, M., SACHS, L., and GINZBURG, B. Z., *J. gen. Physiol.*, **55**, 187 (1970)
16. MITCHELL, P., *Biochem. Soc. Symp.*, No. 16, 73 (1959)
17. EPSTEIN, W., and SCHULTZ, S. G., *J. gen. Physiol.*, **49**, 221 (1965)
18. WEIBULL, C., *J. Bacteriol.*, **66**, 688 (1953)
19. EDDY, A. A., and WILLIAMSON, D. H., *Nature, Lond.*, **179**, 1252 (1957)
20. ARMSTRONG, W. MCD., *Arch. Biochem. Biophys.*, **71**, 137 (1957)
21. ARMSTRONG, W. MCD., *Arch. Biochem. Biophys.*, **73**, 153 (1958)
22. PASSOW, H., ROTHSTEIN, A., and LOEWENSTEIN, B., *J. gen. Physiol.*, **43**, 97 (1959)
23. PASSOW, H., and ROTHSTEIN, A., *J. gen. Physiol.*, **43**, 621 (1960)
24. HITAOKA, J., and TAKADA, H., *J. Inst. Polytechn., Osaka City Univ.*, Ser. D, **8**, 79 (1957)
25. ARMSTRONG, W. MCD., *Arch. Biochem. Biophys.*, **102**, 210 (1963)
26. ROTHSTEIN, A., and HAYES, A. D., *Arch. Biochem. Biophys.*, **63**, 87 (1956)
27. WILLS, E. D., *Biochem. J.*, **57**, 109 (1954)
28. ARMSTRONG, W. MCD., *Arch. Biochem. Biophys.*, **74**, 427 (1958)
29. KOVÁČ, L., *Folia microbiol. Praha*, **6**, 311 (1961)
30. KOVÁČ, L., *Folia microbiol. Praha*, **6**, 319 (1961)
31. HEVESY, G., and NIELSEN, N., *Acta physiol. Scand.*, **2**, 347 (1941)
32. CONWAY, E. J., RYAN, H., and CARTON, E., *Biochem. J.*, **58**, 158 (1954)
33. CONWAY, E. J., and O'MALLEY, E., *Biochem. J.*, **40**, 59 (1946)
34. ROTHSTEIN, A., and ENNS, L. H., *J. cell. comp. Physiol.*, **28**, 231 (1946)
35. ROTHSTEIN, A., and BRUCE, M., *J. cell. comp. Physiol.*, **51**, 145 (1958)
36. ØRSKOV, S. L., *Acta physiol. scand.*, **20**, 62 (1950)
37. CONWAY, E. J., and BRADY, T. G., *Biochem. J.*, **47**, 360 (1950)
38. CONWAY, E. J., and DOWNEY, M., *Biochem. J.*, **47**, 355 (1950)
39. KOTYK, A., *Folia microbiol., Praha*, **8**, 27 (1963)

443

40. ARMSTRONG, W. MCD., *Nature, Lond.*, **192**, 65 (1962)
41. ROSENBERG, T., *Acta chem. scand.*, **2**, 14 (1948)
42. FOULKES, E. C., *J. gen. Physiol.*, **39**, 687 (1956)
43. CONWAY, E. J., and DUGGAN, F., *Biochem. J.*, **69**, 265 (1958)
44. ARMSTRONG, W. MCD., and ROTHSTEIN, A., *J. gen. Physiol.*, **48**, 61 (1964)
45. CONWAY, E. J., and MOORE, P. T., *Biochem. J.*, **57**, 523 (1954)
46. KOTYK, A., and KLEINZELLER, A., *J. gen. Microbiol.*, **20**, 197 (1959)
47. CONWAY, E. J., DUGGAN, P. F., and KERNAN, R. P., *Proc. R. Irish Acad.*, **63B**, 93 (1963)
48. ARMSTRONG, W. MCD., and ROTHSTEIN, A., *J. gen. Physiol.*, **50**, 967 (1967)
49. CONWAY, E. J., and BEARY, M. E., *Biochem. J.*, **69**, 275 (1958)
50. ROTHSTEIN, A., and BRUCE, M., *J. cell. comp. Physiol.*, **51**, 439 (1958)
51. WEBB, J. L., *Enzyme and Metabolic Inhibitors*, Academic, New York, **1**, 55 (1963)
52. CONWAY, E. J., and KERNAN, R. P., *Biochem. J.*, **61**, 32 (1955)
53. GOODMAN, J., and ROTHSTEIN, A., *J. gen. Physiol.*, **40**, 915 (1957)
54. KAMEN, M. D., and SPIEGELMAN, S., *Cold Spring Harbor Symp. quant. Biol.*, **13**, 151 (1948)
55. SCHMIDT, G., HECHT, L., and THANNHAUSER, S. J., *J. biol. Chem.*, **178**, 733 (1949)
56. ROTHSTEIN, A., HAYES, A., JENNINGS, D., and HOOPER, D., *J. gen. Physiol.*, **41**, 585 (1958)
57. CONWAY, E. J., and GAFFNEY, H. M., *Biochem. J.*, **101**, 385 (1966)
58. ROTHSTEIN, A., HAYES, A., JENNINGS, D., and HOOPER, D., *J. gen. Physiol.*, **41**, 585 (1958)
59. FUHRMANN, G. F., and ROTHSTEIN, A., *Biochim. biophys. Acta*, **163**, 325 (1968)
60. FUHRMANN, G. F., and ROTHSTEIN, A., *Biochim. biophys. Acta*, **163**, 331 (1968)
61. ROTHSTEIN, A., and DONOVON, K., *J. gen. Physiol.*, **47**, 1075 (1963)
62. WARBURG, O., and CHRISTIAN, W., *Biochem. J.*, **303**, 40 (1939)
63. KATZ, J., and HASSID, W. Z., *Arch. Biochem. Biophys.*, **30**, 272 (1951)
64. JENNINGS, D. H., HOOPER, D. C., and ROTHSTEIN, A., *J. gen. Physiol.*, **41**, 1019 (1958)
65. SCATCHARD, G., *Ann. N.Y. Acad. Sci.*, **51**, 660
66. ROTHSTEIN, A., and VAN STEVENINCK, J., *Ann. N.Y. Acad. Sci.*, **137**, 606 (1966)
67. ROTHSTEIN, A., in *Effects of Metals on Cells, Subcellular Elements, and Macromolecules* (eds. J. MANILOFF, J. R. COLEMAN, and M. W. MILLER), Thomas, Springfield, Ill., 365 (1970)
68. PASSOW, H., ROTHSTEIN, A., and CLARKSON, T. W., *Pharmacol. Rev.*, **13**, 185 (1961)
69. NEWEY, H., SANFORD, P. A., and SMYTH, D. H., *J. Physiol.*, **186**, 493 (1966)
70. SANDOW, A. J., and ISAACSON, A., *J. gen. Physiol.*, **49**, 937 (1966)
71. HAGIWARA, S., and TAKAHASHI, K., *J. gen. Physiol.*, **50**, 583 (1967)
72. NAYLOR, W. G., and PRICE, J. M., *Amer. J. Physiol.*, **213**, 1459 (1967)
73. DEMIS, D. J., ROTHSTEIN, A., and MEIER, R., *Arch. biochem. Biophys.*, **48**, 55 (1954)
74. ROTHSTEIN, A., MEIER, R., and HURWITZ, L., *J. cell. comp. Physiol.*, **37**, 57 (1951)
75. ROTHSTEIN, A., and MEIER, R., *J. cell. comp. Physiol.*, **38**, 245 (1951)
76. VAN STEVENINCK, J., *Biochim. biophys. Acta*, **163**, 386 (1968)
77. VAN STEVENINCK, J., and BOOIJ, H. L., *J. gen. Physiol.*, **48**, 43 (1964)
78. VAN STEVENINCK, J., *Arch. Biochem. and Biophys.*, **130**, 244 (1969)
79. MYRBACK, K., and WILSTAEDT, E., *Ark. Kemi*, **8**, 53 (1955)
80. CIRILLO, V. P., *A. Rev. Microbiol.*, **15**, 197 (1961)
81. KOTYK, A., *Folia microbiol., Praha*, **10**, 30 (1965)
82. VAN STEVENINCK, J., and ROTHSTEIN, A., *J. gen. Physiol.*, **49**, 235 (1965)
83. WILBRANDT, W., and ROSENBERG, T., *Pharmacol. Revs.*, **13**, 109 (1961)
84. COWIE, D. B., ROBERTS, R. B., and ROBERTS, I. Z., *J. cell. comp. Physiol.*, **34**, 243 (1949)
85. KREBS, H. A., WHITTAM, R., and HEMS, R., *Biochem. J.*, **66**, 53 (1957)

86. MACLEOD, R. A., and ONOFREY, E., *J. cell. comp. Physiol.*, **50**, 389 (1957)
87. SCHULTZ, S. G., and SOLOMON, A. K., *J. gen. Physiol.*, **45**, 355 (1961)
88. SCHULTZ, S. G., WILSON, N. L., and EPSTEIN, W., *J. gen. Physiol.*, **46**, 159 (1962)
89. SCHULTZ, S. G., EPSTEIN, W., and GOLDSTEIN, D. A., *J. gen. Physiol.*, **46**, 343 (1962)
90. SCHULTZ, S. G., EPSTEIN, W., and SOLOMON, A. K., *J. gen. Physiol.*, **47**, 329 (1963)
91. EDDY, A. A., and HINSHELWOOD, C., *Proc. R. Soc. London* (B), **136**, 544 (1950)
92. WISSEMAN, C. L., SMADEL, J. E., HAHN, F. E., and HOPPS, H. E., *J. Bacteriol.*, **67**, 662 (1954)
93. WEIDEN, P. L., EPSTEIN, W., and SCHULTZ, S. G., *J. gen. Physiol.*, **50**, 1641 (1967)
94. LUBIN, M., and KESSEL, D., *Biochem. biophys. Res. Commun.*, **2**, 249 (1960)
95. ENNIS, H. L., and LUBIN, M., *Biochim. biophys. Acta*, **50**, 399 (1961)
96. LUBIN, M., in *The Cellular Functions of Membrane Transport* (ed. J. F. HOFFMAN), Prentice-Hall, Englewood Cliffs, N.J., 193 (1964)
97. LUBIN, M., and ENNIS, H. L., *Biochim. biophys. Acta*, **80**, 614 (1964)
98. LUBIN, M., *Biochim. biophys. Acta*, **72**, 345 (1963)
99. CHRISTIAN, J. H. B., personal communication
100. HINKE, J. A. M., *J. gen. Physiol.*, **56**, 521 (1970)
101. LING, G. N., *A Physical Theory of the Living State: The Association–Induction Hypothesis*, Blaisdell, New York (1962)
102. LING, G. N., and COPE, F. W., *Science*, **163**, 1335 (1969)
103. ARMSTRONG, W. MCD., and LEE, C. O., *Science*, **171**, 413 (1971)
104. CHRISTIAN, J. H. B., and INGRAM, M., *J. gen. Microbiol.*, **20**, 27 (1969)
105. SALTON, M. R. J., *The Bacterial Cell Wall*, Elsevier, Amsterdam (1964)
106. LARSEN, H., *Adv. Microbiol. Physiol.*, **1**, 97 (1967)
107. GINZBURG, M., *Biochim. biophys. Acta*, **173**, 370 (1969)
108. KABACK, H. R., in *The Molecular Basis of Membrane Function* (ed. D. C. TOSTESON), Prentice-Hall, Englewood Cliffs, N.J., 421 (1969)
109. KUNDIG, W., GHOSH, S., and ROSEMAN, S., *Proc. Nat. Acad. Sci., U.S.A.*, **52**, 1067 (1964)
110. KUNDIG, W., KUNDIG, F. D., ANDERSON, B., and ROSEMAN, S., *J. biol. Chem.*, **241**, 3243 (1966)
111. NEU, H. C., and HEPPEL, L. A., *J. biol. Chem.*, **240**, 3685 (1965)
112. TANAKA, S., FRAENKEL, D. G., and LIN, E. C. C., *Biochem. biophys. Res. Commun.*, **27**, 63 (1967)
113. TANAKA, S., and LIN, E. C. C., *Proc. Natn. Acad. Sci., U.S.A.*, **52**, 913 (1967)
114. SIMONI, R. D., LEVINTHAL, M., KUNDIG, W., ANDERSON, B., HARTMAN, P. E., and ROSEMAN, S., *Proc. Natn. Acad. Sci., U.S.A.*, **58**, 1963 (1967)
115. KABACK, H. R., *J. biol. Chem.*, **243**, 3711 (1968)
116. KABACK, H. R., and STADTMAN, E. R., *Proc. Natn. Acad. Sci., U.S.A.*, **55**, 920 (1966)
117. LUZZATI, V., and HUSSON, F., *J. Cell Biol.*, **12**, 207 (1962)
118. FISCHER, A., *Vorlesungen über Bakterien*, 2nd edn, Fischer, Jena, 20 (1903)
119. KNAYSI, G., in *Elements of Bacterial Cytology*, 2nd edn, Comstock, Ithaca, N.Y., 155 (1951)
120. ØRSKOV, S. L., *Acta path. microbiol. scand.*, **25**, 277 (1948)
121. JONES, W. B. G., ROTHSTEIN, A., SHERMAN, E., and STANNARD, J. N., *Biochim. biophys. Acta*, **104**, 310 (1965)

INDEX